PARISH BOUNDARIES

HISTORICAL STUDIES OF URBAN AMERICA
A series edited by James R. Grossman and Kathleen N. Conzen

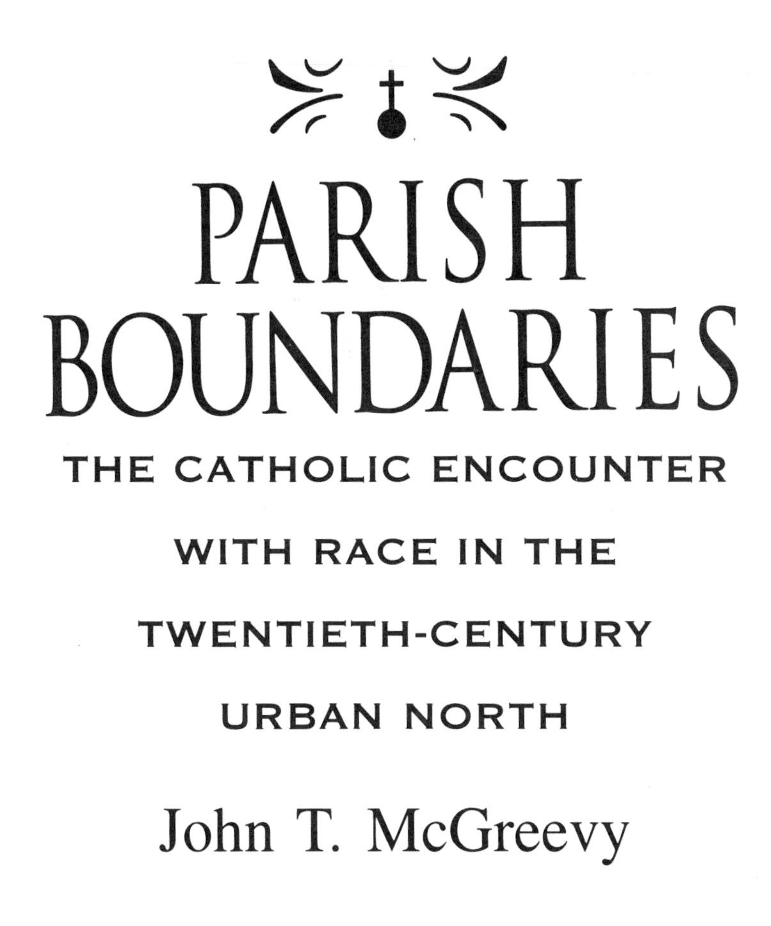

PARISH BOUNDARIES

THE CATHOLIC ENCOUNTER WITH RACE IN THE TWENTIETH-CENTURY URBAN NORTH

John T. McGreevy

THE UNIVERSITY OF CHICAGO PRESS / CHICAGO AND LONDON

John T. McGreevy is an assistant professor of history at Harvard University.

The University of Chicago Press, Chicago 60637
The University of Chicago Press, Ltd., London

Printed in the United States of America
05 04 03 02 01 00 99 98 97 96 1 2 3 4 5
ISBN: 0-226-55873-8 (cloth)

Library of Congress Cataloging-in-Publication Data

McGreevy, John T.
Parish boundaries : the Catholic encounter with race in the twentieth-century urban North / John T. McGreevy.
p. cm. — (Historical studies of urban America)
Includes bibliographical references (p.) and index.
1. Afro-Americans—Northeastern States—Social conditions. 2. Discrimination in housing—Northeastern States—History—20th century. 3. Race relations—Religious aspects—Catholic Church. 4. Community. 5. Parishes—Northeastern States—History—20th century. 6. Northeastern States—Race relations. I. Title. II. Series.
E185.912.M38 1996
305.6′2074—dc20 95-36746

♾ The paper used in this publication meets the minimum requirements of the American National Standard for Information Sciences—Permanence of Paper for Printed Library Materials, ANSI Z39.48-1984.

CONTENTS

INTRODUCTION

I

Riots swept across much of Chicago during the final days of July 1919. Within a week, thirty-eight Chicagoans (twenty-three African-Americans and fifteen whites) were dead and 537 injured. Accounts of the violence are sobering. A mob of several hundred whites ripped a streetcar off its tracks and then chased the African-American passengers, killing one man with a piece of wood. Another white crowd stormed into a restaurant, pursuing and then fatally wounding an African-American patron. Four African-American youths pitched stones at an Italian peddler and concluded their assault by fatally stabbing him.[1]

Charged with explaining the upheaval, members of the Chicago Commission on Race Relations—a group of philanthropists, settlement workers, businessmen, and African-American ministers—pointed to the area of the city where "the largest number of riot clashes occurred," the neighborhoods running parallel to the city's Black Belt along Wentworth from 22nd to 63rd street. A young Langston Hughes, upon arriving in Chicago in 1918, had naively taken a stroll across Wentworth. He returned with bruises administered by an Irish-American street gang "who said they didn't allow niggers in that neighborhood." Little had changed. "Not only," the Commission noted, "[do Negroes] find it impossible to live there, but [they] expose themselves to danger even by passing through." The commission also pointed out that an incompetent local police force (or more ominously, one hostile to African-Americans) had thwarted decisive action during the early stages of the conflict. Fortunately, the commission concluded, city police had been replaced by a state militia, "all-American born."[2]

As use of the term "all-American born" suggests, analysts assigned responsibility for the violence to the immigrant populations clustered along the edge of the African-American ghetto. University of Chicago Settlement Director Mary McDowell argued that "The Ragen Club [one of the "athletic" clubs charged with instigating much of the violence] is mostly

Irish-American. The others are from the second generation of many nationalities.'' The foreman of the grand jury charged with investigating the riots expressed skepticism concerning Mr. Ragen's testimony that ''Father Brian, who had charge of these boys, [simply] taught them how to box and how to hold themselves up physically, and they were doing a most noble work.''[3]

That tensions festered after the riot is also evident. In the early evening of September 20, 1920, an Irish-American named Thomas Barret exchanged words with three African-Americans on the corner of 47th and Halsted, in the heart of Chicago's stockyard section. (Barret had a reputation, according to the city's African-American newspaper, of ''insulting men of our Race.'') The encounter ended with Barret bleeding to death on the street from a knife wound as the three African-Americans dodged through alleyways in an attempt to return to the city's Black Belt. Ducking into St. Gabriel's Catholic church, one man fled up the stairs to the belfry, a second hid under a pile of cassocks and surplices in the sacristry, and another dived into a confessional. Within minutes a crowd estimated at from one to four thousand encircled the church, chanting ''Lynch him, kill him!'' Peering inside the windows of the church, members of the crowd could now see the three men trembling in prayer before the altar. At this point, Father Thomas Burke, the pastor of St. Gabriel's, emerged from the main door clad in a black cassock. Father Burke scolded the crowd, warning them not to ''wreak summary vengeance.'' Even as police smuggled the three men out of the church through a side door and into a waiting patrol wagon, men within Father Burke's eyesight ''were observed to hide their weapons behind their backs.'' Particularly effective, one reporter noted, was Father Burke's gesture of holding his arms out for attention, since ''many had seen him make that gesture several times every Sunday during his sermon.''[4]

II

This study traces the threads connecting religion, race, and community in the nation's northern cities—connections suggested by the image of Father Burke calming his agitated parishioners as well as by the 1919 riots. Investigations of these issues have been infrequent.[5] The prevailing framework in the vast literature on twentieth-century American race relations is biracial; ''white'' groups are presumed to have been ''racist''

(*facing page*) St. Gabriel's Catholic Church, Chicago. Photo courtesy of Donald Hoffmann.

in essentially the same way.[6] Distinctions between the various "white" populations—in terms of culture and contact with African-Americans—are subsumed beneath a "racism" stretching back to the antebellum era. Once placed in this framework, "race" and "racism" become explanations—consequences of diversity—instead of an ongoing process requiring scrutiny.[7]

A frequent corollary to this line of reasoning is to attribute racial disputes in the urban North to "working-class" patterns of behavior. Here the driving force is labor competition, which began in the colonial era and left a legacy of conflict, at the same time as it encircled successive waves of immigrants into a protective "white" embrace.[8] Such analyses do recognize one form of difference within the "white" population, but they also simplify complex matters. The underlying argument—that consciousness formed as a laborer is more important than consciousness developed in the home—is rarely made explicit.

Such a position also fails to explain the location of racial tension in twentieth-century northern cities. The real, if always contested, gains made by African-Americans in the industrial workplace, and the emergence of a powerful African-American middle class, are striking when compared to the violence and tension resulting from even tentative efforts at neighborhood integration.[9] Racial violence in the North centered on housing and not, for the most part, on access to public space, employment issues, or voting rights. Indeed, through most of the twentieth century, neighborhoods in the northern cities were significantly more segregated (in terms of African-American and "white") than their southern counterparts.[10] The more appropriate question is this: what prevented the extension of an occasionally integrated public culture and industrial workplace into the residential communities of the urban North?[11]

Answering this question entails exploring how city residents understood the neighborhoods they created. And a central claim of this book is that American Catholics frequently defined their surroundings in religious terms. Unfortunately, historians of modern America give matters of faith and belief only fleeting attention. Religion frequently ends up at the bottom of a list of variables presumed to shape individual identity, as an ethical afterthought to presumably more serious matters of class, gender, and ethnicity. Churches as institutions—along with local stores, schools, and recreational facilities—receive an occasional acknowledgment, but the emphasis is on organization, not on how theological traditions help believers interpret their surroundings. The role of religion in the literature on modern American race relations is especially circumscribed. Otherwise shrewd analyses of "white" racial formation scrutinize the trajectories of

Irish, Polish, and Italian identity while barely acknowledging the role of religious belief.[12]

Taking religion seriously, to borrow Robert Wuthnow's phrase, blurs customary distinctions between sacred and secular.[13] This strategy is particularly appropriate for twentieth-century American Catholicism. During a century when Catholics made up approximately 20 to 70 percent of the total population in the northern cities, as well as an increasing share of the "white" population, the Catholic Church played a crucial role in molding the world of its communicants.[14] At no point before or since have the connections between the Church and its members been as dense; at no point the Catholic culture so cohesive. As they came into contact with African-Americans, Euro-American Catholics brought with them a religious perspective on community and neighborhood that altered the trajectory of twentieth-century race relations. Catholics did not respond to these African-Americans as simply "workers" or "ethnics"—they also responded in ways that reflected the manner in which religion had structured their lives.[15]

The same issues illuminate the transformation of American Catholicism. This study emphasizes the period between World War I and the early 1970s, when the Catholic system of parishes and schools first expanded into every section of the northern cities, and then, within the last quarter century, began a retreat from what now seemed institutional hubris. It encompasses the journey of Catholics from city alleyways to suburban patios, from ethnic isolation to a religious subculture, from the working to the middle class. In the 1960s, two simultaneous events—the civil rights movement and the Second Vatican Council—combined to place the Catholic struggle over race and religion at the center of the nation's cultural turmoil. By the decade's conclusion, theological discussion had combined with civil rights marches to produce a church almost unrecognizable to longtime communicants.

The story is alternately hopeful and discouraging. Parish boundaries in the urban North served to foster communities of the sort admired by contemporary intellectuals at one historical moment, but proved unable to separate "community" from racial mythology at another. Parochial institutions strengthened individuals while occasionally becoming rallying points for bigotry. The extant literature on religion and race sidesteps this complexity. It enumerates examples of racist behavior that fail to meet standards set by formal doctrine, without attempting to analyze the behavior on its own terms. A guiding principle of this study has been to understand Catholic racism, not simply to catalog it.

ONE
A CATHOLIC WORLD IN AMERICA

I

Echoes of the Chicago riots reverberated far from the shores of Lake Michigan. Just prior to the September 1919 meeting of the American bishops, Cardinal Pietro Gasparri cabled the apostolic delegation in Washington from Rome, noting that "it would be opportune that in the imminent meeting of the episcopacy there be treated the problems of the black population and that there be deplored the recent killings." A similar message made its way to individual bishops, suggesting that Pope Benedict XV was disturbed by both the violence and the feebleness of Catholic work in the African-American community.[1]

This concern was not unusual. For a generation, Vatican officials had pestered American prelates with questions about these matters. The bishops, at Roman instigation, had established a "Commission for Catholic Missions Among the Colored People and the Indians" which distributed the proceeds from an annual collection to priests and sisters working in African-American and Native-American communities. Twenty years later, Pope Pius X asked "all Catholics to be friendly to Negroes, who are called no less than other men to share in all the great benefits of the redemption."[2] Vatican officials had also encouraged Mother Katherine Drexel, the founder of the Sisters of Blessed Sacrament, to use the millions that she inherited from her Philadelphia family to work not in Africa but among Native- and African-Americans. In response to Drexel's entreaties, the bishops in 1906 established another collection, this one entitled "The Catholic Board for Mission Work Among the Colored People." Neither collection proved sufficient. The few priests and sisters working in the African-American community continued to scrape by on the pennies in the collection basket, a paltry annual check from the mission collections, and occasional support from the bishops. Only 2 percent of the nation's African-Americans claimed membership within the church.[3]

In part, these difficulties resulted from the severe strains placed on the Catholic institutional structure by successive waves of European immigrants. These immigrants created a church largely of the North, while the vast majority of African-Americans lived in the South. In 1920, Catholics in Chicago could worship at 228 Catholic parishes, Catholics in Atlanta at 5. Buffalo counted 69 churches, Nashville, 7. Louisiana was heavily Catholic, and home to over half of the nation's roughly 200,000 African-American Catholics, but the South more generally was alien and—during the 1920s—hostile territory. ''It is a well known fact,'' concluded the first historian of the Church's work among African-Americans, ''that these southern states, with the exceptions noted, are not only non-Catholic but are anti-Catholic.''[4]

Attributing the paucity of African-American Catholics to inadequate resources, of course, elided alternative explanations. Speaking informally with the bishops at the 1919 meetings, and offering another perspective on these matters, was Dr. Thomas Wyatt Turner, the leader of a recently formed group called the Committee for the Advancement of Colored Catholics. A biology professor at Howard University and member of the NAACP from its inception, Turner became the most prominent in a group of African-American Catholics who urged Catholics to ''eradicate any and every obstacle which tends to prevent colored men and women from enjoying the full temporal graces of the Church.''[5] Beginning with the desegregation of Knights of Columbus units during the war, Turner moved from his Washington base to organize African-American Catholics throughout the country. Declaring that racism ''is too flagrant a wrong to be tolerated for a moment by the Church after its attention has been sufficiently called to it,'' he issued letters under the committee's name, demanding that the bishops open Catholic institutions to African-Americans.[6]

The American hierarchy politely ignored Turner's missives. Catholicism in the South was essentially a Jim Crow church, with parishes, schools, church societies, seminaries, and even Catholic universities usually segregated. Few southern bishops dared allow an African-American priest or sister to work in the region, and African-American Catholics remained on the margins of what was already a minority religion. Baltimore bishop Michael Curley was typical in his view that only segregation could prevent turmoil. As he testily explained in response to a Vatican query, the Holy See must understand the ''keen race distinction which the Catholic Church in America has not made and cannot solve.''[7]

Persistent questions from Rome, however, suggested that Curley's position was viewed with skepticism. In 1919, Benedict XV issued *Maximum Illud,* the apostolic letter that one historian has called ''the charter for the Catholic missionary movement of modern times.'' Using themes that would be echoed by his successors, Benedict mourned that "there are regions in which the Catholic faith has been introduced for centuries, without indigenous clergy being as yet to be found there" and scolded missionaries for placing national values above those of religion. In 1926, Pius XI reemphasized this theme in *Rerum Ecclesiae,* along with ordaining large numbers of native clergy. "Anyone who looks upon these natives as members of an inferior race," Pius XI warned, "makes a grievous mistake."[8] Papal statements were primarily directed toward church officials in Africa and China, but African-American Catholics, as well as Catholic liberals generally, understandably viewed them as applicable to the American situation. (The phrase "spirit of the Encyclical" was used by one enthusiast for an African-American clergy.)[9] Such hopes were supported by the publication of articles on African-American Catholics in the Vatican newspaper and support from the apostolic delegate.[10]

Indeed, Vatican officials, disturbed by the lack of African-American clergy and reports of discrimination, were already prodding southern bishops into funding an African-American seminary, as well as discussing "bishops for the colored people." American bishops reacted strongly to what one prelate termed "African Cahenslyism"—a pointed reference to the fierce late nineteenth-century dispute over whether German Catholics, led by Peter Cahensly, would receive a separate hierarchical structure. The American (largely Irish) episcopacy had thwarted these German efforts, just as they energetically used their influence to defeat similar appeals to Rome by Polish-American and African-American Catholics. Cardinal Dougherty of Philadelphia made the link explicit by warning his brethren in 1920 that if the Poles received their own bishops, the Indians, African-Americans, French-Canadians and Italians would quickly form a line.[11]

II

That neither African-Americans nor Poles received their own hierarchical structure is less interesting than the window these arguments open onto Catholic thinking on matters of group identity. Crucially, the primary "race" problem for American Catholics before the 1940s was the physical and cultural integration of the various Euro-American groups into the parishes and neighborhoods of the urban North, not conflicts between "blacks" and "whites." These broad notions of "race" are evident in documents ranging from a 1920 Carnegie Foundation report describing how "the great mass of [Catholic] immigrants belong to racial churches of their own" to the 1943 conclusion of a Boston diocesan historian that "the relations between the various racial groups in the Archdiocese have remained singularly harmonious" (even as he entitled a section of the work, "newer Catholic races").[12] The language used by pastor Luigi Giambastiani of Chicago's St. Philip Benizi parish in a 1922 parish bulletin is typical. "It is true that some *idealists* dream of an American millennium when all races will be found fused into one new American race—but in the meantime it is good that each one think of his own. . . . Italians be united to your churches . . . give your offering to the Italian churches who need it . . . the Irish, Polish, and Germans work for their own churches, do the same yourself . . . the Italian church ought to be not only a symbol of glory for you, but a symbol of faith and race."[13]

The city's Back of the Yards area physically exemplified Father Giambastiani's vision. There residents could choose between eleven Catholic churches in the space of little more than a square mile—two Polish, one Lithuanian, one Italian, two German, one Slovak, one Croatian, two Irish, and one Bohemian. Together, the church buildings soared over the frame houses and muddy streets of the impoverished neighborhood in a triumphant display of architectural and theological certitude.[14]

Each parish was a small planet whirling through its orbit, oblivious to the rest of the ecclesiastical solar system. The two Irish churches were the "territorial" parishes—theoretically responsible for all Catholics in the area. As a practical matter, however, all churches—formally territorial or not—tended to attract parishioners of the same national background. The very presence of the church and school buildings encouraged parishioners to purchase homes nearby, helping to create Polish, Bohemian, Irish, and Lithuanian enclaves within the larger neighborhood.[15]

The situation hardly fostered neighborhood unity. Seventy percent of area residents, according to one estimate, were Catholic, but when activist Saul Alinsky began organizing the area in the late 1930s he observed that the various clergy "had nothing but scorn for their fellow priests." A *Washington Post* reporter agreed: "the Lithuanians favored the Poles as enemies, the Slovaks were anti-Bohemian. The Germans were suspected by all four nationalities. The Jews were generally abominated and the Irish called everyone else a 'foreigner.'"[16]

Most of the parishes also included a parochial school staffed by an order of nuns of the same ethnicity as the parish in which they served. The predominantly Irish Sisters of Mercy staffed the parish school at St. Gabriel's, the Polish Felician sisters ran Sacred Heart, and the Sisters of St. Francis of the Immaculate Heart of Mary taught in the school financed by the Bohemian parish of SS. Cyril and Methodius. As the Irish and German population drifted away toward new parishes, Eastern European newcomers resolutely maintained their own schools instead of filling the vacant slots in once booming Irish or German schools.[17]

This Chicago patchwork reflected the complexities of transplanting a European institution, the parish, to American soil. From the Council of Trent in the sixteenth century onward, canon law stressed that the parish served all of the souls living within its boundaries—a conception that implicitly assumed all of those souls were Catholic, as might be the case in rural France, Ireland, Italy, and Poland. The term had a geographical as well as religious meaning. Technically, a parish could even exist without parishioners, so long as its boundaries were clear.[18]

The Council of Trent also recognized, however, that a "large community of distinct national or racial character" needed priests who could administer the sacraments in an intelligible manner.[19] In America, these national parishes became the conscious mechanism used to maintain a religious hold on non-English-speaking immigrants even as the faith lost ground among the European working class. Communicants heard the gospel in their native tongue, worshiped with other immigrants and melded European with American customs.

This grudging acceptance of diversity made the experience in Chicago's Back of the Yards area different from that of the rest of the urban north in degree, not kind. A 1916 U.S. census survey revealed 2,230 Catholic parishes using only a foreign language in their services, while another 2,535 alternated between English and the parishioners' native tongue. Towns as small as Bristol, Rhode Island—population 11,159 in 1940—formally divided its Catholic population into Irish, Italian, and Portuguese parishes. Detroit's Bishop Michael Gallagher, himself the son of Irish immigrants, authorized the founding of 32 national parishes (out of a total of 98) as late as the 1918-1929 period. In 1933, Detroit Catholics could hear the gospel preached in twenty-two different languages.[20]

Indeed, recent writing suggests that episcopal attempts to quash national parishes, schools, and societies only strengthened national identities by creating a sense of shared victimization.[21] A furious battle in the late 1920s between a Providence bishop committed to a centralization of parish finances and instruction in English, and French-Canadian Catholics dedicated to the survival of their culture, resulted in court cases, protests, and fervent appeals to the apostolic delegate.[22] The pattern was similar in northeastern Pennsylvania. When an Irish-American bishop attempted to prohibit the Christmas midnight mass during the 1930s (due to what he perceived as unseemly drinking by Polish groups before the liturgy), thirty-four delegations of Poles immediately complained to the bishop and the apostolic delegate. As one participant in the revolt noted, such quick actions "[gave] proof that we will not permit anyone to destroy a national dignity, pride and traditions."[23]

Rather than face outright revolt, bishops working with national groups generally assigned an auxiliary bishop or senior cleric to handle pastoral appointments and mediate intramural disputes.[24] At times even this was insufficient. Polish priests, for example, formed national federations and periodically sent complaints about discrimination to authorities in Rome. One 1915 statement warned of ominous consequences if Poles were to be "deprived of the care of a Bishop from among our own race." The most serious episode resulted in a demand for greater autonomy for Polish parishes, Polish curricula in diocesan seminaries, and Polish bishops as a counterweight to the power of "Americanizing bishops." One section of the appeal was entitled, "Serious Misunderstandings Between Catholic Poles and the American Clergy."[25]

Relationships between the hierarchy and Italian Catholics were equally strained. American Catholic leaders viewed Italian Catholics, in contrast to the fervent, if troublesome, Poles, as negligent communicants. The "Italian problem" became a topic in clerical journals and frustrated bishops alternated between subsidizing Italian parishes and schools and issuing edicts for more financial support. In Chicago, thirteen times more Polish children than Italian children were enrolled in parochial schools by 1930, even though Poles outnumbered Italians by only two to one.[26] The New York City pastor of Nativity Church, Father Bernard Reilly, informed his archbishop in 1917 that "the Italians are not a sensitive people like our own. When they are told that they are about the worst Catholics that ever came to this country, they don't resent it or deny it. . . . The Italians are callous as regards religion."[27]

As Father Reilly's comments suggest, however, one of the main obstacles to Italian participation in the American Catholic Church was the same national hostility experienced by the Poles. For the first Italian immigrants, special masses were often held in the basement of the parish church. At Transfiguration parish in lower Manhattan, Father Feretti celebrated three Sunday masses for Italians in the basement even as Father McLoughlin ran the main church upstairs. The parish history reports that "Father McLoughlin did his best to make the two races coalesce, by compelling the Italians to attend services in the upper church, but found that far better results could be obtained by having the two peoples worship separately." Father James Groppi, growing up as one of twelve children across from Immaculate Conception parish in Milwaukee recalled watching the Irish parishioners troop into Sunday mass while the Italians gathered with an Italian priest in a shoemaker's shop across the street. An Italian-American interviewed by a social worker in Pittsburgh refused to leave her neighborhood because she would then have to attend an "American" church. "I want my own religion for the children," she explained.[28]

Given these difficulties with Poles, Italians, and other national groups, the few steps taken toward weakening national identities were understandably tentative ones. Only in 1921 did the National Catholic Welfare Conference form a bureau "to rank with and be recognized as the equal of other national immigrant aid organizations." The formation of the National Conference of Catholic Charities produced less a national organization than a federation of wary partners. "How often," lamented Reverend John O'Grady in 1931, "is it said that this organization or that institution is 'Irish' or 'German'?"[29] One non-Catholic researcher commented disparagingly in 1938 upon the proliferation of children's homes in Pittsburgh, arguing that the "uneven balancing of needs among Polish, Italian, Irish-American, German, and other national groups in the Catholic community" resulted in inadequate funds for the most basic care. Fifty-five percent of Catholics in Chicago worshiped at national parishes in 1936, only a slight decrease from 65 percent in 1916. In addition, over 80 percent of the clergy (100 percent of the Poles and Lithuanians) received assignments in parishes matching their own national background.[30]

III

The temptation when reading the heated exchanges endemic to these intra-Catholic disputes is to assume that the combatants had nothing in common. And, in fact, the cultural sensibilities within these parishes, ranging from Irish novenas to Italian *feste,* were disparate. Nonetheless, a specifically Catholic style of merging neighborhood and religion organized life in large sections of the northern cities by the 1920s and 1930s. Even Italian-Americans were, as Protestant missionaries discovered, defiantly Catholic, and eager participants in baptisms, first communions, weddings, and funerals. (Protestant outreach efforts, according to one source, occasionally "ended [with] an angry crowd throwing missiles at the missionary.")[31]

Virtually all of the Catholic immigrant groups were within two generations of immigration, and all placed enormous financial, social, and cultural weight on the parish church as an organizer of local life. In the 1950s, a Detroit study found 70 percent of the city's Catholics claiming to attend services once a week, as opposed to 33 percent of the city's white Protestants and 12 percent of the city's Jews. "Those whose experience of Church influence has been confined to Protestant bodies," concluded one study of Poles in Philadelphia, "will have exceedingly little idea of the extent of the Church's power in a Roman Catholic community." Or as one Newark neighborhood survey explained, the Catholic churches, whether they were Polish, Italian, Portuguese or Irish, simply "dominate[d] the life and activities of the community" with "quite popular and well-attended programs."[32]

An obvious parallel was the role of the local clergy. Typically one of the most educated men in the community, a priest welcomed children into the Church at baptism, gave couples his blessings at weddings, and officiated at funerals. Within the confessional, he listened to parishioners detail the deepest pains in their lives. Along with nuns affiliated with the parish, he directed the school where many of the parishioners' children were educated. Yale sociologists investigating Newburyport, Massachusetts, in the 1930s professed amazement at the ability of priests to "define norms of everyday social behavior for the church's members."[33]

One local historian commenting on an Italian neighborhood in St. Louis noted that "the center for all activities is St. Ambrose. The priest indeed directs most of the activities of this section, and it is only rarely that something is done without his advice." At St. Sabina's parish in Chicago, one man remembered "getting on a bike and riding up and down the streets," upon hearing of the longtime pastor's death, "shouting the news like we were newspaper carriers or Paul Revere . . . and the church bells were tolling. When the church bells tolled, everybody came out in front of their houses to find out what's wrong or what's going on . . . and we yelled, 'Monsignor Egan's died, Monsignor Egan's died.' "[34] Attending the funeral of one prominent Buffalo pastor were the mayor and what the local paper termed "a large delegation of city officials." An estimated thirty thousand parishioners and friends filed past the body. Hundreds of mourners unable to squeeze into the church for the service stood in a one-block radius around the building, listening to the music and catching whiffs of the incense pouring from the windows.[35]

Some pastors stayed in particular areas for long periods, creating reputations that extended over several generations. Monsignor Charles Dauray, of Most Precious Blood parish in Woonsocket, Rhode Island, spent fifty-six years constructing an ecclesiastical empire. By the end of his tenure, Dauray supervised the parish and the various religious societies, a grammar school, a high school, and an orphanage. All institutions were staffed by women religious and priests imported by Dauray from Quebec.[36] In Lackawanna, New York, an industrial suburb on the edge of Buffalo, the priests managing Our Lady of Victory parish, high school, and hospital were the city's second largest employers, second only to Bethlehem

Steel.[37] In Chicago's Back of the Yards, Father Karabasz manned the pulpit at Sacred Heart for 44 years, Father Cholewinski at St. Joseph's for 55 years, and Father Bobal for a remarkable 62 years at SS. Cyril and Methodius Bohemian parish.[38]

The Catholic world supervised by these priests was disciplined and local. Pastors were notorious for refusing to cooperate with (or even visit) neighboring parishes—all attention was devoted to one's own institution. Many parishes sponsored enormous neighborhood carnivals each year (with local politicians making appearances and local businesses donating supplies). Most parishes also contained a large number of formal organizations—including youth groups, mothers' clubs, parish choirs, and fraternal organizations—each with a priest-moderator, the requisite fundraisers, and group masses. Parish sports teams for even the youngest boys shaped parish identity, with fierce (and to outsiders incongruous) rivalries developing in Catholic sports leagues between Immaculate Conception and Sacred Heart, St. Mary's and Little Flower.[39]

These dense social networks centered themselves around an institutional structure of enormous magnitude. Virtually every parish in the northern cities included a church (often of remarkable scale), a parochial school, a convent, a rectory, and occasionally, ancillary gymnasiums or auditoriums. Even hostile observers such as Professor John R. Commons professed admiration for the "marvelous organization and discipline of the Roman Catholic Church [which] has carefully provided every precinct, ward, or district with chapels, cathedrals and priests," while bemoaning those Protestant churches that had yet "to awaken to a serious problem confronting them." Writer Alfred Kazin described Our Lady of Loreto's church and school buildings as a red-bricked "fortress" marking the border between Italian and Jewish sections of Brooklyn's Brownsville neighborhood.[40]

Brooklyn, alone, contained one hundred and twenty-nine parishes and over one hundred Catholic elementary schools (map 1). In New York City more generally, forty-five orders of religious men, ranging from the Jesuits to the Passionist Fathers, lived in community homes. Nuns managed twenty-five hospitals. The clergy and members of religious orders supervised over a hundred high schools, as well as elementary schools that enrolled 214,000 students. The list of summer camps, colleges and universities, retreat centers, retirement homes, seminaries, and orphanages was daunting.[41]

Catholics used the parish to map out—both physically and culturally—space within all of the northern cities. Experts encouraged priests in newly established parishes to turn to precinct voting records for their initial mail-

Aerial view of Holy Cross parish plant, Brooklyn, New York, 1948.
Archives of the Diocese of Brooklyn.

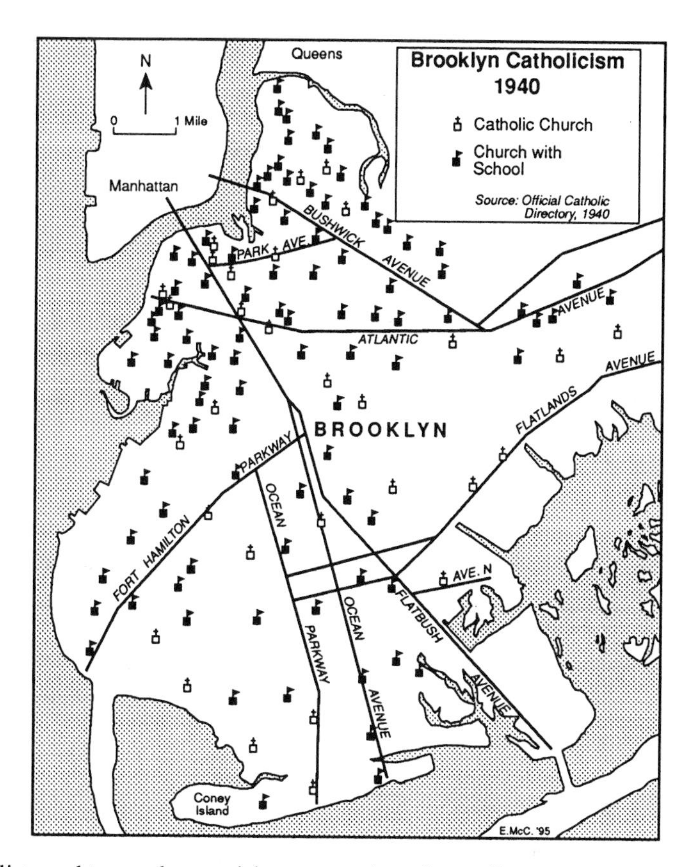

ing list, and to conduct parish censuses in order to discover "unchurched" Catholics. During a parish census, priests visited specific blocks each week (boundaries were announced during mass), inquiring at each door whether a Catholic family lived inside, and if so, whether members of the family were registered in the parish and receiving the sacraments.[42]

Initially, masses were held in any available building—one strategist recommended secretly purchasing any vacant "Protestant temple of error"—while clergy and laypeople planned the construction of a church, rectory, convent, and school. The completed church building itself, with saints' relics lodged in the marble altars, towering spires, and stone walls, also suggested permanence. Following the final payment on long-held mortgages, especially during or after the trials of the Depression, pastors organized religious ceremonies in which the congregation observed the burning of the deed upon the altar.[43]

When African-Americans first began moving north in large numbers, their encounters with the "white" world were filtered through a distinctly Catholic focus on parish and place. One of the fundamental insights of

Chicago sociologist Robert Park was that ''social relations are . . . frequently and . . . inevitably correlated with spacial relations.'' Put another way, the distinctly Catholic heritage of many of the communities surrounding the African-American ghetto made the ghetto's expansion far from a random process. Or as another Chicago sociologist, Ernest Burgess, saw in 1928, ''the relative resistance of different immigrant groups in determining the direction of the movement of Negro population'' would play a crucial role in American urban development.[44]

Part of the explanation lay in rates of homeownership. A cluster of studies of late nineteenth- and early twentieth-century America demonstrated that Catholic immigrant groups invested an inordinate amount of their savings in property. Working-class immigrants were often more likely than middle-class native Americans to own their homes in the urban North, suggesting that home ownership was less an American than an immigrant dream. Sociologists in South Philadelphia during the 1930s found that ''One of the chief objectives of the Italian in this country is to own his home,'' and Stephan Thernstrom's study of Newburyport suggested that Irish workers sacrificed occupational mobility for home ownership. In the heavily Catholic Back of the Yards area, a startling 57 percent of the homes in 1919 were owned by residents, and of these homeowners 90 percent were foreign born.[45]

By contrast, American Jews were often reluctant to purchase homes even though income and education levels for Jews were already beginning to soar above national averages. In Chicago, for example, 13.5 percent of foreign-born whites categorized as from the USSR in 1930—a heavily Jewish population—owned their own homes compared to 43.1 percent of those from the Irish Free State, 42.2 percent from Poland and 39.1 percent from Italy.[46] In the Bronx, Jews flocked to the large apartment buildings on the Grand Concourse, while Irish and Italian Catholics favored single or two-family homes.[47]

Given that homeownership restricted residential mobility, the logical step was to comment, as Burgess did, that ''notably in New York and Chicago, Negroes have pushed forward in the wake of retreating Jews.'' (Burgess added that ''no instance has been noted in the literature where a Negro invasion succeeded in displacing the Irish in possession of a community.'')[48]

Simply citing homeownership rates, however, begs the question. Why did Catholic immigrants so desire these homes? Analysts have stressed peasant origins and the conscious attempt to recreate old patterns of community in a new environment. An often ignored part of this urban environment was the Catholic parish, and the religious meanings the parish extended to heavily Catholic neighborhoods. Parish histories report with numbing regularity pastors commanding parishioners to purchase homes within the parish. One study of a northeastern parish in the late 1930s noted that "In St. Patrick's parish the pastor who immediately preceded the present incumbent was described as frequently urging his parishioners in no uncertain terms, both from the pulpit and in private conversation, to buy their own homes." Another priest proudly announced in 1924 that 81 percent of his parishioners were homeowners, a statistic he perceived as a crucial first step toward a stable, active parish life. A Philadelphia pastor's obituary read that "the priest advised his parishioners, almost entirely of the working class, to strive sacrificially to buy their homes as their greatest step toward security."[49]

Crucially, the parish was immovable. Where Jewish synagogues and Protestant churches could sell their buildings both to recover their equity and to relocate away from the expanding African-American ghetto, Catholic parishes and their property were registered in the name of the diocese and by definition served the people living within the parish boundaries. Even Catholic parishes defined by nationality tended to have geographical boundaries as well, and worked on the assumption that the vast majority of church members would live in the immediate neighborhood.[50]

Studies of white Protestant churches, by contrast, repeatedly found over half of the parishioners living outside the immediate neighborhood—a Newark study even placed 31 percent of parishioners outside the city itself.[51] By the early 1930s, Protestant writers treated the change from the "old neighborhood" model to that of the "interest group" as an accomplished fact. Leading Protestant analysts assumed an inevitable cycle of growth and decay for urban neighborhood churches. (The most widely publicized investigation of the matter was entitled "The Behavior Sequence of a Dying Church.")[52]

When examining the "splendidly organized" system constructed by the Roman Catholics, Protestant analysts bemoaned the "parochial chaos" and the "fragmentation of . . . membership which the Protestant groups have experienced."[53] As one Detroit study emphasized, "The general Protestant lack of the geographical parish makes it impossible to know who should be responsible or to hold anyone responsible for churching any given area."[54] African-American Protestants such as James Weldon Johnson contrasted the rapid multiplication of African-American churches with the Catholic genius for systematic organization and "conservation of power."[55]

Synagogues faced similar dilemmas. Most synagogues drew members from a broad area and competed with neighboring synagogues in terms of ritual and programs. And in comparison with Catholics, the percentage of Jews attending services each week was also relatively small. Orthodox synagogues were more rooted—in part because the prohibition of driving on the day of the Sabbath encouraged members to walk to services. Nonetheless, even Orthodox synagogues abandoned racially changing neighborhoods; only the pace of departure varied.[56]

The mobility of Protestant churches and Jewish synagogues, then, accelerated neighborhood transition by pulling communicants away from particular areas, while the permanence of Catholic parishes anchored Catholics to particular neighborhoods. The lack of census data makes definitive statements impossible, but a scattering of evidence from a variety of cities and the pathbreaking research of Gerald Gamm in Boston convincingly demonstrate that urban Catholics were significantly more likely to remain in particular neighborhoods than non-Catholics.[57] An early description of Boston's Mission Catholic Church is suggestive on this point. An observer noted how the church building occupied an entire block, adding that the building's "resounding chimes, with its immense throngs of worshippers, with its great tower so built that illumined it reveals by the night the outlines of the cross" helped define the area. He also commented that many of the "brewery workers, city employees, and factory hands . . . are no doubt held here by the proximity of the Mission Church."[58]

Put another way, Catholic neighborhoods were created, not found. Most historians lump Euro-American neighborhoods into a broad "white" category, using census data on ethnicity to demonstrate the diverse character of heavily Catholic areas. And in fact Euro-American urban neighborhoods were heterogeneous, although streets around Catholic church buildings were often overwhelmingly Catholic.[59] The argument made here is that the Catholic parish itself, because of its size and community base, helped define what neighborhood would mean. For the parishioners, the neighborhood *was* all-Catholic, given the cultural ghetto constructed by the parish. Invisible to census takers (and later historians) concerned solely with ethnicity and class, Catholics enacted this religiously informed neighborhood identity through both ritual and physical presence.[60]

The result of these Catholic efforts was a merger of educational, religious, and social communities. Significantly, studies as late as the 1960s and 1970s found Catholics unusually apt to form friendship and social networks based upon religious ties, as opposed to ethnic or occupational connections. Catholics stayed longer in urban neighborhoods, counted more friends within walking distance, and were more likely to be involved in neighborhood institutions, especially the local parish church. One Pittsburgh study concluded that perhaps the best way to ensure neighborhood stability was to place a Catholic church in the center of the area.[61]

Even more intriguing, through the 1950s, advertisements in Philadelphia, Buffalo, and Chicago newspapers often listed available apartments and homes by *parish*—''Holy Redeemer 2 Flat'' or ''Little Flower Bungalow''—instead of using community names. (Catholic papers might add ''terrific pastor'' or ''big, new [parochial] school.'')[62] Turn-of-the-century Philadelphia builders emphasized the proximity of newly built rowhouses to particular parishes—''a few minutes walk from St. Ann's.'' By 1930, thirty-seven Philadelphia building and loan associations—such as the St. Columba Building and Loan Association and the St. Donato Building and Loan Association—included parish names in their title, a subtle (and often unauthorized) testament to links between parish and neighborhood.[63]

Non-Catholics also acknowledged this equation. A study of the Buffalo area concluded that in heavily Catholic areas (or ''adverse environment[s]'') Protestant churches rarely developed vital programs. The report summarized one district: ''A very notable concentration of conspicuous Catholic institutions dominates the situation and the Protestant churches are relatively feeble and few.''[64] A 1934 study noted that Catholics rarely faced discrimination in real estate transactions. ''As a matter of fact, real-estate agents welcome the coming of a Catholic church into a community, for it is regarded as an evidence of permanence, and almost invariably it tends to increase the value of the neighboring property.'' The analysis continued:

> Catholics are generally loyal in church attendance; the church itself is as a rule an architectural asset; the coming of the church means the building of a parochial school sooner or later, and the presence of the parochial school is an attraction for other Catholic families. This is so generally recognized that comity commissions in Protestant federations of churches sometimes advise against the building of a new Protestant church in the neighborhood of an established Catholic church.[65]

A powerful indicator of the importance of the Catholic parish is found in the answer of Catholics (and some non-Catholics) to the question ''Where are you from?'' Throughout the urban North, American Catholics answered the question with *parish* names—Visitation, Resurrection, St. Lucy's—a distant echo of a rural Europe where village and parish identities assumed primary importance.[66] Parents in St. Paul, Minnesota, for example, referred to daughters attending dances with boys ''from Nativity'' or ''from St. Mark's.'' John Gregory Dunne recalls that in Hartford ''not to answer 'Immaculate' (for Immaculate Conception), or 'Our Lady of Sorrows,' or. . . 'St. Lawrence O'Toole's' was to be revealed as a heathen at best, a Yank at worst.'' In Cleveland, Slovak Catholics responded ''St. Benedict's'' and never the ''East Side'' or the 29th ward. Holyoke, Massachusetts, Catholics often asked realtors for ''a home in Holy Cross parish.'' Catholics in the Bronx answered ''Holy Cross,'' ''St. Barnabas,'' or ''Christ the King.''[67] In interviews conducted during the 1980s, older Catholics still described cities as clusters of parishes instead of simply listing residential areas.[68]

For all Euro-American Catholic groups, neighborhood, parish, and religion were constantly intertwined. Outsiders frequently remarked on the Catholic habit of genuflecting in front of the advent displays decorating local storefronts in the weeks before Christmas. Also revealing was the custom of arranging for special statues of Mary to stay in the homes of parishioners for one or two weeks at a time. The host family and friends said a rosary in front of the statue each night, and, slowly, the statue moved from house to house and from block to block. Small statues of Mary or local saints appeared in neighborhood yards while crosses and religious artifacts decorated individual rooms. Many clergy promoted the practice of visiting each home in the parish for an annual blessing.[69]

Catholic parishes also routinely sponsored parades and processions through the streets of the parish, claiming both the parish and its inhabitants as sacred ground.[70] Just four years after the founding of St. Sabina's parish in Chicago, for example, the parish organized a massive 186-float parade through the parish boundaries to celebrate a parish carnival. In New Bedford, Massachusetts, members of the local Portuguese parishes organized an elaborate procession in honor of the Blessed Sacrament. Priests announced members of the various festival committees from the pulpit, giving religious sanction to the event. As the eucharistic host was carried through the streets of the neighborhood, parishioners fell on their knees. Tourists who failed to genuflect received glares and caustic remarks.[71]

Local institutions—especially the neighborhood and the school—became layered with religious meanings. A widely used curriculum for Catholic elementary schools included a unit on the parish as a way to ''integrat[e]. . . . religion and social studies,'' emphasizing that ''some people have come to live in this neighborhood to be close to the parish church and send their children to the parish school.''[72] Writers for clerical

May Day procession, early 1920s, Holy Redeemer parish, East Boston.
Photo #1250, Archives of the Archdiocese of Boston.

magazines suggested homilies on "neighborhood ethics" or referred to "the spirit of neighborliness."[73]

Underlying this institutional and local sensibility were particular conceptions of both the human person and the sacred. Crucially, the natural law tradition so central to Catholic thought in the modern period described humans as fundamentally social. As Leo XIII flatly stated, "Man is born to live in society; alone he can neither procure what is necessary or useful for life, nor acquire perfection of mind and heart."[74] A 1943 American Jesuit publication was equally blunt: "Religion is necessarily social. Catholicism at its best and its deepest is social."[75]

The isolated individual was a tragic figure since proper habits could be learned only in the context of specific moral traditions. Formally church-sanctioned institutions—notably marriage and the family—were crucial to inculcating virtue, but so too were local groups ranging from church societies to trade unions and political parties. Such structures were essential to creating a civil society, one capable of resisting either an overreaching state or an unchecked market economy.[76]

Imbued with these views, Catholic leaders deliberately created a Catholic counterpart for virtually every secular organization. In contrast to their European coreligionists, American Catholics failed to sponsor confessional political parties (although in certain areas, the local Democratic party paralleled parish organizations), but the separation of Catholics from the rest of society in other activities was habitual. The assumption was that the Catholic faith could not flourish independent of a Catholic milieu; the schools, parish societies, and religious organizations were seen as pieces of a larger cultural project. "The parish must make every effort to be become a real center of attraction in the lives of the parishioners," one priest advised, "it must become the hub around which a large number of their interests revolve." A Boston social worker commented that "[I]t is only the Catholic Church that has had the vision to see that the very fact of neighborhood organization offers a re-enforcement to every sort of communal effort which is at once both unique and compelling."[77]

And yet Catholic parishes were more than the sum of their organizational parts. Catholic practice depended upon Catholic theology, and more specifically a theological belief that the individual came to know God, and the community came to be church, within a particular, geographically defined space. Communities with distinct physical boundaries—as opposed to communities defined by occupation or gender—actually *became* Church in the context of the liturgy, just as Christ became specific, and corporeal, in the celebration of the Eucharist.[78]

This theological emphasis on the local is also evident in the Catholic tradition of seeing in the neighborhood and in society more generally evidence of God's presence. Where both Jews and Protestants emphasized the reading of texts, Catholics developed multiple routes to the sacred. Theologians describe this as a "sacramental" imagination, willing to endow seemingly mundane daily events with the possibility of grace, always stressing how God became human or Word became flesh.[79]

One telling marker of this sensibility was the answer to a crucial question in the catechism memorized by successive generations of Catholic children. When asked "Where is God?" Catholic children responded "everywhere." God was most visible, of course, during the mass, when the parish community shared Christ's body and blood. But God was also visible in the saints lining the walls of the church, the shrines dotting the yards of Catholic homes, the statue of Mary carted from house to house, the local businesses shutting their doors on the afternoon of Good Friday, the cross on the church steeple looming above neighborhood rowhouses, the priest blessing individual homes, the nuns watching pupils on the school playground while silently reciting the rosary, the religious processions through the streets, and the bells of the church ringing each day over the length of the parish. Catholic versions of the Dick and Jane readers even included stories in which young children gazed at a statue of Mary while riding a city bus.[80]

As the site of both American Catholic spirituality and American Catholic social structure, then, parishes played an indispensable role. With little fear of disagreement, Father Henry Freiburg, O.F.M., of Our Lady of the Angels in Cleveland, could make the following announcement to his congregation in 1947. "Your parish is the most important society to which you as laymen can belong. Your parish speaks not only with the voice of your pastor and your Bishop, but with the voice of the Holy Father, who speaks to you the words of God."[81] The authors of an early history of St. Peter's in Boston chose the revealing title, *A Catholic Stronghold and Its Making*.[82]

IV

This philosophical, theological, and social significance helps explain the centrality of the parish in American Catholic life. As Irish settlements dating from the famine migration broke up under the pressures of upward mobility and new immigrants, Irish-Americans ventured into unknown and largely white Protestant territory clutching the one institution that could order their world. St. Sabina's in Chicago is a typical example. The parish was founded in 1916 upon requests by Irish-Americans whose baptismal records place them in the Back of the Yards area around the turn of the century. The male members of the 7,000-member parish were mostly policemen, streetcar operators, lower management persons, and teachers. Within the tenure of the first pastor the parish erected a church (costing $600,000 and contracted to members of the parish to provide jobs during the Depression), school, convent, and rectory as well as founding a staggering array of athletic, religious, and social organizations. By 1937, the parish plant also included a community center with a full basketball court that seated 1,800 people. Attendance at parish roller-skating shows often climbed over 10,000. Parishioners packed the church and hall for eleven separate Sunday masses, and ushers organized large crowds at multiple Friday novena services.[83]

The vital role of the religious community was also evident among the Poles. According to Thomas and Znaniecki's 1918 study, "the parish is, indeed, simply the old primary community reorganized and concentrated." Where many of the Irish were already two generations away from immigration, the Poles used the parish as both an organizing structure for recent immigrants and a marker for national identity. "I was interested," noted one researcher investigating St. Michael's parish on the South Side of Chicago, "in the word 'Jew' used by the people in defining other nationalities. When I asked them what do they mean by that, I was told that 'everyone who isn't a Pole is a Jew.' " The researcher also discovered that as soon as a parishioner married a non-Pole (even if the spouse was Catholic) the rule was that "he must move out to a non-Polish Catholic parish." The 1,725 children in the parochial school (at tuition of sixty cents a month) split their history classes into thirty minutes on American history and thirty minutes on Polish history. Parish priests required weekly confession and devised an elaborate system of tickets to ensure compliance. Laggards risked hearing their names from the pulpit.[84]

Polish pastors consciously urged parishioners to purchase homes near the unfinished towers of the church building. (One Pittsburgh cleric mistakenly located his church away from the main Polish population, but the lure of the church "brought large numbers of Poles" to the area.) In Brooklyn, priests occupied conspicuous positions on the boards of Polish savings and loan associations, and encouraged workers to invest their savings in local real estate.[85] Thomas and Znaniecki described how this parochialism shaped neighborhood change. "The original population of the district is slowly but ceaselessly driven away, for an Irish, German or Italian tenant or houseowner who sees Polish families take the place of his former neighbors and knows that they have come to stay near their parish-center soon moves to a more congenial neighborhood." The

Clergy procession in Roxbury, Massachusetts, 1954.
Photo #1296, Archives of the Archdiocese of Boston.

parish, they concluded, helped create a Polish-American community which in many ways ''did not consciously exist'' prior to its establishment.[86]

Initially, the integration of religion and neighborhood in Italian communities proceeded apart from the parish church, a reflection of anticlericalism in southern Italy and a popular theology centered around home and family. By the middle of the twentieth century, however, events such as the Good Friday procession at Sacred Heart and St. Stephen's in Brooklyn's heavily Italian Red Hook section were common. Up to 10,000 people followed a lighted ten-foot cross through streets of the parish, with nuns, parochial schoolchildren, two bands, choir boys, and priests marching in distinct groups. Along the route, parishioners placed candles and lamps on windowsills, the tops of doorways, and in front yards.[87] In East Harlem, thousands of New York Italians carrying candles dripping with hot wax followed the statue of the Madonna each summer as it wound its way around the streets of Our Lady Of Mount Carmel parish, concluding at the church building.[88]

The East Harlem *feste,* as Robert Orsi points out, came to be seen as a way to claim the neighborhood in full view of other residents. One resident interviewed in the 1930s recalled that, "In those days we Italians were allowed to worship only in the basement part of the church, a fact which was altogether not to our liking. But the neighborhood became more and more Italian—now Our Lady of Mount Carmel is our very own." In fact, the parish often became the focus of an Italian, as opposed to regional, identity. Promoters of parish schools lauded the opportunity for children to "learn that there does exist in this world a land called Italy, mother of every present civilization and center of Christianity," and parish priests became ardent Italian nationalists. Over time Italian Catholics became more "Irish" in their devotional patterns, and the number of Italian priests, nuns, and parishes steadily increased throughout the period.[89]

Superficially, nothing could be less alike than the devotions to the Madonna of 115th street, the feverish rush by Polish Catholics to create a cluster of institutions, or the Irish-American networks on the outskirts of developing cities. Certainly the bewildered Irish Catholic policemen keeping order while women threw themselves at the bobbing statue of the Madonna in East Harlem saw few similarities. Still, in each instance a society no more than two or three generations away from a culture on the margins of industrial Europe used the parish to define community in a new environment. Over thirty years ago Herbert Gans coined the term "urban villagers" to describe the Italian working-class in Boston. Not just Italians, however, but a whole cluster of European Catholic immigrants built urban villages in the early twentieth century, using the Catholic parish as a focal point. How these villagers, and the Catholic Church they created, reacted to a new, non-Catholic immigrant is the story to which we now turn.

TWO
"RACE" AND THE IMMIGRANT CHURCH

I

Already by the 1920s southern missionaries described an exodus of African-Americans from the region. A Savannah, Georgia, priest wrote that his parishioners "go to New York and to Chicago to what they think is a better standard of social life." From Pass Christian, Mississippi, came word that "many young folks and some old ones leave for the larger cities in the North, especially Chicago." Three-quarters of one Harlem parochial school class raised their hands when asked if they had been born in the South, and one New York pastor estimated that "about 75 percent of our Negro [Catholic] population is American, principally from the seaboard states, the other 25 percent is from the islands and West Indies."[1] A Josephite priest summarized the situation in 1929. "[T]he sudden and abnormal increase in the Negro population creates a new situation demanding the immediate attention of Church authorities."[2]

What was to become of these migrants once they reached their destination? Early treatments of the question in Catholic circles framed it as a traditional immigration matter. African-American Catholics, like Poles, Italians, and other Euro-American groups, were expected to worship in their own parishes, receive the ministrations of religious-order priests specially trained for work in their community, and learn from nuns who were devoted to working in their parochial schools. Supporting the logic of this ethnic segregation was the fact that in cities such as New York and Boston a significant percentage of "black" Catholics were West Indian immigrants, as new to American shores as the latest arrivals from Galway or Galicia.

Also making these analogies plausible was the relative handful of African-Americans generally, let alone African-American Catholics, living in the northern cities. Following a series of violent clashes in the 1850s and 1860s, especially in New York City, residential contacts between Euro-American Catholics and African-Americans dwindled over the next fifty years, as the densely populated pedestrian city of the antebellum era developed into the modern metropolis. Packed into neighboring tenements in 1850, African-Americans and immigrant Catholics were more likely seventy years later to live in separate neighborhoods made possible by improved transportation and lower-density housing.[3]

And in these extended cities, African-Americans made up only a tiny percentage of the total population. In 1920, for example, African-Americans constituted just 2.7 percent of the population in New York and 4.1 percent in Chicago. In heavily Catholic cities such as Buffalo and Milwaukee, African-Americans did not reach 1 percent of the total. By contrast, first- and second-generation European immigrants made up over half of the population in almost all northern cities.[4]

Lines between "race" and what is now considered "ethnicity" were unclear. The assumption that Celtic, Polish, and German races existed was common; as was the belief that differences between these races and the "Anglo-Saxon" race were deep and enduring.[5] The Knights of Columbus funded a monograph series in the early 1920s aimed at publicizing the "racial contributions of the various groups that make up the American people." Subjects included the Irish, Germans, Italians, and African-Americans.[6] At the 1926 Eucharistic Congress in Chicago, the cardinals present agreed to meet with the "thousands of Chicagoans of Foreign birth or extraction." Among the meetings listed were those of German, Italian, Polish, French, Portuguese, and African-American Catholics. A Detroit newspaper list of the Catholic parishes placed the one African-American parish under the "strictly-foreign speaking" category.[7] Philadelphia's Cardinal Denis Dougherty informed diocesan clergy in 1934 that both African-American and immigrant children must be allowed to attend the parochial school nearest them instead of being shunted off to the nearest "colored" or "national" parish. The impetus for Dougherty's ruling, according to one account, was as much to stop Irish pastors from turning away Italian children as it was to lessen discrimination against African-Americans.[8]

Even the 1919 Chicago riot suggested the complexity of these divisions. Irish-American gangs led the attacks on African-Americans, but most Poles and Lithuanians refused to participate, and indeed disparaged the Irish for doing so. (A handful of African-Americans, in fact, lived peaceably within the Polish section of the area.) In a marker of the European racial vocabulary possessed by many immigrants, one Polish priest urged parishioners not to join the "black pogrom." And some Poles during the riot referred to "whites" as a separate group, distinct from Eastern Europeans.[9]

A 1932 study is especially revealing. Members of a Catholic peace group asked sixty Catholic schools to provide "specific instances of Racial Differences in children." Significantly, the organizers of the study equated "racial" problems with both "Negroes" and "foreign children."[10] Responses typically discussed differences within the European tradition. A Pennsylvania principal wrote that "Our Polish and Lithuanian students expressed great surprise that their Catholic associates seemed to dislike them because of their race," while another respondent discussed the effort of the Altar Boys society "to interest the Italian boys to make them feel at home with a group predominantly Irish." One Wisconsin teacher noted the complexities of the situation. "There seemed to be some racial prejudices in our school due to the slowness of the Polish students in grasping situations. . . . Those of Irish descent felt that they were a bit superior to German students. The two races were solid in their disapproval of some Italians who were really bright."[11] The student head of the Chicago Catholic Student Conference made a similar observation as late as 1935, arguing that resolutions opposing discrimination would be too controversial for his organization. Either objections would be made to the motions or "a wag would offer to amend it by adding after the word 'Negro' the words 'Irish, German, English, Polish,' etc . . . "[12]

African-American Catholic activists rarely criticized this racialist language. When the indefatigable Thomas Wyatt Turner formed the Federated Colored Catholics (FCC) in 1925, the group's constitution stated that the organization's "sole purpose is to weld all others [African-American Catholics] into a solid unit for race betterment." In a 1926 letter to the Catholic bishops noting the lack of progress since 1919, the FCC explained that "Our effort is to have every Catholic Negro organization in the country affiliated with the central body which will act as a clearinghouse for Negro Catholic opinion."[13]

Indeed, the separation of African-American Catholics was in part voluntary. When the James family moved to Chicago's South Side from St. Louis in 1929, son David recalled fears that without a "Negro" parish African-American children would not obtain positions as altar boys and that teenagers would date members of another race.[14] Representatives for Cleveland's African-American Catholics made much the same point. Without "opportunity and encouragement," they warned, the "holy aspirations which surge within the hearts of our boys to serve as acolytes," as well as the "virgin longing that inflames the souls of our girls for the loving devoted service in the various activities of a Catholic parish" would wither away.[15]

In Boston, a group of African-American and West Indian Catholics petitioned Archbishop William O'Connell in 1920 for a separate parish, even though outside observers had only recently applauded the integration of worshippers at the local cathedral. The group's spokesman, John Wooten, emphasized that "Many of our Catholic young men and women coming from the South are neglecting their faith because there is no special one in charge of Negro interests in Boston. Of course we are aware of the fact that we might attend any Catholic Church but still like all other races we like our own. For an instance, Irish parishes are interested in the Irish question whilst we are deeply interested in the Negro question."[16] A Cleveland African-American Catholic expressed similar views. "Among the many diverse nationalities that go to make up the Catholic body of this large cosmopolitan city," he informed his bishop, "there are about two hundred persons of Negro birth. . . . The Church in her wisdom, although looking ever toward the day when all nations shall be as one, nevertheless realizes that under earthly conditions religion and piety are best developed along ethnological lines and She therefore tolerates the division into racial groups." Brooklyn African-American Catholics asked for "the establishment of a church among us, and the defense and propagation of our holy faith within our race."[17]

Some Euro-American clerics made parallel arguments. An Omaha pastor argued that "I do not say they are inferior—from man to man all are equal before God—but they feel oppressed, and so they prefer to have a church of their own . . . where they feel themselves on an equality with their fellow worshippers. . . . Have they not the same right to this privilege as Poles, Italians and others?" Or as a member of the Board of Indian and Negro missions put it, "most of our people do not care to go to resorts notably frequented by the Jews, or even in a less pronounced way such a place would not be ferquented [sic] by the Poles if it were known that Italians largely predominated there."[18] Another priest suggested treating the African-American migrant as a "sort of foreigner" while a seminarian maintained that "it seems like a fallacy to conclude from Common Brotherhood in Christ to the essential injustice of racial distinctions; *class* distinctions are natural—are *race* distinctions necessarily unjust?"[19]

American Catholic scholars echoed this racialist ideology. To be sure, theologians pointed out the wrongs of discrimination, discounting any claim of a biblical curse on African-Americans and asserting that African-Americans were equal in all mental capabilities. Discussion of segregation, however, was more nuanced. All citizens, argued Father Francis J. Gilligan, should meet "as equal in street cars, in high-schools, in libraries, in factories, in offices, and in professional societies."[20] And yet, Gilligan added, in a canon law thesis directed by the foremost Catholic liberal, Monsignor John Ryan, "social separation is not seriously uncharitable because some members of the colored group are offended by the practice." Indeed, "social" segregation did not imply inferiority. It was important to distinguish "ideal[s] of life from the commandments which were imposed under the penalty of sin." He continued, "If this Christian tradition is applied to the specific phase of racial contacts under consideration it appears evident that many of the actions involved in social separation are not immoral." Understandably, then,

> It is proper that in every section where large groups of Negroes dwell that separate parishes should be established for them. Many of the colored just as the French, the Germans, and the Polish prefer to worship by themselves. When racial groups in a nation differ strikingly the Church has always permitted them to form separate congregations.[21]

This broad "racial" vocabulary was not limited to Catholics. A primary task of the era's scientific community had been to establish a set of "racial" taxonomies, many of which emphasized differences among European groups.[22] What made Catholics particularly fond of this terminology was its applicability. Although less committed to the essentialist hierarchies pervasive in the scientific literature, American Catholics were accustomed to viewing the world as a series of loosely connected enclaves, with the various "racial" groups staking out claims to different sections of the city and different parochial institutions. One priest discussing the complexities of the American parish system emphasized the "distinct limits" and boundaries placed upon Polish national parishes, casually adding that "the same is true for German parishes, Italian parishes, negro parishes and so forth."[23]

At the same time, efforts by various Catholic "racial" groups to establish genuinely independent sources of authority—especially lines of communication to Rome—met resolute opposition. From the episcopal perspective, the marked similarities between the various Catholic groups, despite obvious "racial" differences, would make parochial distinctions less vivid in the long run. Such assimilation would make the balancing task performed by Catholic leaders more manageable. "The clannishness that made the language parish possible," concluded one cleric in 1936, was being transformed into "mild dislike." Chicago's Cardinal Mundelein optimistically referred to this process as the "fading out of racial lines in our own midst."[24]

II

What became controversial was not Catholic racialism in theory but Catholic racialism in practice. Superimposed upon the Catholic racialist model of a series of geographically and culturally distinct parish neighborhoods was an alternative system of racial organization. Deeply embedded within the America that Catholic immigrants wished to claim as their own, this racial system emphasized distinctions between ''black'' and ''white.''

All Catholic immigrant groups encountered this ideology. Irish immigrants distanced themselves from disparaging references to the ''Celtic Race'' and equations of Irish with ''nigger'' during the years after the famine migration, Poles warded off stereotypes about the ''Slavic Races'' and southern Italians feared that their skin color might place them on the ''colored'' side of the American racial divide.[25] The prejudice against African-Americans resulting in part from these anxieties caused one liberal cleric to complain about ''the infection of Catholic life in this country.''[26]

For Catholics to bracket African-Americans as simply another ''immigrant'' group, in other words, ignored the way in which Euro-Americans claimed a shared identity. The same priest proposing that African-Americans be viewed as ''foreigners'' could also assume the existence of ''our Caucasian society'' and lump Irish, Italians, and Poles into one ''white'' race.[27]

The logic of segregated African-American institutions in this context was difficult to elaborate. By the 1930s, students from various Euro-American backgrounds were accepted at most Catholic schools and parishes. Only African-American Catholics were frequently denied admission. As one pastor of an African-American parish in Philadelphia noted in 1936, ''finally the various National Catholic Immigrant groups—Polish, German, Irish, Italian etc. are slowly merging and are gradually losing their spirit of European parochialism.'' Why then, he wondered, were African-American Catholics still encouraged to form separate parishes? African-American Catholic Gustave Aldrich concisely outlined the situation. ''They [African-American churches] do not stand upon the same plane with the churches set apart by request for the foreign speaking element. Father Cassilly himself states that the second and third generations are absorbed into the general parish work; their separate churches become English speaking or disappear. Not so with the segregated church for colored people.''[28]

Placed in this biracial context, many African-American Catholics viewed separate institutions as a slur. After announcing a policy of ''national'' parishes for Chicago's African-Americans, Archbishop Mundelein received a petition from 81 African-American Catholics stunned that ''they alone shall enter into the sanctuary over whose portals is written in blazing letters of shame, 'SEGREGATED.' '' Cleveland bishop Joseph Schrembs explained to the apostolic delegate that after the opening of a parish for

African-Americans "the greatest opposition to the establishment of this church came from the Negroes themselves. They claimed that by establishing a Negro church I was favoring the policy of so-called segregation. They wanted equal rights with the white people and, therefore, opposed the establishment of a Negro church."[29] The most damning criticism came from W. E. B. DuBois. "Because Catholicism has so much that is splendid in its past . . . it is the greater shame that 'nigger' haters clothed in its episcopal robes should do to black Americans in exclusion [and] segregation . . . all that the Ku Klux Klan ever asked."[30]

Also calling into question the Catholic vision of a city as a collection of parish enclaves was the process of neighborhood transition in American cities. The tension between an immovable parish and a mobile, ethnically diverse American populace had been evident since the late nineteenth century, as members of the original Irish and German parishes spread out across the northern cities. One German pastor in Chicago warned his congregation as early as 1903 that "It is important for our community that those who live within the parish stay there. Don't leave your mother church; don't leave the holy St. Boniface parish." Forty years later a mournful lamentation by one Brooklyn parish historian—"Nobody could forsee the changes of fluctuating populations in these American cities. Those pioneers who clung tenaciously to the homeland traditions, built for the duration of the ages as their ancestors had done in Europe"—also referred to the gradual movement of Poles into a previously German Catholic area.[31]

And yet the early experience of "racial" transition, while hardly painless, still seemed manageable since new groups were usually Catholic. At St. Boniface in Chicago, priests fluent in both German and Polish were assigned to welcome new Polish families; at St. Finbar's in Brooklyn, Irish-American clergy and nuns learned Italian. In such situations, as one Catholic sociologist pointed out, "diocesan authorities have had simply to shift personnel in maintaining parish continuity."[32]

Change from a Euro-American Catholic to an African-American neighborhood moved to a different rhythm. Wholesale abandonment by the Catholic population, often after violent resistance to neighborhood change, replaced gradual transition. James Farrell's *Studs Lonigan* trilogy emphasizes these distinctions. Studs and his Irish-American friends continually grouse about the threat posed by all newcomers to their parish neighborhood, with one friend arguing that "The Polacks and Dagoes, and niggers are the same, only the niggers are the lowest." Studs' father, Patrick Lonigan, notes in a characteristic verbal barrage that "the jiggs have got on Wabash Avenue, and a lot of Polacks and Wops have come in along the southwestern edge of the parish,'' but he adds that ''still I wouldn't be so pessimistic . . . particularly since Father Gilhooley is going to build the new church.''[33]

The most persistent fears of the Lonigans, however, are connected to the possibility of African-American ''invasion.'' Because so few African-Americans are Catholic, only the presence of African-Americans on neighborhood side streets (and in the back pews of St. Patrick's) signals parish collapse. Father Gilhooley, shattered by the exodus of his parishioners, is eventually transferred from the area and replaced by members of a religious order. St. Patrick's maintains its school, but as one former resident puts it, ''the pupils are all jiggabooes and the parish is very poor now.'' Another parishioner laments, ''If we had a pastor like Father Shannon, instead of Gilly, that mightn't have happened. He wouldn't be the kind to build a beautiful new church, and then let the parish go to the dogs. He'd have seen to it that the good parishioners stayed, and that the niggers were kept out. He'd have organized things like vigilante committees to prevent it.''[34]

Great emphasis is placed within the Farrell novels on whiteness. Friends of the Lonigans lurch in conversation between first identifying themselves as Irish, and then in comparison with African-Americans, as ''white.'' When the neighborhood gang encounters a Jewish orator arguing that all races are equal, he is said to be ''trying to prove that a Jew was a white man.''[35]

A 1938 evaluation by one Brooklyn pastor also juxtaposed these two racial languages. When asked to comment on changes within his parish, the pastor replied that ''what was once an Irish German congregation is composed at present of six thousand white people and four hundred Catholic Negroes. . . . The presence of Negroes contributes to the rapid exodus of the Catholic Whites.'' Irish and Germans, in other words, became ''Catholic whites'' only in the context of African-Americans moving in large numbers to a particular area. Ethnicity was flattened into race.[36]

These developments—neighborhood change, a growing distinction between ''black'' and ''white'' and a blurring of lines between Euro-American Catholic groups—were slowly occurring across the urban North. The parish described by Farrell in the *Studs Lonigan* trilogy, for example, had its real-life analogue. St. Anselm's in Chicago admitted African-Americans only after a long struggle and the failure of a new $350,000 church building to ward off neighborhood change. In 1932, the parish became officially designated as an African-American parish and Society of the Divine Word fathers replaced diocesan priests.[37] Hints of similar turmoil could be found in other parts of the city's South Side. The pastor of one Congregationalist church confided to an interviewer in 1928 his satisfaction with the Irish Catholic domination of his neighborhood. "If the restriction [through legal covenants] of the district can't keep them [African-Americans] out," he argued, "the Irish will." The archdiocesan chancellor wrote to a developer planning to rent to African-Americans across from Holy Angels parish and suggested a plan where the Holy Angels pastor would make "every effort to assist you in renting these apartments to Catholic people."[38]

In Detroit, city officials in the mid-1930s referred to "serious resistance" to neighborhood change on the city's heavily Polish East Side. One analyst attributed tension to "[h]omogenous national and religious habits . . . heavy investments in church properties, parochial schools and community places."[39] Detroit businessmen appealed to the local archbishop to establish an Italian national parish in one area, hoping that "an Italian Catholic Church, and possibly a Catholic school would keep away the Negroes."[40] Another pastor, Monsignor Ciarrocchi of Santa Maria parish, informed the archbishop that "I am just in time to stop a property deal which concerns a piece of land and one house, just across our Church door. Negroes would get it." A parish jubilee album published the next year noted that "the many houses we own, are now occupied by our own people; and thus the approach to the Church and to the Parish activities has remained unmolested."[41]

Discerning patterns in these scattered incidents is easier in retrospect. Into the 1940s, church officials, like scholars and politicians, assumed that the pace of the African-American migration, and thus of neighborhood change, would remain steady or even diminish. The northern cities still seemed riven by conflicts between Catholic immigrant groups—a 1939 scholarly account of Boston's Irish and Italians described their relationship as a "race war."[42] African-Americans—dwarfed in population by European immigrants and their descendants—continued to live on the periphery of this Catholic world. No more than a handful of Chicago's over 400 parishes were located in heavily African-American neighborhoods, even on the city's South Side (map 2). One investigation of Chicago's St. Brendan's parish, only a couple of miles west from the African-American ghetto, calmly asserted that "There is small danger that the Negroes will attempt to move in this portion of West Englewood." Even Studs Lonigan, after his family moves two miles south to a new neighborhood, expresses a quiet satisfaction.

> To know that nearly everyone on this street was Catholic
> gave him a different kind of feeling than what he often had

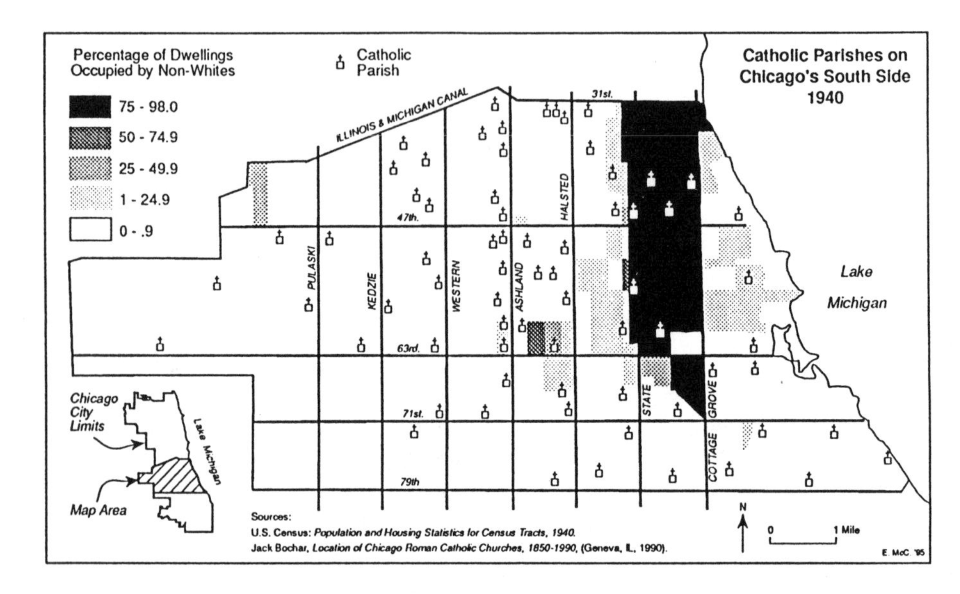

just walking along any street where the people on it were all going about to do any number of things. He felt that he had something in common here and he knew that much about them. They were all on the same side of the fence.[43]

III

The first serious challenge from outside the African-American Catholic community to Catholic thinking on racial issues came from an unlikely pair of Jesuits, John LaFarge and William Markoe. Born in 1880 and named after his father, the Gilded Age artist and confidant of Henry Adams, LaFarge was raised in the rarefied social climes of Newport, Rhode Island. Impressed by the piety of the family's Irish servants, LaFarge decided upon the priesthood as a child. Newport society, unsurprisingly, was "utterly uncongenial" to the idea, but upon graduation from Harvard in 1900, LaFarge, with the support of family friend Theodore Roosevelt, moved to Innsbruck to begin his studies at the Jesuit seminary. In the cosmopolitan atmosphere of the university community, LaFarge quickly demonstrated his linguistic and academic skills. He later recalled with fondness the "friendly rivalry and harmony" of the various national groups studying in the seminary, but he also noted his distaste for the national tensions tormenting both the university and the larger European community. Here too, through conversations with classmates, LaFarge made his initial acquaintance with the developing social doctrine of the

Church—teachings heavily indebted to the scholarly efforts of German Jesuits.[44]

In 1905 LaFarge returned to America a priest. After a few years of teaching, his career path shifted abruptly in 1911 when for health reasons he found himself transferred to a Jesuit parish in Leonardtown, on Maryland's isolated Chesapeake peninsula. Since, as LaFarge put it, "Irish and Germans among them were so rare as to be practically non-existent," he quickly realized what a different Catholic world he was entering. His parishioners were either English Catholics descended from the days of Lord Baltimore or African-Americans (at times even descendants of slaves once owned by the Jesuits) worshiping in small chapels. For fifteen years, he maneuvered his horse and buggy down the dirt paths and across the swollen creeks of the region, and it was during this time that he began to "ponder about the possible applications of the social doctrines of the Church, of which I had heard so much during my Innsbruck years."[45]

Moved by the piety and warmth of his African-American parishioners, LaFarge began tapping family friends for the funds to construct new parishes and schools. His most ambitious project, the Cardinal Gibbons Institute, was built as a Catholic version of Booker T. Washington's school at Tuskegee. LaFarge reasoned that such an institution was the only way to create African-American Catholic leaders, since "the colored man is deeply influenced by those of his own race."[46]

LaFarge's work with the school brought him into contact with Thomas Turner and the Federated Colored Catholics. Turner sat on the board of the Institute and used his contacts to help staff the school and promote it among African-American Catholics. Also working with the FCC was another Jesuit, William Markoe. As a seminarian in Florissant, Missouri, Markoe dispensed food to impoverished African-Americans who frequently appeared at the seminary doors. He soon discovered that little assistance had been given to what one contemporary called the "numerous families of Negroes . . . found scattered in the environs of the seminary," and he set to work.[47]

Markoe dedicated himself to this community with unusual zeal. Catechizing families, encouraging them to attend the local church (despite protests from the white parishioners), and chopping the trees himself for a log chapel, he quickly became a notable figure in the local African-American community. An admirer of St. Peter Claver, the seventeenth century Jesuit honored for his work among newly enslaved Africans, Markoe, along with two seminary classmates (including his brother John) vowed in 1917 "to give and dedicate our whole lives and all our energies,

Father John LaFarge, S.J. John LaFarge papers, Special Collections Division, Georgetown University Library, Washington, D.C.

as far as we are able . . . for the work of the salvation of the Negroes in the United States.''[48]

While LaFarge drummed up support for the Cardinal Gibbons Institute, Markoe published a series of articles on race relations in *America*, the Jesuit weekly, and established a church for African-Americans in the neighborhood surrounding St. Louis University. Markoe warned Catholics that African-Americans would never join a church marked by the segregation they were fleeing in the South. (He wrote and spoke with such urgency that one admiring African-American Catholic noted that Markoe understood the issues ''as much as if he were actually one of us.'') ''A new flood of immigrants,'' Markoe noted in 1920, ''is entering the region by a land route, and there can be no doubt of their color.'' He added, ''When, like the Italians, Poles, or others, the Negroes prefer separation . . . then by all means let the races be treated apart. . . . But even in these cases, as in those of other peoples who have their own churches and schools, the distinction is not meant to be of iron-bound rigidity.''[49]

Much more than the circumspect LaFarge, Markoe aimed directly at Catholic hypocrisy. ''In the North,'' he thundered, ''They [African-Americans] see our great churches, hospitals and schools, they hear in various ways of the wonders of Catholicism, but are . . . certainly not made to realize that it is their sacred inheritance.'' Why is it, Markoe asked, ''that my Negroes believe that the Irish are proverbially hostile to them?'' Even more, some Catholic colleges, unlike Harvard, Northwestern, or Brown, refused to admit African-American students. ''In spite of its name and its Divine Commission to teach all nations,'' he concluded, the Church ''is doing practically nothing.''[50]

Both Markoe and LaFarge became active supporters of the FCC. By the end of the 1920s, however, they had engineered a major shift in the organization's direction. Instead of a race-conscious push for equality, Markoe and LaFarge came to favor an interracial doctrine that eliminated racial categories. Thomas Turner's vision of an African-American Catholic organization comparable to the Ancient Order of the Hibernians or the Polish Roman Catholic Union—hoping to ''perpetuate and increase the national solidarity as an aid in advancing their civic and spiritual lives''—fell before a new vision of interracialism as part of Catholic social action.[51]

This interracialist program was linked to broader intellectual developments. Since the turn of the century, Leo XIII's encouragement of Thomistic philosophy had fostered a distinctive Catholic worldview. Modern problems, Catholic scholars contended, needed to be approached through

Father William Markoe, S.J., with children from St. Elizabeth's parish, St. Louis, Missouri. (c. 1935). Photograph, from the Jesuit Missouri Province Archives, St. Louis, Missouri, was taken by Howard Earl Day.

a rationality dependent on divine law. The code words for this approach, as Philip Gleason and William Halsey have pointed out, were "unity," "integration," and "wholeness," all of which neo-scholastic philosophers used to criticize the disarray they judged endemic to modern society. Especially in the wake of the skepticism and disillusionment resulting from World War I, these Catholics perceived an unusual opportunity to press the claims of an intellectually vigorous faith.[52]

This quest for synthesis extended beyond philosophical journals. Along with a unified intellectual landscape, Catholic intellectuals and leaders desired a more ordered community. The metaphorical model was the medieval world. The corporate life of the Middle Ages—with the various classes linked by mutual obligations—seemed an appealing alternative to the individualism more characteristic of modern life. The "concept of an 'organic' versus an 'atomic' . . . society is the unique contribution," one Catholic sociologist argued, "of Catholic social thought to the American social system." Catholicism could prevent "[the] anarchy of ideas, riot of philosophical systems and breakdown of spiritual forces."[53] The foundation document for such thinking was Leo XIII's 1891 encyclical, *Rerum Novarum,* which asserted the rights of workers in a conscious attempt to win back a European working-class lost to the Church. Specifically referring to medieval guilds, Leo argued that "a dictate of natural justice more imperious and ancient than any bargain between man and man" should order the modern world.[54]

This set of social and intellectual beliefs, which John LaFarge first absorbed in Innsbruck, slowly reshaped American Catholic thinking. As suggested by the 1919 episcopal plea for a "reform in the spirit of labor and capital" Catholic leaders attempted to carve out a middle ground between an individualistic lassez-faire and socialism.[55] The jolt of the global depression and Pius XI's 1931 encyclical, *Quadragesimo Anno,* marking the fortieth anniversary of *Rerum Novarum,* reinforced these trends. Pius XI reiterated that "the common good of all society will be kept inviolate," and he urged Catholics to form Catholic Action organizations dedicated to establishing a more just society.[56]

This emphasis on philosophical and social unity also spurred theological reflection. A notable development of the period in both Europe and America was renewed interest in the Pauline notion of the Mystical Body of Christ. The central tenet of this strand of Catholic theology was that the Church was Christ's Mystical Body on earth. All Catholics, therefore, were united through both this Mystical Body and Christ's literal body and blood as present in the Eucharist. In one popular account, Monsignor Fulton Sheen emphasized that, although made up of many Catholics, "the

Mystical Body is one . . . because all [Catholics are] nourished by the same Eucharist.''[57] Another treatise concluded that ''Catholic Action has its foundation, its deepest reason, in the brotherly communion of the Mystical Body.''[58]

Writings about the Mystical Body also stressed the transcendent bonds existing between all members of the Church. ''The Eucharist,'' Fulton Sheen concluded, ''dissolves all boundaries, nationalities and races into a supernatural fellowship where all men are brothers of the Divine Son and adopted sons of the Heavenly Father.''[59] Milwaukee archbishop Samuel Stritch declared in 1938 that ''The Holy Eucharist among the Christian Sacraments is *par excellence* a great social sacrament, the symbol of true social unity in the Mystical Body of Christ.''[60]

For those interested in race relations, the cumulative meaning of these philosophical and theological developments was obvious. Racialist ideologies destroyed hopes for a genuinely corporate community, one united through faith. During the Middle Ages, according to this view, ''when racial boundaries were minimized by the doctrine of the brotherhood of man and the international character of the Catholic Church, consciousness of racial differences almost disappeared.'' The nationalism dashing hopes for peace in postwar Europe, as well as the science purporting to identify superior and inferior races, were logical consequences of a society lacking authentic sources of consensus. Catholic intellectuals in both Europe and the United States used a strategy similar to that of international socialists, promoting a universalist ideology as a mechanism for disentangling race, nation, and state.[61]

Not coincidentally, the foremost American scholar investigating nationalism during the period, Columbia historian Carlton Hayes, was a Catholic convert and interracial activist. Hayes alternated between pleading for Catholics to emphasize religion over nationalism—''It is a grievous abuse of religion to have it confused with any subnational partisan clique, whether it be German-American, Polish-American, or Irish-American''—and identifying the American race problem as a heightened version of nationalist heresy. In a statement frequently quoted by Catholic interracialists, he termed ''Intolerance toward negroes in the United States . . . perhaps the acme of the racial intolerance of modern nationalism.''[62]

Barriers to unity within the Church itself were intolerable. One Catholic college student, after urging her friends to read John LaFarge's treatment of racial issues, declared that ''Catholic Action will not be limited by the colored line.'' Or as Chicago Bishop Bernard Sheil announced, ''Jim Crowism in the Mystical Body of Christ is a disgraceful anomaly.''[63] A Catholic novel of the period pointedly included a scene in which a ''part

Negro" woman donated blood (a eucharistic symbol) to save the life of a main character.[64]

The story of the FCC must be read in light of this evolving intellectual context. Both LaFarge and Markoe used their influence to shift the emphasis away from a race-consciousness and toward policies consistent with an emerging Catholic Action. In 1930, LaFarge confidently stated that the "practical remedy of [the African-American's] condition, [is] found in the religious and social program of the Church, embodied in Catholic Action."[65]

Instead of training African-American Catholic leaders, LaFarge now emphasized that the "most encouraging feature of the [Cardinal Gibbons] Institute's program has been the opportunity that it has afforded for interracial action." Editorials urged "so-called Negro parishes" to become "bases of interracial Catholic action as opposed to the purely racial parish idea which often aggravates interracial harmony." William Markoe urged the organization to change its name in an article entitled "Our Jim Crow Federation." The name Federated Colored Catholics, he contended, "emphasizes at least by implication, the idea of separation, but implies nothing specifically as regards our real positive objective, which is closer union between all Catholics irrespective of race." At the 1932 convention in New York, one observer noted, "pressure was made upon the group to change their organization to an Inter-racial Federation." By 1933, Markoe, LaFarge, and their allies had ousted Turner from the presidency of the organization and changed the name of the organization's journal to the *Interracial Review*.[66]

Much of the tension that wracked the FCC was between laypeople and clerics. In part because of discriminatory admission policies, counterparts to the Polish, Irish, and Italian clergy were unavailable to African-American Catholics—a serious handicap in a deeply clerical church. While Markoe-edited publications lauded the "important part played by white clergy," Thomas Turner argued at one point in the controversy that "Our aim is to keep the organization strictly in the hands of laymen."[67] W. E. B. DuBois also framed the argument in this manner, warning of a "tragic effort to smash Negro leadership in the Catholic Church."[68]

To the priests, however, as well as their African-American supporters, the real issue was whether race could still be an acceptable Catholic category. Significantly, LaFarge began using the term "un-Catholic" to describe Turner's hostility to the changes.[69] (Advocates of change also favored the phrase "unity instead of separatism.")[70] "I only object," Markoe wrote in a letter to Turner, "to a Catholic organization being racial instead of Catholic, because it is a contradiction in terms." One

Thomas Turner, leader of the Federated Colored Catholics.
Photograph courtesy of Marilyn Nickels.

African-American Catholic refuted Turner's claim that there was a move to "dethrone Negro leadership." Instead, he attacked "Dr. Turner's efforts to perpetuate a Jim-Crow segregated organization which has no place in the Catholic Church."[71]

Turner's protests against what he called the foolish attempt "to put effort primarily on interracial activity than upon increasing racial solidarity and racial improvement" were to no avail. Both the FCC and its successors failed to achieve organizational stability. Markoe continued to attack superiors for permitting discriminatory practices. The more prudent LaFarge—he once warned African-Americans that "we may have to counsel patience and some degree of silence" concerning discrimination—stayed in ecclesiastical good graces. Using his position as an editor at *America* and his contacts across the country, he began in 1934, with the founding of the first Catholic Interracial Council, in New York, to construct an interracial movement based on the model of Catholic Action.[72]

All but invisible to the outside world, the debate over the direction of the Federated Colored Catholics remains instructive. Most important, it marks the first significant break in the racialism undergirding so much of American Catholicism in the first part of the twentieth century. Both Thomas Turner and John LaFarge opposed discrimination. The debate was how to end it. In his first book-length treatment of the subject, LaFarge argued that the "totally inadequate concept of race" could not satisfy a universal church. One of his associates put the matter even more strongly in his memoirs. "The Federated Colored Catholics were a group identified by race. The stronger they became, the more likely were they to increase race tensions and promote race hatred." African-Americans would best be helped by the "combating of race prejudice" through education and the establishment of interracial groups as a "part of Catholic social action." While LaFarge admitted that "certain inherited tendencies present in an ethnic group," might serve a positive function, the force of his argument pushed toward commonalities, not distinctions.[73]

IV

The initial impact of these debates on the larger Catholic community was minimal. Catholic intellectuals spent the Depression years debating economic issues, organizing labor schools, and ultimately, supporting Franklin Roosevelt's New Deal.[74] LaFarge's New York Catholic Interracial Council (NYCIC) worked on the margins of the nascent civil rights movement. More patrician scholar than organizer, LaFarge never understood, according to one longtime African-American Catholic supporter, "the forces that could be mobilized in any particular community." Perennially unstable finances allowed for only a skeleton staff.[75]

Successes were meager. With the backing of NYCIC members Manhattanville college students adopted what came to be called the Manhattanville Principles in 1933, acknowledging that African-American Catholics were members of the Mystical Body and pledging "to give liberally on the Sundays of the year when the collections are devoted to the heroic missionaries laboring among the Negro group." When the Sacred Heart nuns running the college actually decided to enroll an African-American applicant, however, they did so in in the face of considerable alumnae protest, and this in a college located near the center of Harlem. The experience of one Chicago interracialist speaking at a parish on the city's south side was also discouraging. After his presentation, the pastor upbraided him for twenty-five minutes, claiming that the speaker's lecture marked a grievous decline in the community's ability to withstand pressure to integrate the parish neighborhood.[76]

Only the looming importance of anti-Semitism, not conflicts with African-Americans, provoked a broad reconsideration of Catholic positions on "racial" issues.[77] Catholic anti-Semitism did not center around neighborhood change, although Jews in heavily Catholic neighborhoods often encountered hostility. Instead, the linking of Jews with both communism and international financial intrigue, especially by radio priest Father Charles Coughlin, provided a convenient scapegoat. That the situation never reached anything comparable to the tensions then shattering Europe should not, according to historian John Higham, obscure a "common rhythm."[78]

Especially in the northeastern cities, Catholic-Jewish conflicts by the late 1930s seemed capable of getting out of control. In New York City, the more militant Christian Front organizations inspired by Father Coughlin picketed Jewish department stores and in one instance plotted to blow up movie theaters owned by Jews. Boston toughs "laid for Jewish kids" coming out of settlement homes and distributed anti-Semitic literature. One reporter commented on the situation in New York: "Where the population is sharply divided between Jews and Catholics, recognized trouble zones have developed. The intersection of the Grand Concourse and Fordham road is an example. It was there that a priest from Fordham University some months ago loudly denounced anti-Coughlin salesmen as 'dirty kikes.' " In Chicago, lights in an abandoned convent sold to the Chicago Hebrew Institute suggested to Catholics in the neighborhood "that saints were objecting to the building being sold to Jews." Pamphlets distributed outside churches during Holy Week in Boston alleged that Jews had "ruined" parts of Dorchester, Roxbury, and Chelsea, prompting writer Wallace Stegner to argue in the *Atlantic Monthly* that "it is high time Catholicism in Boston made very clear that it is not in sympathy with the prophets of dissension."[79]

Most bishops hesitated to attack the Christian Front groups, correctly arguing that only a small percentage of Catholics were involved but less persuasively that little could be done about the situation. Bishop Francis A. McIntyre of New York expressed his belief that " 'the chalk doodlings' of children have been used by paid publicity agents to conjure up the 'phantom of anti-Semitic hate' to injure the Catholic population of New York." Local priests were as conspicuously absent from interfaith committees as they were prominent in their support for some Christian Front activities. Future historians, concluded the editor of one Catholic labor paper in Chicago, "will be further shocked to learn that pious, church-going Catholics took part in this abominable movement in great numbers."[80]

Among Catholic liberals, what John LaFarge termed this "all-important question" generated a flurry of activity. Social activist Father John Ryan's pamphlet, *American Democracy v. Racism, Communism,* for example, treated "racism" *only* in the context of anti-Semitism.[81] Ryan publicly emphasized the papal rejection of anti-Semitism while privately acknowledging that the attitudes of "a great many Catholics, both lay and clerical" made him "terribly discouraged over the present situation."[82] Catholic Worker supporters formed a "Committee of Catholics to Fight Anti-Semitism" and a Detroit Catholic paper concluded that while Christian Front members might be Catholic, the principles of the movement were not. "It feeds upon Anti-Semitism, [and] shows an alarming disposition to imitate the methods and ideology of German racists and daily converts fresh numbers of our Catholic people."[83]

Signals of concern also came from Washington. President Roosevelt made clear to Vatican representatives his uneasiness with Father Coughlin's anti-Semitic broadcasts, also drawing Vatican attention to reports of Catholic anti-Semitism in Baltimore, Brooklyn, and Detroit. Roosevelt aide Benjamin Cohen, working in conjunction with members of Chicago's Jewish community and Chicago bishop Bernard Sheil, carefully agreed to assist the Vatican in aiding European refugees with a $125,000 donation. As Sheil made clear to colleagues in Rome, however, the Chicago Jews made such a generous offer in part because "the fear of an increase of anti-semitic propaganda [in the United States] was worrisome." More specifically, "Frightened because of the struggle against the Jews in Germany, they [American Jews] sense the advantages [which would result from] a more intimate relationship with the Catholic church."[84]

The German analogy was crucial. Among liberal intellectuals on both sides of the Atlantic, biological assumptions about "race" had moved out of intellectual fashion, discredited in part by National Socialist racial policies.[85] For American Catholic liberals, burdened with a Church structure frequently defined by notions of national and "racial" difference, the process of eliminating traditional categories was painstaking. John LaFarge hinted at this difficulty in a 1939 letter to Columbia anthropologist Franz Boas. "Abroad, particularly in France," LaFarge explained, "this wider use of the word 'race' just as a designation of a cultural or national group does not seem to bring about the same confusion of mind that it does here." He added that "there will be need of more clarification as to the term 'minority group'. There seems to be a confusion in the minds of many between the minority group as a social phenomenon and the minority group as a political entity existing in Europe but not here."[86]

As LaFarge was well aware, however, events in Europe were already presaging a new era of Catholic thinking about racial issues. The greatest spur was the tightening grip of Hitler and the Nazi party on independent institutions within Germany. In response, Vatican representatives grew increasingly concerned about the embattled condition of Catholic schools and youth groups, the persistent harassment of clergy, and the growing hostility toward questions concerning Nazi views on Aryan superiority. After meeting with German bishops, Pius XI authorized the publication of an encyclical, *Mit brennender Sorge* ("With Burning Concern"), addressing the situation.[87]

Smuggled into Germany for a mandatory reading at all pulpits on Palm Sunday, 1937, the encyclical denied theories of racial superiority. "Whoever exalts race," Pius XI argued, "is far from the true faith in God and from the concept of life which that faith upholds."[88] Catholic intellectuals echoed this analysis. In France, Jacques Maritain observed that racialism "to an unimaginable degree degrades and humiliates reason, thought, science and art, which are henceforth subservient to flesh and blood and stripped of their natural 'catholicity.'" John LaFarge gave the statement an American spin. "Let us not flatter ourselves," he warned, "that the extreme forms of German racialism may not penetrate to the United States." American racism, he concluded, was a "pale but venomous older cousin" of the beliefs sweeping Germany.[89] In 1938, LaFarge reminded the Clergy Conference of Negro Welfare, a group of priests working in African-American parishes, of the price of disunity. "Are you upset by what is going on in Europe?" he queried in a form letter, "Are racial and national differences going to embroil this country?"[90]

As conditions in Germany worsened, Pius XI planned more ambitious attacks. In 1938, he ordered Catholic universities and the hierarchy "to forge the intellectual weapons" needed in the battle against racist and totalitarian dogmas. The two volumes resulting from this project reiterated Vatican arguments that "passionless, objective science" eliminated race as a meaningful explanation for behavior. Any individual meant more than categories such as race, nation, or class. "Truth," commented one contributor, "is neither racial nor local, but the same for all races and nations, all times and all places."[91]

Pius XI emphasized his concern more concretely by calling John La-Farge to Rome. In a June 22, 1938, private audience, Pius outlined his own thinking on racial issues and demonstrated his familiarity with La-Farge's own recently published volume on interracial justice. "I found," LaFarge later recalled, that "he wished to talk to me on the question of racialism, which had now become a burning issue." Asking him to "Say simply what you would say if you yourself were Pope," Pius secretly ordered LaFarge and Gustav Gundlach, a German Jesuit, to draft an encyclical "on the topic which is considered most burning at the present time." One month later, Pius hinted at the themes of the proposed encyclical in an address to Roman seminarians, stressing the "all-embracing" character of Catholicism and the fundamental unity of the human race.[92]

For the rest of the summer, LaFarge and Gundlach remained in Paris working furiously on *Humani Generis Unitas* ("On the Unity of the Human Race"). LaFarge seems to have drafted the more pastoral sections of the encyclical, drawing upon his American experience, while Gundlach, already a distinguished Catholic social philosopher, grounded the document in Thomistic principles. The encyclical itself was never issued due to Pius XI's death in February of 1939 and the apparent preference of his successor, Pius XII, and the Jesuit superior general, Wladimir Ledochowski, for a less combative document. Unpublished drafts, however, indicate that the authors viewed the American and German situations through the same unforgiving lens. The "baneful influence [of racism] upon certain regions of the American continent," according to one draft, prevented the American church from becoming a "house of God . . . for all races" even as the European "struggle for racial purity ends by being uniquely the struggle against the Jews."[93]

Despite his cautious approach to the immediate European crisis, Pius XII spurred liberal hopes in 1939 by urging Catholics to give African-Americans "special care and comfort" as well as ordaining bishops from twelve different national groups in Rome. (One observer called the ceremony a "living demonstration by the Church that race does not count in the truly Christian order.") More important, Pius XII released a major encyclical in 1943 endorsing the organic definition of community implicit in Mystical Body theology. Catholic interracialists immediately applied the pope's words to American race relations. "The stupendous Encyclical Letter of His Holiness on the 'Mystical Body of Christ' left me weak and ever so happy," noted one activist, "For now none who will read it, will be able to justify any kind of prejudice against the Negroes, or Interracial Justice." Another editorialist observed that "If any one tenet of the Church may be called all-inclusive, it is this doctrine of the Mystical Body. And in its perfected application it has no common ground with racial discrimination of any sort."[94]

To be sure, the distance between Catholic parish life and papal encyclicals remained vast. Letter writers to New York Catholic newspapers still referred to Italians and Irish as "racial Catholics." Within walking distance of the Catholic Interracial Council office, specific Irish and Italian blocks, churches, and parochial schools marked separate Catholic worlds on New York's lower West Side. The most diverse parochial school counted only 25 out of 700 students as products of "mixed" (meaning Irish Catholic and Italian Catholic) marriages. A study of longshoremen living in the area concluded both that Irish and Italians still treated each other with barely concealed scorn, and that "their social and cultural life like that of the entire neighborhood centered around the Roman Catholic Church."[95]

The necessity of the "national" parish model for African-Americans was still asserted. "The Catholic Church is primarily a religious institution to teach men how to save their souls," Josephite priest Father John Gillard concluded, "and only secondarily a social reform or social institution." In the larger cities, he added, African-Americans "felt freer in congregations composed predominantly of colored Catholics."[96] In 1940, a group of African-American Catholics requested the administrative committee of the bishop's National Catholic Welfare Conference to eliminate racially defined schools and parishes. The committee's response was blunt: "It was the general belief that the petition was promoted by a small but unrepresentative and irresponsible white group, and that experience proved beyond doubt that Negroes by and large went [sic] separate schools and churches." Brooklyn Bishop Thomas Molloy made a similar argument in a letter to John LaFarge:

> While I recognize that this segregation program emphasizes among many colored people that even the Catholic Church

> does not treat colored people as normal human beings, still, on the contrary, I have been informed by priests experienced with the work among the colored people that many of them object to white people attending a church building functioning for colored people.[97]

Still, neither traditional racialist ideologies nor the church that supported them would survive the 1940s unscathed. The easing of Catholic-Jewish tensions, in part because of the economic boom created by the war, would make exhortations about Catholic anti-Semitism less vital. The same wartime prosperity, however, would lure unprecedented numbers of African-Americans toward the heavily Catholic cities of the North. Catholic intellectuals would turn denunciations of "racialism" in the context of anti-Semitism into pleas for African-American inclusion within the Mystical Body of Christ.[98]

Upon their arrival in the northern cities, African-Americans would confront a set of Catholic peoples accustomed to discrimination and sympathetic to the notion that the segregation of "races" was inevitable and natural. Yet even as Catholic leaders hesitantly began the process of eliminating ethnic groupings altogether, the social history of the wartime experience would make Euro-American Catholics more conscious of their identity as whites. Catholics still responded with a parish name to the question "Where are you from?" but they would soon do so in a nation and a church transformed.

THREE

CATHOLICS AND THE SECOND WORLD WAR

I

The most obvious consequence of the war was the sudden surge in African-American migrants searching for good wages and a better life in the booming industries of the urban north. Almost 1.5 million African-Americans abandoned the South (14.6 percent of the region's African-American population) in the 1940s and a slightly smaller proportion during the following decade. By contrast, only 398,000 African-Americans had left the South during the 1930s.[1] A small number of these migrants were Catholics, a partial explanation for the sharp jump in the total number of African-American Catholics in the northern cities. "Almost all of our missions report vast losses of good parishioners due to a wholesale exodus of Negroes from the South," wrote one priest even as another cleric urged northern Catholics to "stop this leakage of Catholic colored migrants" by greeting them with "welcome and solicitude."[2]

"Welcome and solicitude" were not always the operative terms. One Brooklyn pastor, Monsignor John Belford of Nativity Church, spent much of the 1920–50 era fulminating against changes in his neighborhood and directing African-American and West Indian communicants away from his church and school.[3] Even as W. E. B. DuBois privately termed Belford's conduct "obnoxious and insulting," Catholic clergy viewed him as perhaps the "most prominent pastor in the diocese."[4]

In the early 1920s, Belford supported the founding of St. Peter Claver parish for African-Americans a short distance from his own Nativity parish, in part to preserve the racial makeup of Nativity. In 1929, Belford's assertion that African-Americans would be excluded from his church if they became "numerous" attracted national publicity. New York's Cardinal Hayes delegated the matter to a Harlem pastor who told reporters that "the church wishes to treat all nationalities and groups alike" and that "you are all equal when you kneel at the foot of God's altar." Belford reluctantly agreed with this pronouncement but noted that African-Americans did have a church and community of their own. "Our people,"

Belford argued, "do not want the Negroes in their church, in their homes, or their neighborhood," a statement he backed up with support for the Gates Avenue Association dedicated to preventing African-Americans from living in the area. When later approached by Catholic Interracial Council representatives, Belford retorted that "I'm not interested in the parish that has been erected within a thousand feet of my parish."[5]

And yet Belford's hostility did not exhaust the range of Catholic responses. Only blocks away, another priest, Monsignor Bernard Quinn, helped organize St. Peter Claver church. Even as Belford warned of neighborhood deterioration, Quinn stood outside the church building warmly encouraging passersby to come inside. Significantly, Quinn put little emphasis on integrated parishes, attacking the "exclusion of worshipers" but also arguing that "people should support their own parish and should not go where they are not welcome." Within a decade, as Belford's parish slowly withered, St. Peter Claver annually produced over two hundred converts, and Quinn had developed a fiercely loyal following. Upon Quinn's death in 1940, seven thousand African-Americans filled the church and the school in tribute.[6]

Monsignor Quinn's labors were part of a national effort. By the mid-1940s, each northern city contained a handful of priests and nuns laboring in what were ironically termed the "Negro Leagues." They achieved remarkable successes. In Harlem, pastor William McCann and his associates counted 8,000 converts in just sixteen years. When a colleague, Father Michael Mulvoy, was transferred away from the area, Roy Wilkins of the NAACP informed Archbishop Spellman that "Father Mulvoy has been so much a part of the Harlem community that this news is nothing less than a tragedy." Priests and nuns in Philadelphia revitalized a dying parish in the shadow of two predominantly African-American housing projects through a vigorous conversion program. Chicago's Father Joseph Eckert, a Society of the Divine Word priest, baptized 144 African-American adults in one day; the visit of an African-American priest to his parish drew a crowd of twelve thousand.[7]

Here the distinction between the congregational Protestant system and the Catholic parish was again crucial. In contrast to white Protestant churches or Jewish synagogues which generally fled the city ahead of their members, Catholic parishes, defined by geography, not by congregation, remained. This stability, as contemporary observers noted, gave priests and nuns the opportunity to fill nearly vacant facilities with African-American converts.[8]

Education became the most important lure. In a pattern that accelerated in the 1950s and 1960s, many African-American parents began to believe "that the parochial school offered a more thorough education in a quieter atmosphere with adequate discipline."[9] As in other Catholic parishes, the use of women religious as instructors kept tuition charges remarkably low. Half the students at Harlem's St. Mark the Evangelist in 1937, for example, paid no fee, and the more prosperous students paid one dollar annually. Crucially, pastors typically required even non-Catholic parents to attend both instruction classes and a weekly children's mass in order to learn the rudiments of the faith their children encountered in school. In short, while not required to convert, clergy and nuns insisted that parents participate in the new cultural and religious world of which the school was only one part.[10]

These efforts forged a new African-American Catholic community. "The school," wrote one enthusiastic Delaware bishop, "continues to be our best missionary, and most effective instrument of propaganda."[11] African-Americans flooded St. Benedict's boarding school in Milwaukee with over 2,000 applications for 400 spots. Three percent of the students arrived as Catholics but ninety percent converted before graduation. A Detroit pastor noted in a 1939 report that "Since the school was started four years ago, the Parish has increased its membership threefold." A Chicago nun recalled how after planning for 150 students, the school was swamped by eager African-American parents. "After every available desk was occupied the enrollment reached 320. The Sisters were obliged to refuse further supplicants until facilities were provided."[12]

Two Chicago priests, Joseph Richards and Martin Farrell, organized the most impressive work. In his second appointment, Richards moved to a parish, Holy Angels, in which a previous pastor had routinely requested his few African-American parishioners to sit in six back pews on the right side "and let the white people who built this church sit in the center." Even as the parochial school at a neighboring African-American parish burst at the seams, the school at Holy Angels remained virtually all-white, with the building's empty classrooms sending an unwelcoming signal.[13]

Richards introduced techniques that he and Farrell had developed in their first assignment on the city's West Side, and which when applied by their protégés, created a veritable African-American Catholic empire within the city. First, he organized priests to spend the summer refurbishing the school building. At the same time, he persuaded the Sisters of Saint Francis to send women religious dedicated to work in the African-American community. Three years later, these nuns, led by Sister Mary Hortensia, O.S.F., formerly principal of an African-American Catholic school in Yazoo City, Mississippi, formed the core of a staff that educated almost 900 children. Priests and teachers informed parents of two rules at

First Communion class at Detroit's St. Peter Claver parish, 1939.
Archives of the Archdiocese of Detroit.

a mandatory family meeting: "They would go to the full course of instructions [and] everyone would have to go to church every Sunday." (A Gary priest using a similar strategy emphasized to parents that "since the child was responsible for homework in Catechism as well as his other subjects a knowledge of Catholicism would enable them to teach the child in this department.") "Instructions" meant a grueling schedule of thirty-four ninety-minute talks in which African-Americans received the outlines of Catholic doctrine. Lecture titles included "Sin," "Christ—His Divinity," and "Church—Catholic Church Having Authority of God." Instructors warned converts not to step inside non-Catholic churches, "even for social affairs."[14]

Despite these daunting requirements, parishes such as Holy Angels achieved remarkable results. One thousand one hundred and twenty-four parents completed the instruction course in the first three years of its operation, with 751 adults and children requesting baptism into the Church. "They were clamoring at the doors," Richards recalled; "black people wanted their own children to be better." Priests avoided direct interdenominational competition by concentrating on "unchurched" residents. (Virtually all African-Americans identified themselves as Christians, but over 80 percent of the eventual Catholic converts were not active in a church.) Richards organized careful parish censuses in which teams of seminarians, sisters, and priests walked through the neighborhood knocking on doors and later comparing notes on possible convert strategies over the rectory dinner table. "The reaction to that was wonderful . . . by God, they'd say, you Catholics are going after us."[15]

The uneasiness of denominational competitors reflected Catholic successes. As early as 1944, the editor of the mainline Protestant journal *The Christian Century* warned readers that "Catholic leaders believe they are confronted with an unprecedented opportunity" as he described Catholic efforts to point African-Americans toward Rome. In Chicago, sociologists reported African-American ministers holding several meetings in the early 1940s to "discuss ways and means of saving Protestant youth from the Catholics." By that time, "an intensive drive by the Roman Catholic Church" had produced three booming African-American Catholic parishes in the city.[16]

In 1959, researchers counted 600,000 African-American Catholics, over 3 percent of the African-American population, and double the total of 1928. Conversions numbered 12,000 a year. In the largest cities, African-American Catholic communities now formed a distinct component of African-American life, and in some parishes, converts numbered over half the congregation.[17] More vivid evidence came from events such as the

Baptism of African-American converts at Baltimore's St. Francis Xavier parish, early 1950s. Josephite Archives.

confirmation of over four hundred converts in Harlem. Enthusiastic crowds lined the street in front of the church as groups of clerics, followed by Cardinal Spellman himself, entered the building. Detailed explanations of the meaning of the ceremony were given from the pulpit for the benefit of the many non-Catholic friends and relatives present. (This practice was also customary at funerals.) Photos reveal long lines of African-American converts, with the women wearing white veils and the men in their best suits, awaiting the episcopal blessing.[18]

How African-American Catholics viewed their church community is more difficult to decipher. Membership in the Catholic Church ensured pointed questions, if not open hostility, from African-American Protestants distrustful of Rome, or disillusioned by contact with racist Catholics. In Cleveland, baffled neighbors asked African-American Catholics why they refused to eat meat on Fridays, or they gathered to stare at the girls dressed in white veils for religious processions. One non-Catholic relative embarrassed friends by "shouting" during a Catholic service.[19]

Indeed, a leading African-American Catholic could still write in 1933 that "[t]he larger part of the colored population regards the Catholic Church as its bitter enemy, and looks upon Catholics as perhaps the most prejudiced group in the United States." Similarly, a prominent African-American editor wondered why "Negroes fall for Catholicism"—part of his speculative answer was that solicitous church officials viewed "Negroes as a definite pawn in furthering their development until they [Catholics] are the majority religious group here." The Reverend Martin Luther King, Sr.'s often-quoted remark that he would support John F. Kennedy in 1960, *despite* his Catholicism, reflected this enduring skepticism.[20]

These criticisms did little to discourage "cradle" Catholics, often from Louisiana or the West Indies, whose claims on the Catholic tradition equaled those made by immigrant Poles or Irish. Those accustomed to a rich African-American Catholic life in Baltimore, New Orleans, and St. Louis—with active parish societies, high school students encouraged to attend Xavier, the nation's one African-American Catholic university, and particular pride taken in local African-American nuns and clergy—often assumed leadership roles in parishes further north of the Mason-Dixon line. (Richards in Chicago recalled how each African-American parish had at least one Catholic proudly carrying a prayerbook blessed by a long-dead cardinal or bishop.) Other African-Americans seem to have been impressed by the obvious importance of the institution in the urban North. An African-American Catholic noted with awe how "while riding on a trolley car through South Boston . . . practically every man in the car (including the conductor) raised his hat" when passing a Catholic church, and members of African-American parishes eagerly participated in large diocesan parades and celebrations.[21]

Converts also stressed the importance of membership in a global church. At Brooklyn's St. Peter Claver, murals depicting Peter Claver as well as the nineteenth-century Ugandan martyrs were erected next to the altar, and emphasis was also placed on the early African church.[22] One African-American Catholic informed a Buffalo audience in 1946 that "The Catholic Church is the only force able to draw all races and all nations together to establish true peace and understanding and friendship." Another writer compared the United States to Latin America, arguing that "race prejudice is virtually unknown in predominantly Catholic countries."[23]

As institutions, African-American Catholic parishes resembled nothing more than their Polish, Irish, and Italian counterparts. Most African-American Catholics (convert or "cradle") seem to have been solidly working-class. One list of African-American applicants for scholarships to Catholic schools included the occupations of the students' parents. The fathers were janitors, mail clerks, porters, and waiters. The mothers were housewives, cleaning ladies, and office workers. Almost 70 percent of the fathers of Catholic schoolchildren in one Cleveland study worked at semiskilled or unskilled jobs.[24] Catholic clergy reared in Euro-American parishes with similar social compositions inevitably replicated traditional policies and grafted them onto African-American congregations. Women religious ran the school, and priests moderated and encouraged the various parish societies and fundraisers.

"Interracial relations" were a low priority. A Newark priest specifically rejected the term "interracial apostolate," warning John LaFarge that "We always want it understood that our work is principally for the Negro."[25] The Midwestern Clergy Conference on Negro Welfare (founded in 1938 because "pastors found themselves involved with the Negro Apostolate owing to the mass migration of the Negroes . . . ") dedicated itself solely to assisting this "vast conversion movement among the Negroes." A typical address at the 1943 annual meeting in Milwaukee emphasized that while "the members of the clergy conference were interested in social equality and the material well-being of the Negro, their primary purpose is the eternal salvation of the Negroes' immortal soul."[26]

This distinction between convert work and social reform was particularly evident in New York and Chicago. Catholic interracialists applauded Francis Spellman's immediate order upon becoming archbishop in 1939 to open all Catholic schools to qualified African-Americans. Nonetheless, archdiocesan officials clearly viewed interracialism with more skepticism. New York auxiliary bishop Francis McIntyre pointedly asked John La-

Farge in 1946 to inform him whether the Catholic Interracial Council's main purpose was "the promotion of religion amongst the Negroes" or "just interracial relations." One archdiocesan report informed a national missionary board that "The policy of the Archdiocese, however, is to concentrate its efforts on the promotion of the religious social interests of the Negro rather than on the inter-racial relations."[27] In Chicago, Cardinal Samuel Stritch discouraged the local Catholic Interracial Council from contacting priests engaged in "mission work" since they "should be free from any sort of other activities which may interfere with their work." Fathers Farrell and Richards thought "the important thing is to educate and train black people," recalled Monsignor Daniel Cantwell, the chaplain for the Chicago Catholic Interracial Council. "[A]ll this interracial stuff, [they argued] that was nonsense." One source remembers Farrell and Richards arguing that "racial segregation was not intrinsically immoral, discrimination was."[28]

The aspirations of their parishioners, however, altered definitions of successful ministry. The more successful pastors spoke for parishioners in the courts, at diocesan offices, and at City Hall. A visitor to one New York parish saw the pastor contact Catholic nurseries about admitting African-American children and arrange with local police to organize a family meeting for a boy from the school taken into custody. A Camden pastor lobbied the local police department to hire African-American officers, and asked the local theater owner to desegregate his auditorium. To ensure a courteous welcome, Chicago priests conspicuously accompanied their African-American parishioners to meetings of Catholic women's groups.[29]

The very ability of African-American Catholic parishes to push students into high school, college, and then more affluent neighborhoods, again much like their Euro-American counterparts, ensured tension within the Catholic community. As early as 1937 a Philadelphia African-American Catholic pointedly observed how irresponsible it would be for a local Catholic college to reject African-American graduates of parochial schools, especially "when there is so much effort being exerted to improve racial relations through the practice of Catholic Action."[30] Frequently, the first African-American student to apply for admission to local Catholic high schools or colleges did so with the full support of the student's pastor.[31]

Discrimination within the church, veteran pastors realized, not doctrinal ignorance among African-Americans, would be the main obstacle to conversion. In a letter to the Cleveland bishop, one clergyman argued that separate parishes "do a wrong job well. . . . it is, per se, wrong for us to have separate Catholic activities for the colored people and they know it.'' The chancellor of the Milwaukee archdiocese concluded that ''The only reason to fear that the Negro will be lost to the Church is the old and bitter evil of segregation. . . . It is clear that little lasting work can be done in a colored mission if the colored converts are not permitted to attend other 'white' churches when they move elsewhere.''[32]

II

Fervent anticommunism also undermined traditional Catholic thinking on racial matters. Since the first World War, Catholic leaders in both Europe and the United States had waged a relentless campaign against what a 1937 papal encyclical termed ''bolshevistic and atheistic Communism'' that ''aims at upsetting the social order and at undermining the very foundations of Christian civilization.''[33] American and European Jesuits were especially likely to view Catholic social thought as an alternative to a Marxism which had already ''infected nearly the whole world.''[34]

Such efforts linked Catholic intellectuals, Vatican officials, parish priests, and lay leaders more than any other issue. John LaFarge recalled his shock at discovering that ''Some of our fine conservative Protestants took up with enthusiasm the revolutionary bandits in Mexico'' while another interracialist explained to a Philadelphia NAACP leader that ''[T]he right thinking Catholic [must oppose] the Spanish massacre of innocent nuns, priests, and other religious and laity for no other reason except for their religion.'' Parochial school children offered daily prayers for the ''conversion of Russia'' while the American bishops warned of the ''audacity of subversive action.''[35]

The Catholic effort to win converts partially derived from this concern. The African-American community, in this view, became part of the battle between ''Moscow's materialism, and Christ's immortal message voiced by His Vicar.''[36] The foremost researcher on African-American Catholics during this period, Father John Gillard, S.S.J., wrote *Christ, Color and Communism* in 1937. Gillard entitled one section ''Communism: a Lie'' and the following section, ''Catholicism: A Solution.'' A Pittsburgh Dominican described competing with communist orators on street corners by handing out pictures of St. Martin de Porres (notable for his ministry on slave ships and ''mixed'' racial background) and singing hymns. ''The communists had helped to draw a crowd,'' he conceded, ''[but] we had soothed souls.''[37] The Omaha archdiocese authorized the construction of a high school for African-American Catholics in 1937. (No public mention was made of the fact that attempts to place African-American students in regular Catholic schools were "received rather coldly by a number of high school principals.") A Jesuit booster of the school argued that "Only such as are actually on the firing line and realize the dire fate of irreligion and communism that is threatening the colored race can appreciate the high merit and excellence of this important project."[38]

Significantly, Catholic liberals used the communist threat as a wedge with which to open the door to a more interracial church. In Philadelphia, one African-American Catholic complained to Cardinal Dougherty about racial barriers at a local Catholic college. "This debarring of Negroes from Catholic Colleges," she warned, "is furnishing the Communists with additional propaganda to lure the Negro from the Catholic Church."[39] John LaFarge urged North American Jesuits in 1935 to move beyond instruction on communism in the schools and toward an end to segregation in Catholic institutions. "It is merely a question of time till all the Negroes go communist unless we do something," he argued. LaFarge later claimed that "*the attitude of our white Catholics*" would determine whether African-Americans would be "won for God and His Church or for Communism and Satan," and in 1945 he applauded the opening of Jesuit seminaries to African-American candidates. "[A]n occasional Negro in our ranks," he noted "is an answer to a very frequent and harmful type of Communist propaganda."[40]

Even the most idealistic Catholic endeavors contained this mixture of anticommunism and racial liberalism. Russian émigré Baroness Catherine de Hueck began her Catholic settlement house movement after reflecting upon her experiences during the 1917 revolution and concluding that communists are "made by hypocritical Christians, Catholics included, who render to Christ lip service." In 1938, de Hueck established in Harlem the first of several Friendship Houses across the country. As the Baroness later put it, "Then came the call to New York to open a place in Harlem where the atheism of the Communists was making deep inroads."[41]

De Hueck viewed the volunteer staff as "young men and women who see Catholic Action as a life vocation, who have heard Communists speak against the priests and nuns." More prosaically, staffers were African-American and white Catholics, often recent college graduates, who viewed interracial justice as integral to the future of both church and nation. "[W]ithout Interracial Justice," a typical editorial in the organization's newsletter concluded, "America will never find its soul." Anticommunism, however, remained a driving theme. Staff at the Chicago Friendship House took particular pride in their location on 43rd St. across from the South Central Committee of the Communist party. An issue of the New York house newsletter carried a letter from one its African-American sup-

Volunteers and workers at Chicago's Friendship House, 1942.
Courtesy of Chicago Historical Society, item ICHi-03614.

porters describing a debate with another soldier going through basic training. "He quoted Marxian tenets," noted Allan A. Archibald, "I quoted what little I knew of scholastic philosophy [in response]."[42]

Catholic anticommunism would play an even more important role in the context of the cold war. One Baltimore Catholic observed that the city needed an interracial council to counteract leftist union members working within the NAACP. What the city needed, he concluded, was "a solid Catholic organization, first to awaken Catholics, including many priests, to a realization of the colored problem, and secondly to stop this inevitable swing of negroes to Communism."[43] John LaFarge received a letter from a woman who knew an African-American Catholic wavering in her faith. My greatest fear, the woman wrote, was that "she may not only neglect her religion but turn to Communism as a comfort." Father John Cronin, S.S., authored a report for the hierarchy on the extent of communist influence in America even as he used contacts within the FBI to sneak reports on Alger Hiss into the hands of Congressman Richard Nixon. At the same time, Cronin urged liberal racial policies. "It is shameful," he concluded in 1950, "that the most vocal proponents of racial justice have been the American Communists. . . . America would indeed be short-sighted were they to overlook the propaganda value for the Soviet Union derived from American discrimination."[44]

III

Rising African-American Catholic expectations and fears of communist gains within the African-American community fused with more sweeping intellectual currents. In 1943, John LaFarge received a letter from a Fort Wayne woman complaining that diocesan clergy disagreed on the proper handling of racial questions. Some of them, she concluded, "are as . . . apart on the teachings of the Church as the Nazis and the Holy Father." LaFarge told the woman not to worry. "You will be surprised," he wrote, "to find how many are thinking along the same lines as yourself. The recent events have stirred people up and they are proceeding in the right direction."[45]

LaFarge's optimism was appropriate. With notable alacrity, American intellectuals became convinced during the 1940s that a democratic society depended upon tolerance, and ultimately, an end to discrimination. Even as John LaFarge claimed interracial justice to be a "basic Christian dogma," Gunnar Myrdal's landmark 1944 study, *The American Dilemma,* also framed the question in distinctly moral terms. Like LaFarge and other Catholic interracialists, Myrdal emphasized education programs and the irrationality of prejudice.The heart of the American race problem, according to Myrdal, was the contradiction between equitable "national and Christian precepts" and discriminatory behavior. "At bottom," Myrdal concluded, "our problem is the moral dilemma of the American."[46]

Myrdal's work also signaled a subtle redefinition of American national identity. Given the need to rally support for the war, how could intellectuals distinguish American values from those of their fascist (and later communist) opponents? Myrdal answered by comparing the United States to his native Sweden. In particular, he commended the propensity of Americans from all walks of life to define the nation in terms of its highest values—the Constitution, the Declaration of Independence—instead of a particular racial or religious heritage. All Americans, Myrdal argued, shared a "Creed" which included "justice" and "fair opportunity for everybody," although "different individuals and groups" necessarily interpreted these values in slightly different ways. Only this fundamental "cultural unity" allowed "discussion between persons and groups" and even "the democratic process."[47]

This explicitly ideological definition of national identity marked a shift from earlier notions. As evidenced in the Jim Crow system in the South, the debate over immigration restriction, and the revival of the Klu Klux Klan, conceptions of an American identity in the first half of the twentieth century often included Protestantism and Anglo-Saxon ancestry. Twenty years later, both wartime propaganda and American intellectuals applauded a calibrated individual diversity—most obviously in the Hollywood war films emphasizing the necessity for teamwork among soldiers from all backgrounds. Those staffing the hundreds of mayoral commissions and tri-faith committees founded to discuss racial issues during the war defined the "American way" as synonymous with the acceptance of individuals from diverse racial and religious groups. "We are the most composite of people, the most multi-group of all societies," concluded one influential scholar, "No way of life depends more than ours on the decent ordering of inter-group relations."[48]

Only a generation beyond Al Smith's 1928 presidential campaign, Catholics responded to this notion with particular warmth. More than at any previous point in American history, church and nation became intertwined. Invariably, large sections of the parish histories that were completed in this era were given over to a listing of parish veterans and elaborate descriptions of the memorials honoring their service. Parishes purchased war bonds, priest-chaplains donned military uniforms, parochial schoolchildren competed to sell the most war stamps, and hundreds of parishes began newsletters specifically for members serving overseas. At Brooklyn's Our Lady of Perpetual Help, a contingent of Resurrectionist Fathers and 5,000 parishioners marched behind a military honor guard to bless the parish honor roll. A New Jersey parish bulletin requested lists of all parishioners then serving in the armed forces "so that in the future we can prove that we Catholics have surely played our part in the struggle for 'Life, Liberty, and the Pursuit of Happiness.'"[49]

This more capacious American identity—one that included Catholics as full citizens while emphasizing "tolerance" as a distinctively American virtue—led to condemnations of discrimination. In 1943, John LaFarge helped organize a ceremony at the Lincoln Memorial which included Protestant ministers (white and black) and a Jewish rabbi in order to emphasize "support of racial justice and the unity of all citizens." To 30,000 listeners at a 1945 New York rally, as well as in magazine articles, Archbishop Francis Spellman attacked both prejudice against African-Americans and anti-Semitism. By contrast, he applauded men in the armed services who "may dislike one another's personalities, attitudes, beliefs, and actions, but nevertheless patriotism lifts them above disunion." A "real American," Spellman concluded, was one who attempted to stop "the spread of bigotry."[50]

Repudiations of discrimination, in turn, led to questions about the segregation so deeply woven into Catholic life. Margaret Mead, for example, deplored "an immigrant community which is both foreign and Catholic attempt[ing] to keep its young people separate from the community . . . isolating them from the mainstream of American life." Catholic liberals used similar language. Editorials expressed concern over bigotry in "largely self-contained" immigrant Catholic communities. The Detroit Catholic Interracial Council endorsed not only "the sublime doctrine of the Mystical Body of Christ" but the belief that "in Christian and Democratic America, we must, among all races, creeds and nationalities, cultivate the ability to live together."[51]

African-Americans Catholics also favored the notion that the ideological nature of the war against Nazi Germany signaled the beginning of a fight for a more equitable American society. After a Washington, D.C., pastor attempted to move a veteran and Xavier graduate to the side aisle during mass, an integrated group of Catholic Interracial Council members and college students staged a silent protest by conspicuously sitting in the side aisles during an almost empty service. An African-American speaker declared before the Brooklyn Catholic Interracial Council in 1945 that the sacrifices of the war necessitated an attack on prejudice, while a pastor in an African-American parish maintained that "Unless the Negro is given his place in church and in the nation, we fight in vain."[52]

These sentiments are also evident in the reaction of some African-American Catholics to Father John Ryan's 1943 address to the faculty and students of Howard University. Ryan vigorously attacked southern segregation but downplayed the need for action on voting rights. "The only moral right possessed by the citizen in the political field is the right to have a government that promotes the common good," he argued, "This end can be obtained without universal suffrage." Some Howard faculty "sharply dissented" from Ryan's position and one student asked how he could square his views with the need for "interracial democracy." A group of African-American Catholics living in Philadelphia also registered their dismay, using the language of Catholic Action and interracialism developed in the previous decade. The group urged the "fullest integration of all their [the bishops'] spiritual subjects, regardless of race in order that the American parish may present an example of Catholic culture in keeping with the Encyclicals of Pius XII exhorting unity in the opposing of world evils and on the Mystical Body of Christ."[53]

"Racial" organizations came under greater scrutiny. John LaFarge warned in 1943 that "a church is not welcomed, no matter how attractive in itself, which is suspected of being a device for discouraging attendance in other churches."[54] Boston archbishop Richard Cushing, after numerous requests from the city's African-American Catholics, agreed in 1945 to found a parish specifically for them, "as the Italians, French, Syrians and other groups have for theirs within the city." Cushing specifically noted, however, that "This new church is not established in accordance with any concept of 'segregation.'" Even that statement did not forestall a bitter retort from the city's African-American newspaper. "Now that the colored people of Boston have won several fights against Jim Crow clubs and places of entertainment in the last few months," one columnist queried, "will this Roman Catholic Church undo this progressive trend by extending segregation?"[55] Members of the Chicago chapter of the Federated Colored Catholics voluntarily dissolved into the local Catholic Interracial Council in 1945 after reaching "the conclusion that the effectiveness of its program lies in its affiliation with Catholics of all races."[56]

Catholic liberals also orchestrated a surge of educational activity. Programs included lectures on the Mystical Body of Christ and summer school classes for priests aimed at securing "the integration of racial groups into the life of the civic and religious community." One course syllabus warned that Catholic action on racial issues was imperative given the "immediate threat to our society and to religion."[57] Catholic colleges hosted debates on topics such as "The Encyclicals and the Negro," with

Catholic faculty outlining the applicability of papal writings to racial questions, and NAACP representatives answering questions from Catholic college students about contemporary racial issues. The shortage of papal writings explicitly addressing the American racial situation proved no obstacle; activists simply drew what they perceived as logical conclusions from the social encyclicals. At one Providence College session, students confidently resolved that "the policies of the Encyclical [*Quadragesimo Anno*], fully carried out, would completely solve the problems confronting the Negro in America."[58]

By the mid-1940s, then, Catholic interracialists and their supporters claimed the moral high ground. Nervous southern bishops might prevent a public endorsement of the Fair Employment Practices Commission, but high-ranking clergy knew that the apostolic delegate, on Vatican instruction, checked the racial views of all candidates for the episcopacy. "The Apostolic Delegate," commented one well-placed cleric, "is 'indignant' over the situation of the Negro and our indifference to it."[59] Following a decade of somnolence, the National Catholic Welfare Conference's Social Action Department sponsored a conference on the issue in 1946. Their conclusion—that only "integration into the life of the community" could solve the problems of African-Americans—echoed the developing orthodoxy.[60]

Most remarkably, Catholic liberals started to term segregation a sin. To be sure, this position marked the edge of the discussion, not yet its center. New Orleans archbishop Joseph Rummel, for example, formerly the pastor of a parish in a racially changing neighborhood in New York, still emphasized in 1945 how "even in the northern sections of our country, where segregation is by no means compulsory, colored groups lead to a considerable extent a life apart, seemingly by preference."[61]

Nonetheless, warnings about "racial clannishness" began appearing in the Catholic press.[62] Jesuit George Dunne's widely reprinted 1945 *Commonweal* article struck the most important blow. Dunne wrote in the immediate aftermath of the integration of St. Louis University, where he resided as a member of the Jesuit community until his advocacy of immediate racial integration provoked a sudden transfer. While the school eventually became the first "white" southern university (Catholic or non-Catholic) to admit significant numbers of African-American students, this development occurred only after a public dispute.[63]

Dunne compared segregation to slavery, arguing "that one hundred years from now the Christian conscience will repudiate with equal decisiveness the whole pattern of racial segregation." Assuming that African-Americans might lower home values in a neighborhood simply because of their race, Dunne maintained, was to ignore Vatican injunctions to "repudiate racism in all forms." Technical discussions of rights obscured a moral imperative. "We can go to hell for sins against charity," he concluded, "as easily for sins against justice, perhaps more easily."[64]

Other Catholics echoed Dunne's conclusions. Detroit Catholic interracialists specifically recommended that "Catholic Negroes should be accorded full and equal (not separate) membership in the Catholic churches of America." The Detroit women's group, after discussing Dunne's article, resolved to begin working to end segregation in local Catholic schools, and African-American parents in Washington, D.C., enclosed copies of the article in letters to the local bishop when they encountered discrimination.[65]

IV

Skirmishes over the precise Catholic position on "race" and segregation were consequential. Perhaps more than at any previous moment, Catholics and their institutions played important roles in all of the northern cities; their response to the African-American migration would help determine the future of the urban North. The problem was not simply neighborhood transition, especially since the virtual cessation of private-home construction during first the Depression and then the war inadvertently stabilized urban neighborhoods. Instead, tension centered around tentative efforts by federal and local government in the early 1940s to house African-Americans streaming into the already overcrowded ghettoes of the northern cities. Possible locations for these housing projects inevitably included heavily Catholic neighborhoods, a situation that sparked a series of public controversies. As early as 1941 federal housing officials were appealing to representatives from the National Catholic Welfare Conference for assistance. One government representative delicately referred to the "growing hostility toward the Church among the Negro people because of the intemperate and unjust attitude of some priests who have led opposition to the housing programs in certain cities." Similarly, African-American scholar Charles S. Johnson emphasized in 1943 the importance of resolving "inter-minority conflicts involving, particularly, Negroes and Poles and Irish Catholics in such northern cities as Detroit, Philadelphia, Chicago and Buffalo."[66]

Incidents occurred across the country. In South Bend, Indiana, a local priest led efforts to prevent an integrated public housing complex from being located on the city's heavily Polish West Side. The pastor of South Bend's one African-American parish wrote to his bishop complaining that

African-American Protestants ridiculed his parishioners for belonging to a church that would tolerate such behavior. The bishop replied that a conversation with a man familiar with such situations, Archbishop Mooney in Detroit, had convinced him that a housing project would lead to property depreciation in the neighborhood. In such a situation, "if the coming of the colored people caused others to move, the pastors would be left with a big problem on their hands."[67]

In Chicago, the African-American press sharply criticized Father Luigi Giambiastiani for leading the (ultimately futile) opposition to the inclusion of African-Americans in the Cabrini homes on the city's North Side. While Father Giambiastiani maintained that "Separation of the two groups while not the ideal theoretical solution, is the only practical road to community brotherhood," the *Chicago Defender* observed that "The church is a power in the community. But not a power for unity, nor for equality."[68]

The situation in Buffalo was unusually instructive. As late as 1927, researchers had commented that the city's Poles merely expressed "indifference" to the city's tiny African-American population.[69] By the early 1940s, however, Poles on the city's east side, led by Monsignor Alexander Pitass, president of the Polish clergy association, had organized successful crusades against a proposed housing project for African-Americans.[70] One newspaper account indirectly highlighted the contribution of parish groups by observing that while most letters came from "individual citizens," protest telegrams from SS. Peter & Paul Church, the Mothers' Club of SS. Peter & Paul, and the Polish-American Clergy Association were treated as representative of the larger community.[71]

An even greater furor was created by a federal proposal to place an African-American housing project in the heart of heavily Irish South Buffalo. Two prominent local priests, pastors of large parishes in the area, led neighborhood rallies and represented the area at emotional City Hall meetings. (At the height of the dispute, Buffalo's mayor conceded that the "town is in an uproar.") Monsignor John Nash of Holy Family Church emphasized that "The Okell Street project threatens to destroy the work of 40 years to build up the South Park district. The right to protect our homes is as sacred as the right to defend our lives." Father David Coughlin of St. Ambrose made a similar argument: "Of course the Negro workers should be taken care of but this is not the place. . . . Execution of such a project would tremendously reduce the value of homes in this neighborhood, homes which represent the life savings of many of our parishioners."[72]

Tensions in Detroit were the most severe. An estimated 60,000 African-Americans moved to Detroit between 1940 and 1946, and the number of

African-Americans in the city increased by 41.5 percent. To help ease the subsequent housing shortage, the federal government planned a housing project for African-Americans, the Sojourner Truth Homes, on Detroit's north side. The fiercest resistance came from the Poles living in nearby St. Louis the King parish.[73] "The conflict," concluded an investigator for the Office of War Information, "could better be called a Polish-Negro conflict, or a European vs. Negro conflict than a white-black conflict." Grocery store notices, handbills, a newsletter, meetings in the parish hall, and Sunday sermons by pastor Constantine Djiuk united the community in opposition to the proposed housing plan. Djiuk and his associates coordinated the formation of a local improvement association and put pressure on local politicians to renounce their support of the project. They were successful. After an emotional meeting at a packed City Hall which included testimony by Djiuk and two other priests, local politicians wired Washington that they were withdrawing their support of the project. One councilman, as he did so, affirmed that he was "a Catholic first, last, and always." The first attempted move-in of twenty-four African-American families resulted in a full-scale riot as hundreds of whites roused by automobile horns confronted equally angry African-Americans. The neighborhood hostility, combined with policemen sympathetic to the rioters, prevented occupation of the units.[74]

St. Louis the King was a prototype of the tightly bound parish community. In a 1940 parish census, Father Djiuk estimated that only 2 percent of parishioners missed mass on a given Sunday. The 224 children in the elementary school received instructions from an order of Polish nuns. At four of the five packed Sunday masses in a church that seated over six hundred people, Djiuk gave the homily in Polish, since, as he put it, "the Poles generally like to hear Polish spoken, even those who are born here and whose parents were born here. And I like to have them keep up with their Polish." Parish albums recall that the "1930s were bleak years. Those were the days of depression, unemployment, material want and consequently, times of poverty and indigency." The benevolently remembered Djiuk visited each family twice a year to check on their religious status and ask for contributions. During the late 1930s, Djiuk repeatedly emphasized the need to "buy and own your own home."[75]

Djiuk's primary argument rested upon the need to protect his community. While he occasionally lapsed into private warnings about the safety of young women near African-American males, Djiuk publicly stressed his willingness to "take up collections for the negroes in his church" and support other locations for the housing project.[76] The key issue, he pointed out in a letter to the Division of Defense Housing coordinator, Charles F.

Palmer, was neighborhood stability. ''This [decision] would mean utter ruin for many people who have mortgaged their homes to the F.H.A. . . . May I feel that I have found a friend in you, Mr. Palmer, and that the many sleepless nights that I have spent in trying to ward off this future danger to my parish and citizens will not prove in vain.''[77]

This fierce protection of community extended beyond opposition to African-Americans. In one ''instruction'' at a Sunday mass attended by a local researcher, a curate used the parable of the Good Samaritan as a launching pad for a discussion of how ''God finally wearied of trying to tell them [Jews] anything and left them to work out their own salvation.'' In an interview, Djiuk commented that ''The whole history of the Jews everywhere has been one of ingratitude for what they receive.'' He continued, ''And now here in Detroit, we have the Jews and Niggers making a combination [in city elections]. . . . they cheat the Niggers worse than they used to cheat the Polish peasants who trusted them.'' The only solution, he concluded, ''is for each family to have and to own its own home.''[78]

Catholic liberals observed these events with dismay. John LaFarge confided to John Ryan that ''Unfortunately there is very painful evidence about the Polish clergy. . . . I have not heard about Italian clergy, but knowing how some of the foreign clergy behave in these neighborhood questions I should be by no means surprised if some of them took an equally unjustified attitude.''[79]

Detroit's Catholic leadership refused to address the issue. In an interview with auxiliary bishop Stephen Woznicki, responsible for Detroit's thirty-five Polish Catholic parishes, researchers learned that ''He likes to have Catholic people live in Catholic communities. These Catholic communities of single family homes must include a Catholic church, a Catholic school, and facilities for recreation that meet Catholic standards. The Bishop added that these are his specifications for the Poles.''[80] Louis Martin, an African-American Catholic newspaper editor, led a delegation to archdiocesan offices to discuss the matter but received little sympathy. The archdiocesan chancellor informed the group that ''priests had responsibility for the financial well-being of their parish as well as the spiritual well-being'' and refused to comment on Martin's assertion of a ''Catholic conspiracy'' due to the united front presented by clergy and Catholic city officials.[81]

NAACP executive secretary Walter White warned of ''a tremendous feeling growing on the part of Negroes against Catholics in Detroit,'' and NAACP pamphlets specifically attacked ''Polish Catholics.'' Even the gentle request of John LaFarge for a statement by the bishops on interracial justice met with minimal success. The administrative board of the National

Catholic Welfare Conference did urge Catholics to "create a neighborhood spirit of justice" in "many of our great industrial centers [where] racial tensions exist," but calls for a more wide-ranging statement were rejected. While supportive of what he called "a positive declaration" in the field of human rights, Archbishop Mooney argued that a more specific statement might "complicate a problem which is by no means as simple as it is sometimes assumed to be."[82]

National rivalries continued to paralyze the church structure. Following a conversation with Archbishop Mooney, LaFarge added that "He seemed somewhat obsessed with the Polish problem, which I suppose is natural enough." Mooney recounted his recent meeting with two representatives of the NAACP. "They were both apparently naive in their belief in the effectiveness of a word from me. . . . Feeling on the subject runs high among the Poles. They are by temperament never phlegmatic, and in fact just now they are heavy-hearted over what has happened in the homeland and in Russia." Mooney added, "I must in conscience consider that any declaration of mine which might have a general apologetic value for the Church among the Negroes would most certainly have a disastrously disturbing effect on the more than two-hundred thousand Polish Catholics who are a large part of my direct responsibility."[83]

V

And yet, the balance of rhetorical power in these disputes was clearly shifting. In Buffalo, Catholic priests and parishioners successfully prevented the construction of defense housing for African-Americans within their parishes. Nonetheless, African-Americans in the city quickly claimed the developing language of interracialism, with one African-American reminding Buffalo priests of "the words of His Holiness Pope Pius XI when he condemned Mussolini for saying that the Italians were superior to the Jews."[84] In 1946, the decision by the local bishop to close the city's two African-American mission parishes (in order to spur integration) sparked triumphant headlines—"Catholic Churches Amalgamated!" and "Catholics Set the Pace"—in the local African-American newspaper.[85] The editor of the paper, while acknowledging that some African-American Catholics were uneasy about the abandonment of their own parishes, applauded the diocesan commitment to "absolute equality in human brotherhood."[86]

The outcome in Detroit, as Dominic Capeci argues, was foregone as soon as the issue moved out of local hands. Eleanor Roosevelt became personally involved, writing Archbishop Spellman of New York and observing that "it [the problem] is laid to the Church very largely. They tell me the same situation has come up in Buffalo." Playing on patriotic symbols, African-Americans in Detroit coordinated an interracial campaign that ensured occupancy of the Sojourner Truth Homes within a few months. ("[T]he Negro," noted one African-American, "was a part of this country before the ancestors of these people ever heard of America.") African-Americans in the Roosevelt administration assisted in an intense publicity campaign which made the logical comparison between tyranny abroad and segregation at home.[87] In Detroit, African-American Catholic Louis Martin lambasted the priests opposing the housing project and their supporters in an editorial entitled "The Ungodly." "These three Nazi-minded priests," Martin wrote, "have damaged the Catholic faith and have struck democracy a foul blow." He added that the "Catholics have abused their religion and resorted to measures which become only the Nazi heathens" and compared them to "greedy Hitlerite stooges."[88]

In addition, liberal Catholics appalled by the behavior of their coreligionists joined the civic coalition. Father John Coogan of the University of Detroit began giving sermons on interracial justice (which at times were "flatly contradicted" by other priests) because "I believe we are looked upon as one of the most bitterly reactionary elements in the Detroit situation, far behind the Protestants and the CIO." The American Catholic Trade Unionists sent representatives to rallies endorsing African-American occupancy, and two members of the Detroit Catholic Worker's group attempted to explain Catholic teaching to residents milling around the besieged housing project.[89]

"Catholic teaching," of course, was precisely at issue. At the Sojourner Truth Homes, Catholic liberals debated opponents of integration who still explained "their position to us as fellow-Catholics." One problem liberals rarely acknowledged, in other words, was their own tendency to view social teachings as immutable. A typical *Interracial Review* editorial lauding Vatican attempts to combat racism declared in 1938 that "In this teaching of the Church there is nothing new, just as there is nothing new in any teaching of the Church."[90]

To the contrary, asking all Catholics to follow appropriate teachings on race relations remained an ambiguous request. The confrontations of the early 1940s are vivid reminders of the power of an older idea of Catholic racialism and community even as a new theology and approach to urban problems took hold. Fourteen years prior to the Sojourner Truth Homes riots, a liberal Catholic theologian had written that "in such matters as the selling and renting of houses considerable freedom is permitted by the theologians." After a few caveats, he added, "A group of citizens might, without violating the rights of the Negro, exhort other neighbors privately not to sell.''[91] This emphasis on the rights of property owners and groups supported the more general notion that a parish marked a sacred space in the city as well as identifying a people.

Upon such a foundation was built the Catholic world of the urban North. By the 1940s, however, the structure was weakening. A full generation had passed since the last wave of European immigration, a time during which the trials of the Depression, often in conjunction with membership in industrial unions and the Democratic party, provided common experiences for people accustomed to defining themselves as distinct.[92] The African-American migration and the intensity with which Euro-American Catholics of all backgrounds would defend parish neighborhoods marked a second stage in this process, providing another common experience while placing a different set of ''racial'' barriers—those separating ''black'' and ''white''—at the front of the Catholic imagination. Wartime service in units composed of soldiers from all European nationalities, with only African-Americans segregated into separate units, sharpened these distinctions for both parishioners and priests serving as chaplains.[93]

When an African-American member of the local CIO steel workers' union stood up at one South Buffalo rally and asserted that as an American citizen he should be allowed to live anywhere in the city, the heavily Catholic audience shouted him down. Presumably members of the crowd also belonged to the steelworkers' union, and certainly many in the audience were blue-collar workers. Their message was that African-Americans at the workplace might be tolerated, but not a housing project that seemed to threaten the most important achievements of a ''white'' Catholic population.[94]

Ironically, from the moment of its full articulation in the 1930s and 1940s, this particular biracial framing of housing issues faced formidable opposition. In Washington, the logic of the fight against the Axis powers enabled liberals to knit together American nationalism and integration, a development that would become even more clear in the context of the cold war. In Rome, and among Catholic liberals more generally, the legacy of the 1930s was the conviction that a universal church must work to eliminate all racial categories, both those dividing European national groups and those separating ''white'' from ''black.'' Much like the African-Americans jamming the train cars north, the church constructed so painstakingly by successive waves of European immigrants found itself in the midst of dramatic change. A more fractured Catholic community had emerged in the midst of a more unified nation.

FOUR
NEIGHBORHOOD TRANSITION IN A CHANGING CHURCH

I

Formally, Catholic life after the war moved along familiar paths. Nuns still prepared boys in bow ties and girls in white dresses for first communion, parishes sponsored dances and dinners, altar boys struggled to learn Latin responses. The Catholic world remained encompassing. Half of all Catholics claimed to live in neighborhoods where 50 percent of the residents were Catholic. Growing up in New York City, author Mary Gordon observed that "Until I went to college I had no genuine contact with anyone who wasn't Catholic. . . . Real life, the friendships, the feuds, the passions of proximate existence, took place in the sectarian compound." Sociologists marveled at the schools and hospitals, associations of Catholic doctors and Catholic lawyers, Catholic war veterans and Catholic philosophers, even Catholic audio-visual educators and Catholic librarians.[1]

The strength of the Catholic subculture, however, concealed a historic transformation in the various Catholic communities. In retrospect, the end of mass immigration in the 1920s and the lack of economic mobility (at least upward) during the Depression imposed a false sense of stasis on Catholic life; a notion that the sense of order imposed on heavily Catholic, urban neighborhoods by priests, nuns, and politicians would endure indefinitely. Following the war, while a world of parallel societies remained, and even flourished, much of what made the Catholic experience of the early twentieth century distinctive faded into the larger American kaleidoscope. What one worried Jesuit sociologist termed the "time and space ignoring solidarity" of American Catholicism seemed suddenly vulnerable as a largely working-class community developed an extended class structure, a highly urban population joined in the exodus to the suburbs, and different national traditions blurred into a denominational religious culture. Symbolized by the election of John F. Kennedy in 1960, these changes were already creating a different American Catholicism before the twin shocks of the Second Vatican Council and the cultural upheavals of the 1960s.[2]

The foundation for change in the various Catholic communities came from the extraordinary growth of the American economy. Perhaps more than any other group, Catholics used the extended boom during and after the war to move into the expanding American middle class. Surveys of religious groups done in the late 1930s and early 1940s showed Catholics (especially Polish and Italian Catholics) lagging behind national income averages. (One study noted how "precisely in stratification" northern Catholics compared to southern Baptists.)[3] Irish Catholics were more secure, but even the Irish, particularly those located in urban enclaves, only gradually climbed the occupational ladder. Stephan Thernstrom notes that as late as 1950 80 percent of Boston's Irish immigrants still worked with their hands, and he points to a "distinctive Catholic mobility pattern, a skidding syndrome in which youths who started with a foothold in the white-collar occupational world were unable to maintain it for long." A study of a largely Irish parish in the northeast in the late 1930s is also revealing. Almost half of the parish members reported some unemployment during the Depression. Only two of the parish's students were reported to have gone on to higher education, and both attended a local Catholic college. Most of the women who worked were either telephone operators or domestic servants; half the men were skilled or unskilled laborers.[4]

The war changed all this. Located in fast-growing metropolitan areas, and prime beneficiaries of both higher wages in industrial unions and the explosion in lower-management positions, Catholics matched or surpassed national income and education averages by the mid-1960s. (In Connecticut, priests interested in labor issues complained of their inability to attract newly affluent workers to meetings, grousing that the workers seemed more interested in vacations and new televisions.) College enrollments across the country skyrocketed after the war, but the percentage of Catholics advancing beyond high school increased faster than the national average. In Catholic colleges alone, 92,000 students registered in 1944–45, and 220,000 three years later.[5]

More subtle alterations centered around the waning of European traditions in a population now often two generations distant from immigration.

(*facing page*) 1945 Homecoming dance at St. Margaret Mary parish, Detroit. Archives of the Archdiocese of Detroit.

Older Catholics recalled heated parental arguments when an Irish boy and Italian girl discussed marriage, but studies now suggested increasing rates of intermarriage within the Catholic tradition. Will Herberg's influential 1955 study, *Protestant, Catholic, Jew,* expanded upon this observation. Focusing on national origins, Herberg argued, made observers miss the "ethnic amalgamation" that helped create a "new type of American Catholic."[6] Priests dropped the final mass in German, while nuns discovered that students found it difficult to memorize catechisms in languages other than English. Pastors who had immigrated with their parishioners in the early part of the twentieth century—and whose sensibilities seemed strangely out of place in mid-century America—retired. Their replacements, trained in American seminaries, quietly made the parish "up to date."[7]

Barriers between various national groups weakened. Community organizers in Chicago's Back of the Yards area recalled how in the late 1930s priests from the various national parishes rarely addressed one another on the street, and refused to cooperate in community affairs. In 1945, all priests in the area made a common retreat. "Only now," commented one cleric, "are we beginning to feel through community cooperation that the word Catholic means catholic, the church universal."[8] Following their extended probe into the folkways of Newburyport, Massachusetts, Lloyd Warner and Leo Srole concluded that "the future of American ethnic groups seems to be limited; it is likely that they will be quickly absorbed." Sociologists forecast the rapid decline in "numerical strength and importance" of the national parish.[9]

Of course the process of assimilation also contained ideological components. As with "racial" parishes for African-Americans, liberals now viewed national parishes as an unfortunate aspect of American Catholic history. Interracialists conceded that "all minority groups tend to segregate themselves" but argued that "this attitude is destructive of the unity which should distinguish all Catholics."[10] Chicago's Archbishop Stritch refused the request of one Italian national parish for increased financial support in 1945. "It is a fact," he wrote, that "national groups are passing phenomena in our cities. In time these groups no longer use their own vernacular and they become part of the general Catholic faith. We should not make the mistake of building chapels not urgently needed which will in time become problems."[11]

The power of this integrationist ideology was most evident in the response to Puerto Rican and Mexican immigrants. Bishops confronted with Latino migrants to the northern cities in the early part of the century typically permitted the formation of national parishes, since experience with the larger Polish and Italian migrations demonstrated the desire of immigrants for distinctive congregations. In the postwar period, however, at the exact moment when the Puerto Rican and Mexican migrations accelerated, bishops and Catholic intellectuals frowned upon this sort of "segregation." One 1955 clergy conference report noted that while "the sentiment of the Puerto Rican people themselves" on the matter remained unclear, "the establishment of new national parishes was not advisable." Much better, the assembled clergy concluded, "to begin the process of integration from the very beginning." A few priests defended the national parish as a way to integrate the faith and culture of "the uprooted," but such arguments failed to sway conference organizers, supported by New York's Cardinal Spellman, who considered integration the ideal.[12] Work in Chicago followed a similar pattern. According to one local organizer, Cardinal Stritch argued that "National Parishes have sometimes in the past prolonged unnecessarily the transition period, thus delaying or even avoiding integration into the territorial parishes." Territorial parishes should attempt to attract Spanish-speaking immigrants "for the purpose of eventual integration."[13]

Integration of the various nationality groups within the Church mirrored the integration of Catholics into the broader society. By the end of the 1940s, the anti-Catholicism that once helped fuel the Ku Klux Klan had largely disappeared from public view. As early as 1948—twenty years after Al Smith's presidential campaign—respected commentators could observe that "religion *by itself* [was not] of crucial importance in provoking the tensions and cleavages manifested in the everyday relationships of American society."[14] Intellectuals still warned of "Catholic Power" but ordinary Catholics rarely encountered hostility. Public images were largely favorable—the pope giving daily blessings to American troops during the war, Bishop Sheen preaching on television, Bing Crosby in *The Bells of Saint Mary's*.[15]

Even more dramatic was the literal integration of American Catholics with the other nine million Americans leaving the northern cities for the suburbs during the first decade after the war. ("More Catholics moved last year to suburbia," complained one Catholic writer, "than arrived from Ireland in any decade of the 19th century.") To be sure, Catholics seem to have been less likely than than their non-Catholic neighbors to leave old neighborhoods, and a Catholic birthrate 40 percent higher than that of white Protestants or Jews in the northern cities masked the changes by keeping many city churches full. In addition, booming parish and school enrollments in areas within city limits that had been relatively undeveloped before the war (Staten Island, Queens, Chicago's Southwest

Side, Northeast Philadelphia) suggested a propensity to make short hops within the city instead of immediately moving into the suburbs.[16]

Still, as Andrew Greeley concluded in 1959, growth in the "somewhat isolated" world of "national parishes and old neighborhoods" lagged far behind the boom outside the cities. Proof of the exodus came from the endless pictures in the diocesan press of bishops holding ceremonial shovelfuls of dirt and blessing the four parishes built in each week across the country during the early 1950s. Boston priests boasted of the seventy-five archdiocesan churches built from 1944 to 1959, many "in small historic communities in the midst of which no church steeple bore the cross."[17]

Changes in Catholic spiritual life echoed this geographical and cultural integration. Especially in the suburbs, parking lots replaced processions and upwardly mobile parents demanded schools before even the church building itself.[18] For the most part, spiritual practices remained local and communal, with the context of the parish shaping formally religious events. A characteristically Catholic spiritual innovation of the late 1940s was the "block rosary" in which neighbors prayed the rosary together each evening, timing their prayers so as not to "conflict with parochial services at church." That the goals of the prayers were world peace and the conversion of Russia, however, suggests international concerns; similarly, the popularity of prayers and rosaries over the radio, as well as the new prominence of "national" shrines, hinted at a waning emphasis on the neighborhood.[19]

II

African-Americans moving to the North thus encountered an institution and set of Catholic peoples in the midst of a momentous transition. What became clear during the 1950s was that while the migration to the suburbs possessed an independent momentum, its connections to racial issues were profound. In fact, the most obvious distinction between suburban Catholicism and parish life within the cities was the relative importance of racial concerns. Given the legal and economic barriers preventing African-Americans from purchasing homes outside the city, racial transition was an abstraction to most suburban Catholics.

For Catholics in city neighborhoods, however, the stakes remained high. To these parishioners, the suburban boom also meant the possibility of expansion for the African-American ghetto as the postwar housing crunch eased. Exactly at the point of triumph, having weathered the Depression, built the school, and finished the church, Catholic parishes in the northern cities confronted the possibility that generations of painstaking work might be rendered obsolete in a few years. As Monsignor John O'Grady of the National Council for Catholic Charities noted in 1955, this fear marked parish life in "all our city parishes, even the parishes in the outlying sections of the city."[20]

Church leaders in the cities receiving the largest African-American migrations also understood by the 1950s how the phenomenon might impoverish the institution as a whole. No other denomination possessed such numbers in the northern cities; no other organization faced such financial risk. Maintenance costs on enormous, recently constructed physical plants in the city continued to soar even as parish membership dropped precipitously. The Buffalo diocese loaned $346,000 to one hard-pressed inner-city parish in less than a decade; a neighboring parish saw annual income fall from $36,000 to $10,000 in twenty years.[21]

At the same time, demand for new facilities placed added pressures upon diocesan coffers. Refugees from racial change in the West Philadelphia parishes of Most Blessed Sacrament, St. Francis de Sales, and Transfiguration, for example, pushed the population of St. Denis parish, in suburban Delaware County, from 1,000 families in 1944 to 2,532 in 1963. Pressure from parishioners to build a new school and church resulted in parish debt leaping from zero to over $300,000 and annual expenditures multiplying eight times. "The new parishes," estimated one city planner, "are now costing two to five times as much as the perfectly satisfactory buildings which are deprived of their value solely through the process of neighborhood deterioration."[22]

Awareness of these figures provoked Buffalo bishop Joseph Burke to distribute a pamphlet on neighborhood conservation to thirty-one city pastors, observing in a cover letter that "certainly it would be to the advantage of the diocese to maintain the Central City parishes rather than to have to build more and more expensive churches and schools in the suburbs."[23] Chicago's Stritch warned his clergy that "As things are going now, we are building a new archdiocese on the perimeter of Chicago." In Detroit, priests bitterly joked that they had built the archdiocese one thousand times.[24]

Concerns about racial change ushered in a new generation of Catholic liberals. Instead of the depression-era priest marching alongside a union organizer, the typical Catholic liberal of the 1950s was an "intergroup relations" professional, often a layperson, and interested in a broad range of social and theological issues. Embarrassed by the reluctance of Church officials to initiate programs on racial matters, these liberals pressed for educational efforts, scholarship programs, and work in "changing" neighborhoods. In Philadelphia, Dennis Clark and John McDermott (later executive secretary of the Chicago Catholic Interracial Council) used their positions with city housing and human relations agencies to advocate a program of community meetings. "Surely," they argued, "pastors who administer parishes that can be gravely affected by neighborhood changes should be alert to techniques such as this one." Despite being "barely tolerated," Hope Brophy began working through the Detroit Archdiocesan Council of Catholic Women to develop education programs, since Catholics were "such a substantial part of Detroit" and "it [had begun] to dawn on me that there was a moral dilemma and issue involved." The Catholic Interracial Council in Chicago received a dramatic boost when Sargent Shriver became chairman of the board in 1954. Interested in theological issues and connected to the city's business community through his position as manager of the Merchandise Mart for the Kennedy family, Shriver and the CIC staff organized well-attended benefit shows, scholarship programs for African-American students, and ceaseless rounds of educational seminars for nuns, priests, and lay leaders.[25]

From this work emerged a national organization, the National Catholic Conference for Interracial Justice (NCCIJ), in 1958. More than 400 delegates representing 36 Catholic Interracial Councils resolved "as Catholics and Americans" to "fully implement the principles of Christian Social Justice and American Democracy in regard to race relations." Conference organizers conceded the "scandalous" persistence of segregation in hospitals, parishes, and schools and pledged to work for its elimination. With Mathew Ahmann, a sociologist and Catholic activist as executive director, supporters of the NCCIJ hoped to develop an organization capable of addressing national racial questions as well as fostering local Catholic leadership.[26]

These same Catholic liberals welcomed a noticeable increase in support for racial integration among clergy and the hierarchy. Into the mid-1940s, moral theologians working from neo-scholastic premises still argued that the "common good" did not necessarily require the integration of all Catholic institutions.[27] In 1945, for example, when queried as to whether Jesuit schools should accept otherwise qualified African-American applicants, theologian John Courtney Murray favored integration as a matter of charity, but expressed skepticism that "the Negro's right to a Catholic education . . . create[s] a right to get it in association with white boys and girls in the same school." Eleven years later, when New Orleans archbishop Joseph Rummel publicly denounced Louisiana parishioners unwilling to allow an African-American priest to celebrate mass, Murray came to different conclusions. The moral sinfulness of segregation, he informed one critic of Rummel's decision, was "entirely clear."

Speakers and delegates at the inaugural meeting of the National Catholic Conference for Interracial Justice. From left, Sargent Shriver, president of the Chicago Catholic Interracial Council; John LaFarge S.J.; Bishop Emmanuel Mabathoana, O.M.I., of Leribe, Basutoland; Archbishop Owen McCann of Capetown, South Africa; Monsignor Daniel Cantwell, chaplain of the Chicago Catholic Interracial Council. Archdiocese of Chicago, Archives and Records Center.

French theologian Yves Congar also argued that "The idea of race . . . has no place in theology, missiology, pastoral theology, or canon law."[28]

Catholic interracialists could count on support in Rome, including front-page editorials in the Vatican newspaper favoring Archbishop Rummel's actions in Louisiana. In turn, interracialists warmly welcomed native African bishops on tours of the United States, and ensured that the the naming of an African-American priest as bishop of Accra, in the Gold Coast, would receive extended coverage in both the Catholic and African-American press.[29] Liberals watched developments in South Africa with especially keen interest. As in the United States, the South African bishops of the 1950s inherited a tradition of racial segregation in church institutions. In South Africa, however, Catholic teaching became seen as flagrantly contradicted by the explicitly racialist policies of the Nationalist government. (In part, this occurred because over 70 percent of the members of the South African Catholic Church were already of African descent.) Beginning in 1952, the South African bishops released a series of increasingly defiant statements attacking apartheid and racialist notions of white supremacy. Like their American counterparts, South African bishops drew upon papal attacks on racialism, and warned of "communism and unbelief" emerging from a refusal to confront the "fundamental opposition between Catholicism and apartheid." In 1957, one Vatican source made the connection between America and South Africa explicit, noting that "the Church is completely and unalterably opposed to all forms of discrimination—in New Orleans as much as in the Union of South Africa."[30]

With increasing frequency, priests dared their communicants to welcome all newcomers to heavily Catholic neighborhoods. Upon learning of the movement of African-American Catholics into heavily Catholic areas, a few priests sympathetic to the interracial cause placed their imprimatur upon integrated neighborhoods by performing highly conspicuous blessings of homes, even as movers unloaded a family's furniture.[31] In 1943, a representative of the Cleveland real estate board could casually ask Bishop Hoban for his assistance in blocking Federal housing for African-Americans, adding "that many of the priests [in the area] have attended meetings that have been held trying to combat the idea of locating such a development there." Ten years later, Cleveland residents attacked the home of the first African-American to move into an area of the city estimated to be 70 percent Catholic. Diocesan authorities quickly published pastor Monsignor John T. Ruffing's expression of disapproval. "I can't help it if they don't like the truth," Ruffing announced, "the parish is rapidly changing, and I knew we'd have to face this thing sooner or later.''[32] In Detroit's St. Leo parish, Fr. Hubert Roberge, inspired by visits to the South during the war, worked to create a genuinely interracial community. Roberge forcefully integrated parish societies, worked with lay women volunteers on a program of home visitation, and brought in African-American priests during church holidays. ''The salvation of this parish,'' Roberge told the archdiocesan chancellor, ''depends on the conversion of the people living within its confines.'' Roberge refused to tolerate discussion of segregation among his parishioners. The authority of the church, he concluded, ''resides in the hierarchy and the clergy. If the authority speaks forthrightly, the mass of the congregation will follow.''[33]

Reaction to student strikes in Gary, Indiana, also reflected new awareness of the need for Catholic support of integration. The first strike, organized in 1945 by white students at Froebel Public High School, resulted from anger at a school board decision ensuring greater numbers of African-American students at the school. Three Catholic parishes were located near the school building, which was situated in a racially changing, but still heavily Catholic, area. (Much of the student and parent disgruntlement stemmed from the fact that schools in more affluent areas would enroll virtually no African-Americans under the new plan.) Local school and NAACP officials immediately blamed Catholic priests for covertly supporting the disorder, hinting at a Catholic plan to boost enrollment in parochial schools as well as what NAACP leader Walter White termed Catholic fears of ''parishioners moving to other sections of the city.''[34] Catholic leaders denied the charges, but only after some hesitation did local priests publicly announce support for nonracial enrollment policies.[35] Two years later, however, following another student strike, Gary priests immediately termed defiance of the school board ''undemocratic and un-Christian.''[36]

Episcopal attitudes also evolved. Despite encouraging all-white neighborhood improvement organizations in the early 1940s, Detroit's Cardinal Mooney eagerly endorsed plans to make St. Leo's parish a model of integration and showered praise upon priests working in the African-American community.[37] In Chicago, Cardinal Stritch placed increasing pressure on Catholic schools to accept African-American applicants and quickly, if privately, eliminated dissent among the clergy. In a 1956 sermon, for example, Father Owen Mattimore compared the forced ''flight of the Hungarians'' to the ''unchristian cold war of nerves'' compelling his long-time parishioners to leave his South Side neighborhood. ''Now, Father,'' Stritch responded, ''you know the teaching of Holy Church and you know your theology. . . . If in the changes which have come, there have come into your parish the Negroes, then this is the work of the parish to bring this population to Christ.'' Stritch forced Mattimore to integrate his parish school with the daughter of an African-American Catholic schoolteacher and encouraged him to consider ''an appointment that will be more agreeable to you.'' Shortly thereafter, Mattimore ''announced from his pulpit that he wanted every Negro Catholic child in his parish to register for his school.''[38]

By the end of the decade, the interracialism first voiced in the 1930s organized all public discussion of racial issues. Without fear of disagreement, John LaFarge could state in 1959 that ''the ideological controversy, the acknowledgement of the principle of interracial justice—has been won. . . . the cause itself admits no further argument.'' Franciscans honoring St. Benedict the Moor explained that they they had dropped the phrase ''friend of the colored race'' in their literature because it implied segregation. Now, they admired St. Benedict because ''He was a Negro accepted without question into an all-white community.'' Following a 1958 fire that destroyed a church for African-American Catholics in Newark, the pastor of the church actually requested that the church *not* be rebuilt. ''Segregation,'' wrote Father Thomas Carey to his archbishop,''is an outmoded social and moral principle which brings untold consequences and injustice. It impedes the conversion of the Negro to Catholicism [and it]. . . . can be an excuse for some pastors to be delinquent in their obligation to the souls of the colored in their own parish.''[39] One southern bishop privately complained that his ''righteous'' northern colleagues now believed—''almost overnight''—that segregation was both sin and heresy.[40]

Proof that interracialism equaled orthodoxy came with the release of a 1958 pastoral letter from the American bishops. After the first NCCIJ conference (which included an address by Capetown, South Africa, archbishop Owen McCann), Father John Cronin, S.S., began lobbying the American bishops for a statement on race issues. In November, the bishops released a statement authored by Cronin entitled ''Discrimination and the Christian Conscience.'' In this, the first extended comment on racial issues by the episcopacy since the war, the bishops declared that ''the heart of the race question is moral and religious.'' Noting that ''no region of our land is immune from strife'' and rejecting compulsory segregation, the bishops argued that African-Americans, like European immigrants, would advance in society once the chains of segregation and prejudice had been lifted.[41]

Up until the last minute, the statement faced opposition from bishops uncertain of its effectiveness in the largely Protestant South. As one southern bishop noted, the problem could not be viewed from "only the theoretical angle" but must include a discussion of "the practice of the Catholic Church herself providing separate schools for the negroes with the permission of the Holy See."[42] Despite such objections, the southern bishops fell into line following the death of one prominent opponent and the secret request of Pius XII that the statement be "issued *at once*" one day before his own death in Rome. (The younger American bishops also expressed their support "in no uncertain terms.")[43]

As Cronin observed, "Even if the worst fears of some bishops were realized, and we lost some Catholics in the South, I still think that we would be far ahead arithmetically in a few years, to say nothing of the worldwide impact of our moral leadership." Even more important, Cronin noted, "Our Catholics in New York, Detroit, Chicago, and other major cities need guidance also. . . . [F]rom my experience, many of them do not know the principles of the Church on this matter."[44]

III

Confusion about the "principles of the Church" was indeed widespread. The release of the bishops' letter and the formation of the NCCIJ in 1958 were significant markers in the evolution of liberal Catholic thought on race, but appeals to universalist theological principles often fell upon rocky soil. Successful convert programs, public rejection of segregation, and declarations that Catholic institutions were open to Catholics of any race marked the formal relationship of the church to African-Americans. Comments such as those by one Michigan Catholic—"Segregation is not wrong just because you say it is"—and resistance to neighborhood change suggested an alternative, and yet still "Catholic," interpretation.[45]

Tension was evident across the country. African-Americans in Pittsburgh complained that "in those areas where there has been the most tension . . . the 'opposition' contained a high percentage of Catholics and in certain instances a few of the less responsible Catholic leaders, including the clergy." Brooklyn Catholics heard warnings from a young Daniel Berrigan against "stirring up racial hatred of a new housing center, to prevent a fine young family from having a decent home because they are Negroes."[46]

Liberal Catholics in Philadelphia moved throughout the city during the 1950s, attempting to prevent anxiety over African-American entrance from exploding into rioting. The reception was rarely enthusiastic. "I spoke to a community group yesterday on racially changing neighborhoods," noted

Dennis Clark, a member of the Philadelphia Catholic Housing Council in his diary, "Silence. Stone cold silence. The people were just intensely and silently aggravated by the movement of Negroes throughout the city. As I left I knew that they despised me because they believe I am 'against them.' "[47]

National attention centered on the Philadelphia area when the first African-American family attempted to move into suburban Levittown. Forty percent of the development was Catholic and close to half of these Catholics had moved directly to the area from inner-city Philadelphia neighborhoods. In an area already tense because of a slack housing market, the news that a home had been sold to an African-American war veteran provoked unruly crowds and boycotts of local merchants sympathetic to the family. The family endured this harassment and remained in the community, but local Catholic churches came under criticism for their silence during the dispute. As a report to Cardinal O'Hara concluded, "Catholics are much too prominent in the anti-Negro forces and conspicuous by their absence in the effort to restore harmony."[48]

This Catholic presence was even more vivid three years later in the city's heavily Irish and working-class Kensington neighborhood. There the mere threat of an African-American move-in brought several hundred area residents into the streets. Rocks quickly crashed through the windows of the alleged residence of the African-American family and neither the assurances of city government officials nor police could persuade the crowd that a move-in would not take place. Finally, a squad car arrived on the scene with Father Charles Mallon, pastor of Ascension parish. Since "many of the local people were members of this parish" the crowd quieted and then cheered as Mallon informed them that no move-in would occur. Only at this point, did the crowd disperse.[49]

The request of city officials to discuss with Cardinal Stritch the "problem of the movement of members of other minority groups into communities foreign-born and Catholic" in 1946 suggests the severity of the problem in Chicago.[50] Although heir to the the nation's most vigorous liberal Catholic tradition, the Chicago church also served as a defining institution for fiercely cohesive and heavily Catholic parish neighborhoods. Publicly, the news remained positive. Diocesan publications touted spiraling convert totals and the slowly increasing number of African-American priests and religious. Occasionally, to the delight of the local Catholic Interracial Council, pastors ensured that African-American newcomers were welcomed into parish life.[51]

Bitterness among both clerics and parishioners, however, was at least as common. One researcher noted the "number of instances in which members of various Property Owners Associations referred to the participation of local priests at least behind the scenes. In several cases, priests in white parishes bordering the main South Side Negro community were quite outspoken in their concern to prevent encroachment on their parish by Negroes.''[52] Pastor F. J. Quinn of St. Ambrose spent much of the late 1940s fulminating against the entrance of African-Americans into the Kenwood area in which the parish was located. According to one report, Quinn announced at eight a.m. mass that ''The Niggers have taken over Corpus Christi Church, Holy Angels and St. Ann's and they are now trying to take over this church; but if it's left to me, they will not. . . . Our forefathers from Ireland came over here and prepared the way for us in this church and the Niggers are not going to run us out.'' Shocked observers immediately warned Cardinal Stritch that ''We feel certain that his approach to the tense and perhaps explosive situation existing in this area is in diametrical opposition to the Holy Father's attitude as expressed in numerous statements.'' Stritch requested Quinn to ''keep strictly to Catholic doctrine'' in his homilies but not before Quinn asked in his parish bulletin ''How did they even find their way into the last beautiful neighborhood in Chicago?. . . . Sometimes we feel like Christ weeping over the City of Jerusalem. If we lose this, we have lost everything. Oh, my people, rally to the cause and save what is left.''[53]

Neighborhoods with higher Catholic populations occasionally witnessed a more organized response. Commenting on a series of racial disturbances that rocked the city's South and West Sides during the late 1940s, one Catholic Interracial Council staffer concluded that ''the mobs are entirely white, many of them Catholics. . . . It seems that it has come to the point where Catholics believe our Church condones and approves segregation.'' Catholic Interracial Council chaplain Father Daniel Cantwell noted with disgust the behavior of citizens in the area near St. Columbanus Church in 1949. After an African-American couple, Mr. and Mrs. Roscoe Johnson, purchased a home in the area, all windows in the house were broken and police protected the couple from a crowd of 2,000. The pastor refused to visit the family because ''he'd talked to the chancery office and they said he shouldn't do it.'' ''I cite this case,'' said Father Cantwell, ''because Mrs. Johnson is a Catholic, because Mrs. Johnson had to go to Mass last Sunday under police protection, and because she has yet to receive a friendly gesture from the priests of her present parish.'' Distraught African-American families complained to one Catholic settlement house worker that ''you're the only one, among the Catholics, who has taken an interest in us.'' The integration of the parochial school at St. Columbanus that fall also stirred controversy in the neighborhood.[54]

The most notorious incidents occurred near Visitation parish in the heart of the city's South Side. Founded in 1886 by Irish-Americans moving south from the Back of the Yards, "Vis" had nurtured three generations of Chicago city employees, priests, and aldermen by the 1950s. Chicago priests called it a "little diocese" because of the tight connections between local priests and politicians, the enormous church building dominating 55th street, the twelve masses each Sunday, the convent housing over sixty Dominican sisters, the five priests working out of the rectory, the community center, the 1,847 children in the grammar school, the 1,062 girls in the high school, and the special summer camp for parishioners.[55]

Parish, neighborhood, and patriotism reinforced each other. In 1942, pastor Daniel Byrnes, born in County Tipperary, placed a huge red thermometer on the side of the high school which counted the number of rosaries said in support of American troops. Each Mother's Day saw the closing of Garfield Boulevard—the main street of the area—and a procession of over 3,000 students, nuns, and priests through the neighborhood. The procession concluded with a series of prayers, the national anthem, and a salute to the flag. Monsignor Byrnes himself took a proprietary interest in the immediate neighborhood around the church—purchasing lots for parish facilities, referring to the main street as "my boulevard" and scheduling massive parish carnivals on side streets near the church building. Monsignor Byrnes's successor as pastor frequently used the phrase "Our Lady of the Boulevard" as shorthand for the notion that Mary watched over the neighborhood—physically through the large statue facing Garfield Boulevard and metaphorically through the parish community.[56]

This set of religious forms and practices facilitated the response to racial "threats." For a number of years, a handful of African-Americans had lived south of 59th Street while the parish borders stopped at the railroad tracks between 58th and 59th. On November 8, 1949, a woman observed eight African-Americans in the home of labor organizer Aaron Bindman near the intersection of 56th and Peoria, less than two blocks from Visitation Church. Quickly the rumor spread that Bindman had sold his home to the visitors. For three consecutive nights, crowds of up to 10,000 milled outside Bindman's home.[57] The fears of the crowd multiplied when word leaked out that Bindman was Jewish, and rumors spread that he was a communist. Small groups of rioters chanted "dirty kikes" and "communists" along with "let's get them out" and "burn the house." Bystanders muttered that a "dirty Jew had sold it [the house] to a 'nigger' for $20,000, to move 'niggers' into the neighborhood." Groups of men stopped unfamiliar faces and accused them of being Jews, University of Chicago stu-

Procession on Garfield Boulevard, Visitation Parish, 1944.
Archdiocese of Chicago, Archives and Record Center.

dents, and communists. Police only halfheartedly restrained gangs of youths from beating up those who failed to satisfy their interrogators. Crucially, one question used to discover "outsiders" was "what parish are you from?" A group of neighborhood residents approached Bindman and asked both him and his guests to leave. Bindman replied that "you people are Catholic and you should speak to the priest as I know he does not believe in such." A member of the crowd then yelled out, "leave religion out of this."[58]

Of course religion had already played a vital role. Arrest records revealed a profusion of Irish surnames and home addresses within twelve blocks of the incident. The most organized response to the crisis came from within the parish itself. With Monsignor Byrnes's quiet encouragement, parish members had earlier organized a Garfield Boulevard Improvement Association dedicated to "keep[ing] Negroes out." (Meetings were occasionally held in a parish hall.)[59] The usual method was for block captains—sometimes precinct workers for the local ward—to go door to door in the neighborhood collecting one-dollar bills in order to "buy out" African-Americans attempting to rent apartments in the area. Word would go out to "tell your non-Catholic neighbors" about the threat. During the Bindman incident, a reporter for *The New Republic* observed residents clustered in a local tavern listening to a radio with a police band and monitoring the movements of both police and "radicals." Indeed, during the rioting, ACLU observers heard "Some people . . . passing the words they attributed to Monseigneur [sic] Burns [sic] . . . to the effect that he would buy any piece of property in the neighborhood before he would allow a Negro to move in." Byrnes did nothing to discourage such sentiments by periodically announcing from the pulpit the parish boundaries or emphasizing the need to clean up the neighborhood. "He had a way of giving a sermon without mentioning race once," observed Father Martin Farrell, "and everybody in the congregation knew what he was talking about. He'd get up there and say, clean up and paint up."[60]

Catholic liberals watched these events helplessly. To their chagrin, the African-American and liberal press quickly attacked the Church. "Like a 'brood of vipers'," as one African-American newspaper put it after the disturbances on Chicago's Peoria Street, "some of them [Irish Catholics] have led rioters . . . and spread discontent throughout Chicago." Protestant minister Homer Jack noted that "Despite many requests, some by liberal Catholics, neither the local parish nor Cardinal Stritch, so far as is known, has in any way condemned the violence."[61] Critics were further inflamed by Judge Joseph H. McGarry's dismissal of charges brought against the rioters. After commending the district's ability to produce ''leaders in civic, social and political life,'' McGarry attributed the disturbances to ''professional agitators and saboteurs bent upon creating and furthering racial and religious incidents.''[62]

During the rioting, representatives from the Catholic Interracial Council visited the rectory (Monsignor Byrnes was on vacation) and spoke to one of the clergy about ''the anti-Semitism taking place and the potential anti-Catholicism which could rise out of it.'' One report for the Catholic Interracial Council plaintively observed that to prospective African-American converts ''things like this do more damage in two days than we can do good in three years. . . . Can't the Mayor, a Catholic, see the Cardinal about doing something about Monsignor Byrnes?''[63]

Events two years later in Cicero, a heavily Bohemian and Slavic working-class suburb west of the city were equally disturbing. Tensions escalated with the movement of an African-American family into a neighborhood apartment building. A crowd of 5,000 encircled the building for several nights until dispersed by the National Guard. Estimates placed the Catholic population of the community at 60 percent, and observers described the mob as largely Catholic, with boys wearing letter sweaters from Catholic high schools and girls fingering crosses around their necks. A man in the crowd was heard to remark that ''I don't want those jigs sitting in the same pew with me.'' A seventeen-year-old woman quickly responded that ''those niggers don't join the Church anyhow.''[64] Walter White reacted by writing Pope Pius XII about ''the seriousness of the situation regarding Catholic-Negro relations'' and publicly attacking area priests unwilling to educate their parishioners.[65]

Representatives from the Catholic Interracial Council quickly formed a suburban unit in conjunction with the American Friends Service Committee. According to press releases, they hoped to inform local residents of ''Catholic dogma and principles directly [related] to this one social problem of our time.'' One Chicago priest also denounced the ''middle-class materialism'' which attempted to ''justify segregation by saying that it produces peace and harmony by keeping separate people who would otherwise be in conflict.''[66]

Local pastors made no public statements during the rioting, and Cardinal Stritch simply sent a letter to area pastors urging sermons on equality. Confidential discussions with pastors in Cicero and Berwyn (a neighboring suburb) revealed despair at the ''biased and bitter'' attitudes of parishioners as well as belief in the ''right to choose their own neighbors.'' A few priests also recalled old feuds between German and Czech Catholics, adding that ''Some Irish families [in the past] had to move because of unfriendly attitudes.'' Several priests cautiously supported Cardinal

Stritch's letter, but others refused to speak on such a "delicate" question or blamed radical organizations for the disturbances.[67]

Even more extraordinary disputes occurred in South Deering, a working-class area on the far southeastern edge of the city. In 1953, housing authorities began moving African-American families into the Trumbull Park Homes for the first time since the construction of the project in 1938. Community reaction was ferocious. For several years, the few African-American families found themselves threatened whenever they moved beyond project walls. At night, homemade bombs and fireworks rocked the project and at times several hundred Chicago policemen were required to protect the families as they shopped, attended church, or traveled to their jobs.[68]

The neighborhood was heavily Catholic. In 1925, parishioners at St. Kevin's parish had erected a new church and school just two blocks from the future site of the Trumbull Park Homes. According to the parish history, "It was hoped that the erection of the school and church would be an incentive for home building and development of this section."[69]

From the beginning of the disturbances in the 1950s, representatives from the Catholic Interracial Council were in "constant communication" with the parish priests. Liberal Catholics denounced the violence and attempted to organize Catholic nationality groups "in order to seek their help in maintaining law and order in [the] community." Members of the council met repeatedly with Cardinal Stritch to discuss the situation and even journeyed to the project to support the African-American Catholic families. Stritch discussed the "scandalous" situation with city and federal officials and conspicuously warned an audience of Catholic schoolchildren that "the Mystical Body of Christ is being torn from limb to limb in many neighborhoods populated by Catholics who practice racial injustice."[70]

This statement produced few results. At St. Kevin's, three African-American Catholic women who dared to attend Sunday mass found parishioners ostentatiously moving themselves away from the pews in which the women were seated. The sound of homemade explosives could be heard outside the church building during the service. Following one mass, crowds lining both sides of the steps leading to the church "hissed at . . . hooted at and assaulted" the terrified women as they walked to the rectory to pick up their church envelopes. One sixty-four-year-old woman parishioner slugged a Catholic Interracial Council member escorting the Trumbull Park residents.[71]

Developments away from the actual church building were also discouraging. One parish volunteer for the local Big Brothers program served as a dubious role model by looking the other way while boys in the program

Woman protests African-American families moving into Chicago's Trumbull Park housing project in 1953. UPI/Bettmann.

vandalized cars owned by African-Americans. A letter from one of the beleaguered supporters of integration in the neighborhood attacked the "Catholic children who go to parochial schools [and] are noted for their intolerance of people who are different than they in religion as well as in color." Priests reported hearing Catholic schoolchildren "bragging about the number of times their fathers and brothers have been put in jail for participating in destructive and violent actions against Negroes."[72]

White supremacists warned Catholics not to trust the CIC. "I tell you in plain direct language the Catholic Interracial Councils are not Catholic in any sense," one letter to the editor argued, "The one thing that they show craftiness about is to hide behind the beautiful name of Catholicism to deceive people into thinking they have the approval of the Catholic church." Fresh from South Africa, Alan Paton traveled to Trumbull Park for *Collier's* magazine where he interviewed a man standing outside the project. "We mean to get the jigs out, that's all. . . . I'm a Catholic . . . and a good one I'd say. But the Church hasn't got a right to tell me who I should live next to. And the Church knows it too, because it hasn't said anything about Trumbull Park."[73]

On November 18, 1955, CIC representatives met with the African-American Catholic families in the project. That night, Cardinal Stritch listened to a tape of the conversation in his home. Following an opening prayer which concluded with a plea for the assistance of St. Martin de Porres (the patron saint of the interracial movement), the families began answering questions. The discussion was heartwrenching. "All of the time we attended St. Kevin's Church we were called names," said one woman, "even during Mass. . . .When we went to the communion rail there were comments. . . . " One woman quit attending church briefly. After local priests asked her to return, policemen had to protect her as a woman from the parish attempted to push her down the stairs leading into the building. "It's so hard to understand that we are living in America," another resident commented. She continued, "I know the Pope said that Negro people should be helped."[74]

Local Catholics still refused to meet CIC representatives "for fear that they might be discovered by the rest of the community." Eavesdroppers hired by the ACLU reported that steelworkers in the local bars believed that the parish priest placed the "blame for the whole trouble" on the African-American families. Two years after the first move-in, hecklers continued to boo African-American Catholic churchgoers and their police escorts. When a priest admonished the parishioners after church one Sunday, the "angry crowd then directed their boos toward him as well as the Negro women."[75] Following a later disturbance in a nearby park,

African-American sociologist St. Clair Drake plaintively asked for "weekly references to papal encyclicals against racialism" in Catholic churches and schools.[76]

IV

This cluster of incidents in the fifteen years after the war illuminates how two perspectives on racial issues bedeviled the Church in the North. Formally, by 1945, a new emphasis on interracialism was apparent in Catholic life. Exams for a class at Fordham University entitled "Community Organizations and Cultural Relations," for example, included a series of questions on Catholic social teaching. In answer to a question on prejudice, Marguerite Mahoney faithfully argued that "By the creation of an 'inferior race' and a 'superior race' we have violated the unity of Christ's Mystical Body." Mariagnes O'Neill added that students who "have learned the true facts" must assist African-Americans in making their "contribution to America's democracy." Instructors also exposed both students to the ideas of John LaFarge, whose primer on Catholics and race relations became a staple in sociology courses taught at Catholic colleges.[77]

Other Catholics invoked different "Catholic" traditions as they persistently refused African-Americans admission to particular neighborhoods, schools, and churches. In Detroit, a priest upon learning that a penitent in the confessional was an African-American, refused him absolution, "informing [him] that he had his [own] church and that he should go there to confession. He also informed him that he need not go to the assistants because they had their instructions not to absolve Negroes."[78] When asked why he ignored "church teaching" and the "general civic effort toward integration" by placing African-Americans in a separate mission parish, one Chicago priest responded "What teaching? . . . The hierarchy knows very well what's going on out here. There's always been this sort of situation in the Church. There's always a Polish Church and a Mexican Church. Nationality churches. This is the same thing. As a matter of fact, when this parish was set up the colored requested it themselves."[79]

Etching these two Catholic positions in bold relief was the trauma of neighborhood change. Here the contrast between the Catholic experience and that of other city residents is telling. The number of white Protestants remaining in the northern cities by the 1950s was already relatively small. In 1952, Protestant church members numbered only 3.6 percent of the white population in the Bronx and 16 percent of the white population in

Philadelphia, compared to Catholic totals of 33 percent and 40.7 percent. The average white Protestant congregation in Philadelphia counted 476 members; the average Catholic parish, 4,926.[80]

Mainline Protestant church officials acknowledged a series of challenges. On the one hand, the proliferation of Protestant churches in a given area tended to weaken each institution, in contrast to more centralized Catholic planning. Alternatively, Protestant churches, as one church researcher put it, "move out when their people move out." This habit, especially in what one author called "dominantly Roman Catholic communities in northern cities" frequently led to the almost complete absence of a white Protestant church presence.[81] Protestant churchgoers were also less predictable. A Newark observer commented that "Catholics were regular in their attendance at masses irrespective of the weather. Even when the big snow came, five masses were celebrated and parishioners were seen by Protestants at an early hour wending their way. Something about the masses compelled them to attend, whereas many Protestants remained home from divine worship."[82]

Jews were far more likely to live in city neighborhoods than white Protestants, but the institutional and social dynamic was similar. Even more than white Protestants, American Jews placed little importance upon weekly attendance at religious services. A New York survey estimated that while 61 percent of the city's Irish and 50 percent of the city's Italians attended church each week, the same was true for only 2.8 percent of the city's Jews.[83]

Similarly, as Gerald Gamm demonstrates, Boston's Jews were more likely—regardless of income—to flee urban neighborhoods than Boston's Catholics. In Philadelphia as well, the more affluent Jewish area of Strawberry Mansion peacefully integrated, and then resegregated, even as bitter conflicts occurred in working-class, more Catholic sections of the city.[84] In New York's Washington Heights area, the mostly Jewish neighborhoods tended to undergo racial transition in advance of the Irish-Catholic Inwood section.[85]

The heavily Jewish area of North Lawndale on Chicago's West Side provides an unusually telling example. What was once the heart of Jewish Chicago changed into an African-American neighborhood with astonishing rapidity. Eighty-seven thousand whites lived in the North Lawndale area in 1950, only 11,000 in 1960. During the same period, 100,000 African-Americans moved into the area. In one year, 15 synagogues out of 48 closed; by 1960 not one of the 48 in the area was still active. *South* Lawndale—an area dominated by Poles, Bohemians, and Italians—remained virtually all-white during precisely the same period. By the early

1960s, the only local clergy available to join a Lawndale neighborhood community organization were white Catholic priests and African-American ministers.[86]

Comparisons between the resistance found in heavily Catholic neighborhoods and more peaceful racial transitions in other sections of the city were inevitable. Social workers noticed the reluctance of Jewish youths in New York to join street gangs, even as Irish and Italian gangs confronted their African-American counterparts. A 1945 report on racial tensions compiled for the Detroit chapter of the NAACP concluded that "In view of the fact that most of the opposition had come from Catholic neighborhoods, it was thought it might be desirable to have a prominent Catholic layman [working with the NAACP] if such a person could be obtained." Two years later one city agency reported that "Improvement Associations" on Detroit's East Side "conform to the bounds of Catholic parish lines rather than subdivisions."[87] In Chicago, African-American realtors steered clients into heavily Jewish areas in order to avoid contact with the Catholic homeowners on the South and Southwest Sides. "I find it difficult to find a proper apology," wrote one Chicago seminarian in 1956, "for the fact that neighborhoods strongly Catholic are the most opposed to integration."[88]

The scale of the African-American migration during and after the war, however, ensured that not only heavily Jewish or white Protestant neighborhoods would change in racial composition. Instead of the piecemeal change of the 1920s and 1930s, the massive wartime migration transformed whole sections of northern cities. The raw numbers are less memorable than the image of literally hundreds of African-Americans arriving at train stations in Chicago, Cleveland, Detroit, Philadelphia, and New York. Diocesan priests in Chicago recalled hearing a speaker implore clergy to move south and minister to impoverished African-Americans. Archbishop Stritch, however, told the priests to stay. "The big work is right here," he announced, "Don't leave Chicago. They're all coming North." Stritch himself stopped at the Illinois Central Station one day in the early 1950s and marveled at the baggage and shouted greetings of the over five hundred African-Americans then pouring into Chicago each week. Occasionally, Stritch would receive letters from African-Americans he had known while growing up in Tennessee, but who now also boasted Chicago addresses.[89]

Typically, the sudden rush to sell homes occurred after a racial incident or when an invisible "tipping point" was reached in the area's population. Hoping to generate neighborhood panic, realtors urged white homeowners to sell immediately, gave heavy publicity to each home sale to an African-American, and even hired groups of African-Americans to walk through neighborhoods near the racial divide. In the stampede, prices often plummeted and realtors bought the homes for resale to African-Americans.[90] Some parishioners in the area surrounding Cleveland's St. Thomas Aquinas parish, for example, received twenty-five visits from local realtors, despite angry condemnations of these tactics by the local pastor, who embarked upon his own program to visit each home in the parish in hopes of stabilizing the area.[91]

Neighborhood densities increased and upkeep on the buildings lagged during and after the initial integration, as several non-white families jammed into what had been single-family homes in order to finance exorbitant monthly payments. Once an area became perceived as unstable, younger white families shied away, ensuring that the majority of potential buyers would be African-Americans. To those residents tied to an area by religion, family, and culture, the process was particularly painful—psychologists investigating the phenomenon compared it to the grief felt upon the death of a family member.[92]

The most severe traumas associated with racial transition—violent incidents, a sudden evacuation of the neighborhood by whites, a climate of paralyzing fear—were limited to encounters between African and Euro-Americans. Contact between Euro-American Catholics and Spanish-speaking immigrants occurred in a different context. Mexican-American leaders resolutely identified themselves as "white" during this period with much the same fervor as Italians or Poles, and of course could easily compare their own experience (arriving in a new land with limited English skills) to that of many Euro-American Catholics.[93]

In Chicago, African-Americans on the city's South Side lumped Mexicans along with Poles, Italians, and Irish as the primary groups clustered along the western border of the African-American ghetto.[94] Near Trumbull Park, Mexican-Americans made up 9 percent of the "foreign-stock" population in 1960, even as the few African-Americans in the area feared to venture out of their homes. The acceptance of a small Mexican-American community in Chicago's Bridgeport neighborhood also contrasted sharply with pervasive hostility toward African-Americans.[95] Puerto Rican immigrants (especially in New York City) faced more barriers, but even Puerto Ricans became less segregated from whites than African-Americans.[96]

A partial explanation for this phenomenon is that predominantly Catholic groups could become integrated into what one sociologist calls the "web of relationships" so critical to the neighborhood structure, one of the most important of which was the Catholic parish.[97] Ties between Spanish-speaking peoples and the parish structure in the postwar period were often tenuous, in part because of the absence of Mexican and Puerto Rican clergy. Even so, neighborhood contacts between Euro-American and Spanish-speaking Catholics at parish activities or in the parochial school were comparatively frequent when compared to contact between Euro-Americans and African-Americans. In fact, the very presence of Mexicans or Puerto Ricans in a parish community might signify to Euro-American Catholics that the new group was not "black," regardless of physical features. Even the practice of changing one weekly service to a "Spanish Mass" implied at least a tentative institutional endorsement.[98]

Shared responsibility for neighborhood institutions, in other words, was plausible. Transition from a heavily Italian area to a predominantly Mexican one guaranteed turmoil, but in contrast to the transition to an overwhelmingly Protestant population, it did not automatically portend a crisis in parish finances or the potential collapse of the parish school. In some cities, at least, Mexican-American Catholics developed loyalties to particular shrines or schools, and even began answering the question "where are you from?" with parish names.[99]

By contrast, Euro-American Catholics constantly referred to African-American "invasion" as a catastrophe. "In the past few years we have seen the colored people almost take over a number of parishes," wrote one parishioner to the editor of the Detroit diocesan newspaper in 1953, describing how members of the parish "where I grew up and where my family still lives, are on pins and needles, feeling that theirs will be the next parish to go."[100] "[E]verything they have been taught to value, as Catholics and Americans," observed one reporter in Cicero, "is perceived as at risk. . . . [T]he churches and schools they built would become empty, the neighborhood priests, if any were left, would be missionaries from some motherhouse that sends off religious to the far East." One Chicago parish included eight priests, twenty-five brothers, and forty-eight nuns as well as 1,755 children in the parish grade school and 2,662 students in two parish high schools. "It disturbs me deeply," wrote the pastor, "to think that the valiant efforts of all these consecrated lives, are, in a certain sense, far less effective than they would be, if it were not for the emaciating effects of creeping blight."[101]

The most prominent rhetorical strategy available to opponents of integration was anticommunism. Betrayal abroad—notably the loss of Eastern Europe to the Soviet Union—became linked to a perceived betrayal at home. In Levittown, Pennsylvania, the pastor of a parish wracked by racial tension believed that "there are unknown groups behind this who are purposely stirring up the trouble." One Catholic liberal concluded that "The Catholics were particularly affected by the rumors about Communist promotion of the episode.'' Members of the neighborhood improvement association near Trumbull Park attacked city officials for tolerating a ''carefully planned Communistic plot''; on the city's West Side the leader of a community group concerned about racial change, and holding its meetings in a Catholic parish hall, warned that ''this breaking of neighborhoods is one of the most successful ways the subversives are working in this city.''[102]

While persuasive in local areas, this analysis failed to sway liberals already using the threat of communism to forward their own social agenda. Like parishioners concerned about racial change, Catholic liberals also believed communists within the United States threatened national security, but they saw the ''Reds seeking to press the issue'' by publicizing America's troubles with racial minorities to the world. ''Interracial Justice,'' John LaFarge declared in 1958, ''as prayed and lived in the Catholic interracial movement in the United States, is a direct answer to Marxist theory and to world-commmunist propaganda.''[103] Robert Kennedy, fresh from assisting Senator Joseph McCarthy's search for communists within the American government, told a New York CIC audience in 1955 that any effort to win ''colonial peoples'' rested on cleaning ''up our own house and persuad[ing] our allies to do the same.'' These sentiments even permeated the essays written by high school students for Catholic Interracial Council essay contests. One student writing on ''What a Student Can Do To Combat Racial Prejudice?'' argued that ''the dissention [sic] and friction caused by prejudice is playing directly into the hands of society's no. 1. enemy, Communism. Their idea is an ancient one; divide and conquer.''[104]

To an extent, the division between liberal and conservative on racial matters by the 1950s mirrored class lines. The parishes most worried about racial change—St. Anthony's of Padua in Detroit where parishioners placed ''This House Is Not for Sale'' stickers in their windows; St. John the Baptist in Pittsburgh where residents watched the city's Urban Redevelopment Gateway plan with foreboding—were often primarily working-class. Most of the male parishioners at St. John the Baptist, for example, were operatives or foremen in heavy industry, with a sprinkling of salesmen.[105] In one survey of the elderly in a racially changing Buffalo parish, only 4 percent of the men and 15 percent of the women had even *attended* high school; two-thirds had lived in the parish for over twenty-five years, and many recalled first communions and weddings in the church building. Over half of the residents owned their own home—some had built it with their own hands—and many of the homes were close to the church building itself. When interviewed, ''about half of the respondents gave spontaneous expression of their attitudes toward the Negro during discussion of housing and relationships with neighbors.''[106]

Simply invoking the talisman of "class" to explain racial tension, however, is inadequate. Even the most working-class Catholic neighborhoods contained a core of physicians, lawyers, city government officials, and police officers, and Catholics were significantly more likely than white Protestants or Jews to structure social networks along religious, and not simply occupational, lines. Particular individuals, in other words, might identify themselves as "workers" on the job, but as Catholics in the parish neighborhood where racial tensions were at their most severe.[107]

This cultural separation of residence from work helps explain the phenomenon noted as early as the 1919 Chicago riot and emphasized later by a wide array of observers—that the same individuals accepting of an African-American presence in the workplace violently resisted neighborhood integration.[108] Catholic interracialists constantly encountered this problem. Educational seminars on race issues held in Detroit parishes, for example, often included union members informing discussion leaders that "you want them in this neighborhood and we don't think that's good for us." Interviews conducted in Chicago's meatpacking plants are also revealing. One butcher working in an integrated plant urged African-Americans to "*stay where they belong and be proud of their race.* My son is a Catholic, and we'll stay that way. I'll stay in my class, my own class and don't mix with others." Another worker claimed that at the workplace he was "friendly with the colored" even as he acknowledged being arrested during the riots near Visitation parish. Catholic workers in Gary expressed support for equal opportunity legislation while in the same breath asking "why should they [African-Americans] move in with us?"[109]

This incongruity—between racial tolerance in one area and opposition to integrated neighborhoods—is also evident in the various surveys and opinions polls of the period. One Chicago woman interviewed in the early 1950s emphasized that "We respect a colored priest . . . because they were sent by God. We have a colored Saint. . . . Our souls are all equal. We pray to the Blessed Martin [de Porres] even though his mother was an Indian and his father colored." When asked whether she favored neighborhood integration, the woman responded "I do object to that! I think they should keep them in their own neighborhood. . . . I wouldn't like it if my neighbor sold to colored." A 1945 survey of one Philadelphia Catholic high school found that 92 percent of the students "realized that that baptized Negroes are members of the Mystical Body of Christ," and 67 percent agreed that African-Americans had a right to attend any Catholic church. To the instructor's dismay, however, 94 percent of the students favored restriction of African-American residents to certain areas of the city.[110]

Nothing better illustrates the complexity of these issues than the efforts of liberal Catholics to persuade fellow communicants to accept integration. They faced formidable obstacles. Generous to more historically "liberal" groups (such as the Quakers), national foundations characteristically viewed Catholic organizations as inherently conservative and tightly controlled by the institutional church.[111] For their part, few of the Catholic dioceses, busy with traditional school and parish programs, expressed interest in funding education programs on such controversial matters. A Detroit lay leader recalled that, despite the 1958 episcopal statement, "the isolated individual who sought to equate race concerns with moral concerns was highly suspect, repeatedly required to explain his position and subjected to counter-efforts from clergy and laymen alike."[112]

Equally important, priests suspicious of lay initiative held the levers of power in each diocese. While African-American and liberal groups applauded the efforts of Catholic Interracial Councils, priests nourished in the powerful clerical subculture of the era equated criticism with disloyalty. "Our opinion," noted one outside observer of the active Catholic Interracial Council in Chicago, "is that it is not so much radical as it is ineffectual and tactless. Too many Catholic churchmen are not certain which side the Council is on, their own, that is the Catholic one, or the 'other'."[113]

Inadequate support, however, only underlined more fundamental difficulties. A persistent refrain in the available records is the frustrating inability to intrude upon the social and liturgical whirl of the local parish. One Philadelphia resident noted the irony of the Catholic Book Club choosing one of John LaFarge's volumes on race relations as its main selection. "I have personal knowledge," he wrote, "that the pastor of my own parish feels that the negro race as a whole has not yet earned the right to live with the white race."[114]

High school study days and scholarship contests usually received cordial welcomes, but the efforts of CIC members to work in communities suffering from racial tension rarely succeeded. Despite their best efforts, they were outsiders—outsiders without a financial and emotional investment in the neighborhood's future. In 1954, a Catholic settlement house worker traveled to the Bronx to discuss housing issues. A group of property owners, according to one source, "simply refused to listen to her when she indicated her connection." One Chicago activist bemoaned the situation. "[W]e ordinarily do not get interested in a neighborhood, get to know the people, until it is in transition. As a result the focus is solely on race; we are seen as an organization crusading . . . for a group and an ideal in which they don't believe; and we . . . are not seen by the people in the neighborhood as at all interested in their problems."[115]

Nonetheless, racial divisions would come under increasing scrutiny. The dispute over civil rights at the 1948 Democratic convention, the Supreme Court ruling outlawing restrictive housing covenants that same year, the 1954 *Brown* decision, and the surge of racial activism in the South pushed many national institutions away from the economic issues so central in the 1930s and toward civil rights disputes. NAACP attorneys replaced union leaders as the keepers of the liberal flame. Prodded by the arguments of interracialists, the Catholic Church followed these same paths, with its leaders knowing full well that the specific interests and prejudices of many of its members stood in opposition to the larger national community of which it was increasingly a part.[116]

This tension is illustrated by the efforts of one Catholic interracialist, a sociology teacher at Chicago's St. Stanislaus Kostka high school. As a class assignment, the instructor asked students in 1959 to comment on the possibility of minority groups attending the school. In papers topped by a JMJ (Jesus, Mary, Joseph), the students wrestled with the tension between Church doctrine (as presented by the instructor) and segregation. Most students cautiously favored integration, with several referring to Catholic teaching that "all races are equal." Others agonized. "I know we are supposed to love everyone because everyone is created equal," wrote one student, "but colored people bring down the standards of the school. They have their own schools which were built especially for them." A classmate referred to the slums she saw when traveling through the city's South Side. "I think that I would probably quit school. Because they have no right coming to our school at all." Or, as another student put it, "one of the two things I am attending St. Stan's for is to study religion, and get the best possible Catholic training I can, and to go to a school were [sic] minority groups are not popular." "The neighborhood in which our school is located," one respondent concluded, "is bad enough without having negroes add to the reputation of it."[117]

None of the students referred to the Polish heritage of the school or the still heavily Polish population in the neighborhood. Indeed, to judge by the surnames, a few of the students in the school were not Polish. To those students opposed to integration, "white" nationalities were acceptable; African-Americans (even African-American Catholics) were not. This opposition to integration remained tied to distinctively Catholic language and forms. Students at St. Stanislaus, along with Catholics across the urban North, still evoked the sacred character of parish neighborhoods, the social ties established through the parish community, and the financial and religious investment in the physical plant.

Early in the twentieth century, these Catholic arguments were logical outgrowths of an institution and a set of peoples committed to a union of faith and culture. Parishes and ethnic identities—ranging from African-American to Polish—were presumed to possess stable, geographical referents. By 1959 this Catholic emphasis on "neighborhood" and "community" remained, but in a less defensible context. Catholic defense of segregation now depended upon skin color, not simply culture, and a clear bifurcation between "black" and "white." But "black" and "white" were unacceptable theological categories. Denying a home to an African-American family simply because whites (of whatever religious or cultural background) might flee the neighborhood placed prudence above morality, and tore at the heart of the Mystical Body of Christ.

How then, could Catholic leaders prevent abandonment of both church facilities and urban neighborhoods without condoning opposition to racial integration? Few church officials could answer the question, and most simply left pastors and parishioners to fend for themselves. John McDermott described the attitude of the typical pastor in 1959. "A feeling of hopelessness is common, due perhaps to the dread of being unable to stem the tide with a consequent fear that nothing can be done or need be done but wait for events to take their course." In Cleveland one prominent pastor was described as in a "panic regarding the Negro question—wants to be moved before the parish collapses." A Gary priest enclosed a map of a local housing project in a nervous letter to his bishop, maintaining that since "the new housing unit is being built two and a half miles from here" his parish, contrary to rumor, would not have a "Negro problem." A Buffalo pastor concluded that "The changes are bad. The colored people are coming in. Coming from St. Mary's, St. Ann's, St. Nicholas next door. That's the greatest change. Eventually it'll ruin the parish. . . . I think it will have the same effect on the neighborhood. Don't know what can be done to stop it."[118]

Parishioners discussed the question incessantly. Genevieve Rafferty of Chicago's Presentation parish noted in 1955 how "the first few Negroes are beginning to be seen around our parish" and "the chief topic of conversation [is] the colored." Rafferty reported rumors that "several Catholic pastors were buying the property for sale in their parishes so it could not be purchased by Negroes. . . . " "Can a Catholic who leaves a neighborhood solely because Colored are moving in," she asked, "actively participate in trying to keep the colored out? . . . What is the practical application of the teaching of Our Lord in our relation to Negroes and what is just idealism?"[119]

FIVE

COMMUNITY ORGANIZATION AND URBAN RENEWAL

I

The most creative attempt to answer Genevieve Rafferty's questions emerged from the work of organizer Saul Alinsky.[1] During the late 1930s and early 1940s, Alinsky organized a Back of the Yards Council (BOYC) in the area immediately south of Chicago's stockyards. To Alinsky, the piecemeal benevolence of social workers obscured the capability of slum-dwellers themselves to form effective organizations, given the proper direction. Within a few years of its founding, the BOYC had mobilized neighborhood residents in support of better housing, more recreation facilities, and improved city services. Journalists and sociologists trumpeted the organization as an example of American democracy and pointed to the various ethnic groups working together as a symbol of national unity.[2]

In an area 90 percent Catholic, church support for the organization was crucial to its success. With the aid of the younger clergy, Alinsky carefully negotiated alliances with the various parishes and ensured clerical support in showdowns with city government and local meatpacking plants. Auxiliary Bishop Bernard J. Sheil proved the most important supporter. Publicly endorsing the council and assisting in fundraising projects, Sheil helped shield the BOYC from accusations of communist influence and attacks by local politicians. By 1949, local Catholic clergy had assumed a variety of leadership roles within the organization. "Acting upon the Catholic belief that all human beings alike reflect the face of God," wrote one avid supporter, "they [the BOYC] are finding a path on which democracy and Catholicism can go forward harmoniously toward mutual enrichment." Another article in a Catholic periodical termed the group "a heartening demonstration of democratic Catholic social action" and proof of "radicalism rooted in religion."[3]

The philosophical underpinnings of Alinsky's project excited Catholic intellectuals. In particular, Alinsky's mentor, Jacques Maritain, applauded the council's achievements. Along with other European Catholic intellectuals, Maritain had been embarrassed by the complicity of the Church in France's Vichy regime, and the hesitancy with which some Catholic leaders supported democratic governments. From this perspective, the success of a Catholic-sponsored community organization confirmed the papal emphasis on subsidiarity (in opposition to state programs) while demonstrating that Catholicism, social justice, and democracy were compatible phenomena. The mission of Catholic scholars would be to convince skeptics of the links between Thomistic philosophy and democracy; the mission for activists was to build local organizations that promised to reinvigorate democracy in a world desperate for alternatives to fascism and communism. Such a program would necessarily have a positive effect on spiritual life. "An internal moral awakening," Maritain emphasized, is linked to "the awakening to the elementary requirements of true political life."[4]

Alinsky attempted to transplant his methods to other areas immediately after his success in Chicago. He established fledgling organizations in St. Paul and Kansas City and in 1945 proudly announced that "fourteen Catholic churches of different nationalities" formed the core of his "People's Council of South Omaha." To Maritain, he wrote that "I thought of you while I was out there and when I was asked to come to the Chancery and address all of the priests in the Diocese on the subject of Christian principles and Social Action."[5]

Outside Chicago, however, these community groups never took root, in part because of the shifting dynamics of "race" and ethnicity. By the 1950s, boasts about uniting Euro-American residents seemed beside the point when compared to the challenge of African-American integration. To Alinsky's embarrassment, one of the most determinedly segregated areas of Chicago remained the Back of the Yards. African-Americans now dominated the stockyards workforce, but as during the 1919 riots, few dared to live in the immediate neighborhood. BOYC officials publicly explained that "Negroes don't have anything in common with the people who live here," but residents privately expressed a blunt determination to prevent African-Americans from moving into the area. Even after a direct request, the leader of the BOYC refused to sponsor the move-in of several African-American families, fearing that such a gesture would destroy both his career and the positive work of the organization. One nun from a Catholic school in the area discouraged a CIC proposal for an interracial day, commenting that "Since we reside in the 'Back of the Yards' neighborhood, we are not in a position to conduct assemblies, etc. In other words, it would be impossible to hold any sort of interracial program at our school."[6]

The challenge, then, was to build a local community organization without succumbing to local attitudes on racial questions. The solution, according to Alinsky and supporters like Monsignor John O'Grady, the longtime head of the National Conference of Catholic Charities, lay in a program of conservation and rehabilitation. Only "a new type of social organization, which will be built around the neighborhood and the parish," could prevent neighborhood decay.[7] "[T]he economic loss directly accruing to the Church by virtue of the deterioration of its holdings," sympathetic priests argued, "presented a very practical justification."[8]

Beginning in Detroit in 1951, O'Grady scheduled a series of seminars on neighborhood conservation in conjunction with the National Conference of Catholic Charities. One year later, Cardinal Stritch, Alinsky, and O'Grady addressed over 150 Chicago priests on the topic, "The Parish in a Changing City." Here Stritch called conservation of property an "essential part of the pastoral ministry" and O'Grady emphasized how "Blight destroys the capital investment made by our city governments and by our Church."[9] Since northern cities already contained many local groups publicly dedicated to conservation and, privately, neighborhood segregation, O'Grady encountered some skepticism. In a 1955 pamphlet, he conceded that any discussion of "conservation of our parishes and neighborhoods opens up the question of the relationship of our people in the cities to the Negroes that are moving in from the South. When people in our parishes talk about conservation, they are talking about Negroes." O'Grady then issued a charge to welcome all newcomers to the neighborhood.[10]

O'Grady and the National Conference of Catholic Charities also proved vital to a new burst of community organizing begun by Alinsky in the mid-1950s.[11] To Alinsky, the connection was inevitable. "We have always worked in industrial settings," he told one interviewer, "It just so happens that the populations which are there to be worked with are Catholic." A handful of clerics funded a series of Alinsky-style projects with similar pragmatism. Their hope was to create citizen organizations capable of reaching beyond the traditional, insular life of the parish in order to shape plans developed in Washington or City Hall. "An organized and an aroused public opinion among the people whose areas are threatened by the spread of blight," argued Monsignor O'Grady, "is the only force that will bring about a genuine law enforcement program."[12]

The first significant attempt to put these ideas into practice outside Chicago occurred in 1956 in Lackawanna, New York.[13] Here, under the auspices of the National Conference of Catholic Charities, Alinsky-trained organizers used an initial $45,300 grant to establish a Citizens Federation of Lackawanna.[14] With the vigorous financial and logistical support of Monsignor Julius Szabo, a Lackawanna pastor and director of the local Catholic Charities, the organization protested corrupt city government, unenforced housing codes, and the domination of local politics by Bethlehem Steel. "We believe," argued Monsignor Szabo, "that this is not just another 'do-gooder' organization."[15]

Simultaneously, with $300,000 in foundation support, Alinsky began work in New York City's Chelsea area. Of the area's 60,000 residents, perhaps one-third were still Irish Catholics clinging to walk-ups between 14th and 34th Streets, and another one-third were Puerto Rican migrants. As part of the grant requirements, Alinsky uncharacteristically agreed to work with an existing organization, the Hudson Guild. While well-established in the area, the Hudson Guild possessed all of the faults Alinsky habitually identified with social service organizations—it enrolled relatively few local members and funding came from outside the neighborhood or from institutions (the YMCA, St. Peter's Episcopal Church, B'Nai B'rith) headquartered in the community but irrelevant to most community residents. Most important, the Hudson Guild had been engaged in a long-running religious skirmish with the neighboring Catholic parishes, distributing birth control literature to low-income families and advocating an Ethical Culture program with the motto "deeds not creed." When the Guild organized boy scout troops or promoted youth activities, Catholic parishes responded in kind.[16]

Alinsky selected Father Robert Dunn of St. Columba's parish as the leader of the new Chelsea Community Council and emphasized to the reluctant Hudson Guild leadership that "the lodestar of this entire program [was] to involve as many of the citizens of the Chelsea Community Council as possible in citizen participation activities." (Instead of working with the large institutions located in the area, Catholics habitually asked the local Democratic club for assistance with problems. In turn, when a politician "wanted to verify his notions of what people were thinking about, he would check with the priest.") Alinsky organized a trip for Dunn to Chicago to meet Catholic leaders and observe the Back of the Yards Council and urged him to integrate Puerto Rican Catholics into both parish life and the work of the council.[17]

The organizations in both Lackawanna and Chelsea were short-lived. In Lackawanna, efforts to organize the community faltered against the ability of local political barons to lure organizers into lucrative patronage jobs. Religious tensions also marred group activities. To be fair, any attempt to bring all faiths together in a common project was unprecedented, and supporters understandably boasted of overcoming "national, racial and religious differences'' in ''cooperative endeavors for the common good.'' Still, as one organizer concluded, ''the gap between the Protestants and Catholics in Lackawanna [was] so great'' that any attempts by Catholics to organize the community raised Protestant suspicions of a Catholic takeover. When Catholics attempted to raise money through a raffle, Protestant ministers (white and African-American) precipitated an organizational crisis by refusing to participate on moral grounds.[18]

Most important, the fear of racial change shaped discussion of every issue. ''The issue [race] is dynamite here,'' concluded Nicholas Von Hoffman. ''The threat of violence is in the air. As much as you would like to postpone talking about it, you cannot, because it is always brought up.''[19] In time, Catholic politicians attacked the Church for ''meddling in politics,'' while residents whispered that the Citizen's Federation planned to integrate white neighborhoods or bring public housing into the area.[20] The rumor that an African-American doctor planned to purchase a home in an Irish section of the city initiated a series of heated private meetings in parish rectories about the proper course of action, even though the rumor later proved false. Father Kowalenski of St. Barbara's feared two things, reported an IAF organizer: ''One being the inability to deal with some of his parishioners and the other the negro threat of moving into his parish.'' Even a priest who resolved to welcome African-Americans and ''service them as a minister of God,'' worried about the impact of neighborhood transition on parish building programs.[21]

The collapse of the Chelsea effort began with the announcement of city plans for an urban renewal project, Penn Station South, in the area. Officials of the Hudson Guild welcomed the project since it promised to eliminate vast tracts of what they termed slum housing and bring a new population into the neighborhood. One supporter explicitly applauded the ''displacing of hundreds of poorer families, mostly Roman Catholic, with a more heterogeneous population of a higher social class.''[22]

Alinsky, Father Dunn, and the community council responded with less enthusiasm. St. Columba's church and school would soon be encircled by high-rise apartment buildings, with current residents having little hope of receiving one of the new units. Father Dunn and the council quickly organized rent strikes, a women's group to protest relocation procedures, and a series of meetings with Mayor Wagner. Priests from the parish also sponsored walking tours of the area, interviewing infuriated residents and publicizing the fact that 20 percent of the parish population had received notices to search for new apartments.[23]

Archdiocesan officials seemed torn between unease with the displacement of Catholic populations and previously established alliances with powerful public officials favoring the project, particularly New York housing and development czar Robert Moses. Moses dashed off a memo to Cardinal Spellman complaining about the Chelsea Community Council and Father Dunn. "In this project," Moses maintained, "special arrangements were made to permit the churches to remain in the area including St. Columba's. No useful purpose can be served by obstructing or delaying the construction of the new housing which has been authorized and is under way." Spellman's response was deferential, although he did note that protests about relocation procedures had also reached his ears.[24]

The efforts of the Chelsea Community Council failed to impress the members of the Hudson Guild. Already, the selection of a Catholic priest as president of the council and fears of a Catholic "power center" had frightened non-Catholic members. (By the same token, Father Dunn's observation that the CCC was his first contact with non-Catholic groups after sixteen years of work in the parish suggests the isolation of Catholics from the rest of the community.)[25] As in Lackawanna, sparks flew when the Catholic parishes promoted a raffle and carnival as a fundraiser, in opposition to non-Catholic preference for "giving as a matter of intelligent concern for others." In June 1959, the Hudson Guild and sixteen related organizations left the council and within a year, bereft of funds, the council collapsed.[26]

Perhaps the most enduring legacy of Alinsky's short-lived New York experiment was a new attempt at community organization on New York's Upper West Side. Closely following events in Chelsea was Father Harry Browne, a New York priest and historian then teaching at a local seminary. Himself a proud product of Hell's Kitchen, Browne sympathized with any attempt to save "low-middle-income people for our neighborhoods, which is to save them also for our parishes." In 1959, Browne attended a community meeting in his own parish, St. Gregory the Great, and watched aged Irish longshoremen berate city officials for abandoning the interests of low-income residents. Soon, Browne had become the catalyst for a neighborhood group, the Stryker's Bay Neighborhood Council, that quickly won the attention of city officials through vocal protests against city plans to eliminate much of the low- and middle-income housing between the Hudson River and Broadway.[27]

Browne argued that his strategy possessed three merits. Like Alinsky, he scorned "paternalistic social agencies" and promised that a local community council would respect the wishes of those whom it served. In addition, parish sponsorship of a community organization would bring the aging, largely Irish, parish membership into contact with the Puerto Rican families rapidly moving into the neighborhood. For too long, he conceded, parish administrators had catered to Irish sensibilities. "[B]y the 1950s," he noted, "the bagpipes in the street on Easter Sunday were awakening a block full of mystified Puerto Ricans." Active participation in affairs vital to the Puerto Rican community—particularly housing issues—might result in more acceptance for the parish church. Finally, community organizations pressuring the city for low-income housing might enable the still sizeable Irish population to stay within the neighborhood. "The American city has up to now been Catholic," he declared in a bulletin distributed to archdiocesan priests, "it is not time to desert the city. We need not argue with the family that chooses the suburbs, but for those who choose to stay where our oldest parishes are, we *can* assist in the choice."[28]

Browne himself received little formal archdiocesan guidance, but Cardinal Spellman by 1961 was sufficiently concerned about these issues to form an Archdiocesan Housing and Urban Renewal Committee. The initial impulse behind the committee came from deputy mayor Paul O'Keefe. In a meeting with Spellman, O'Keefe argued that the combination of luxury housing and large Jewish-sponsored, middle-income housing developments, particularly Coop City in the Bronx and Penn Station South in Chelsea, combined to exclude middle- and low-income Catholics from the city. "You have to get into this housing field," he said, "to keep a proper balance in the city."[29]

The committee's chairman, Monsignor Harry Byrne, agreed with these sentiments. Since the early 1950s, from the vantage point of St. Joseph's parish on East 87th Street, Byrne had watched developers in the once heavily Catholic area replace modest two- and three-story structures with luxury high-rises. Already, a number of the fifteen Catholic churches north of midtown and east of Park Avenue were virtually empty on Sundays, the parishioners gone to Queens or Staten Island. In a memo to members of the urban renewal committee, Byrne argued that "Catholics weren't up to other groups in the community in this area [housing]." By 1962, Byrne had begun laying the groundwork for a number of housing projects, all geared at allowing middle-income residents an opportunity to remain in the city.[30]

Other Catholics also recognized the importance of Church participation in urban renewal programs. In Syracuse, Bishop Foery urged the involvement of priests with local urban renewal plans. Only such involvement, he argued, could "maintain the value of our properties and the properties of our neighborhood." A special issue of the *Interracial Review* encouraged Catholics to wade into neighborhood and city-planning disputes. One priest with a planning degree from MIT remarked that "So long as we stick to hymns, holy water and innocuous invocations at civic banquets, we will be accepted, even applauded." Another planner dismissed as obsolete a theology of the parish that assumed a "politically docile and residentially static community." A church historian, noting the range of Catholic institutions in northern cities, commented that Catholics had "a potential for leadership in every kind of community that is limited only by their own felt sense of responsibility."[31]

Despite the disappointments in Chelsea and Lackawanna, Alinsky also rejoined the fray. In 1959 testimony before the U.S. Civil Rights Commission, Alinsky declared that "No white Chicago community wants Negroes." What white communities, including the Back of the Yards where he had begun his career, had seen of "integration"—a decrease in home values, overcrowding of the neighborhood, rapid transition—mocked appeals to higher principles. Alluding to the vital role played by the Catholic Church, Alinsky compared the situation to the problem faced by a "foreign language national parish established to serve one particular group [and] faced with a fearful dilemma when the prospect of a radical change in the population's composition arises." (In a separate interview Alinsky emphasized that "When a community changes from white to Negro, the Catholic church is in a different position than Protestant and Jewish churches. It has a bigger real investment.")[32]

The solution, Alinsky argued, was simple. Few whites were naive enough to believe that their community could prevent the entrance of African-Americans for eternity. Neighborhoods should establish minority quotas, perhaps 5 percent, for any particular area. Giving community organizations in each neighborhood the right to limit African-American penetration would prevent the devastating cycle of resistance, abandonment, and resegregation. "This kind of organization," Alinsky claimed, "would be based on their issues of vital concern and framed by their self-interest; an enlightened self-interest which accepts the previously mentioned facts of life."[33]

Alinsky delighted in the furor created by his proposal. "I recently stepped hard on a great many morally swollen and pompous toes in this town," he told one friend.[34] Since an Alinsky aide also wrote the testimony of Chicago's Archbishop Meyer before the same committee, Alinsky particularly enjoyed the similarities between his own statement and the archbishop's plea for "community organization[s] to ensure that Negroes do gain access to our communities, but not to the degree that we merely extend the boundaries of the racial ghetto."[35] (Liberal Catholic groups quickly attempted to decouple the Chicago archdiocese from discussions of quotas. "The Archbishop," the Catholic Interracial Council decided, "was discussing moral principles and moral goals, and not discussing or advocating particular strategies or techniques.")[36]

Following Alinsky's testimony, civil rights commission member and Notre Dame president Father Theodore Hesburgh asked him whether extant community organizations could serve his purposes. Alinsky replied that he was "aching to answer you" but declined for "organizational reasons." In fact, Alinsky, his ally Monsignor John J. Egan, the head of the Cardinal's Conservation Committee, and Archbishop Meyer had that spring launched an organizational project on the city's heavily Catholic South Side.[37] Here residents of the mainly Catholic neighborhoods west of the African-American ghetto had watched the ghetto expand in spurts for a generation, first meeting a wall of resistance, then conquering it, and again moving west. "Predominantly Catholic communities," according to one IAF report, "which have nurtured and supported large parishes, have disappeared overnight, to be replaced by overwhelmingly Protestant and non-Catholic populations." By the late 1950s, most of the South Side neighborhoods and parishes (notably Visitation, where rioting had occurred a decade before) were still virtually all-white, but parishioners viewed racial change with extraordinary foreboding.[38]

Upon Monsignor Egan's urging, Alinsky agreed to build a community organization in the area. Remarkably, Alinsky received support from Chicago's new archbishop, Albert Meyer. The shy, cautious Meyer provided a marked contrast to the flamboyant Alinsky, but Meyer came to Chicago committed to integration and eager to stop the population turnover and violence common to neighborhoods near the color line. In the spring of 1959, Meyer gathered the pastors of ten Catholic churches on the southwest side and allowed Egan and Alinsky to address them on the virtues of the new Organization for a Southwest Community (OSC) as a method of stable integration. Following the speeches, Meyer ordered the assembled pastors to fund the ecumenical organization and endorsed it in the strongest terms. The next year, in a conference for diocesan clergy, Meyer referred to the OSC and encouraged pastors "to work out the problems of community change on a community-wide basis . . . bearing in mind that change and tensions are not just a Catholic problem."[39]

As Meyer's comment suggests, organizers attempted to include all denominational groups. To Protestant ministers fearful of advocating racial integration—with the possibility that the congregation might then oust the pastor—IAF organizers stressed the safety of group action. To Catholic audiences, Alinsky stressed the organization's practicality. "When the Roman Catholic Church, . . . or any other organization, attempts to cope with problems outside of its particular specific functions," he argued, "it must then join forces with every other possible community resource." In meetings with priests, Alinsky emphasized the "analogy of vaccination" or "the idea of being inoculated with a certain quota of Negroes for white immunity."[40] Organizers also attempted to prevent harassment of African-American visitors to the neighborhood, and the almost ritualistic vandalism of the first African-American home on a particular block by gangs such as "The Saints."[41]

The most important institutions in the area were Catholic parishes. A cluster of second- and third-generation, heavily Irish-American parishes, with the requisite grade schools, high schools, convents (and at one parish even a football stadium), enrolled the majority of the neighborhood's residents. To the delight of organizers, Monsignor John McMahon of St. Sabina's gave the OSC his complete support. Fresh from an assignment as pastor in another Chicago parish that had undergone a painful racial transition, McMahon in the early 1950s began a parish conservation program using techniques developed by Alinsky and Monsignor O'Grady. Meeting with small groups in the parish, he made members of parish devotional societies leaders of block clubs to watch for illegal home-conversions. McMahon also assisted in the formation of a parish credit union to enable parishioners to obtain home loans. Each night, he walked the neighborhood, greeting parishioners and peering down alleys to check on trash collection. Since he viewed "our most serious problem [as] hysteria," he refused to discuss racial issues in the pulpit. Instead, he urged parishioners to have the attitude that "nowhere else can you get the service and conveniences that our particular parish can give you."[42]

Other pastors were less supportive. Monsignor Stephen McMahon, pastor of Little Flower parish since 1925, refused to participate, and other pastors suspicious of integration were known "for their disapproval of what O.S.C. tries to do." Despite Archbishop Meyer's endorsement, many clergy and parishioners, ill with what Monsignor Egan termed the "virus of anti-communism," instinctively equated integration with communist sympathies.[43] The situation was particularly tense at St. Leo's. As in so many of the Catholic parishes of the era, church and community in the heavily Catholic neighborhood had become synonymous. Bulletins informed parishioners that a collection for an addition to the parish grade school was "part of a large program that assures you of a neighborhood worth living in, and worth sacrificing for." Local newspapers assured readers that the vice of materialism could be avoided only "through the community wherein is rooted an active parish such as St. Leo."[44]

Directly in the path of the expanding African-American community, Monsignor Patrick Molloy and the parishioners at St. Leo's had feared racial transition for a number of years. Molloy himself embodied all of the stereotypes of a big-city Catholic pastor. Connected to the city's politicians through friendships with local aldermen and the park commissioner, he enjoyed speeding down to Comiskey Park for a White Sox game and easing into box seats just behind the Mayor.[45]

Molloy also possessed a well-deserved reputation for bigotry. In his previous assignment, he had informed an aspiring African-American seminarian that "we've been doing well enough without you and we'll keep on doing the same thing." At St. Leo's, he occasionally ordered "burrheads" out of the church, watched area real estate transactions, and kept the school lily-white through the 1950s. African-American newspapers printed articles alleging discrimination in parish activities and hostility to African-American Catholics. When one African-American woman appeared at the parish door to register in 1960, Molloy bluntly informed her that no room existed for "you people." He added that "I don't give a damn about him [Catholic school superintendent Monsignor William McManus] or [Archbishop] Meyer. Nobody's gonna push me around."[46]

At the same time, however, a desperate tone crept into parish bulletins. Commenting on urban renewal plans, Molloy pleaded that "we must take care of the people that we dispossess and they should be taken care of in the neighborhood in which they were born and raised." Molloy constantly reiterated the virtues of city living and scorned suburban migrants. "It is a foregone conclusion," one bulletin noted, "that undesirables will be everywhere, not only here in our city but in the suburbs. Their birth rate is practically four to one hence there is no use being panicky and immediately moving out, because it makes no difference where you go, you will find more undesirables." Molloy urged parishioners not to "sell your birthright to some real estate man and then lament your loss throughout your lives! Clean up, paint up and be proud of your Auburn Park. It is YOUR home, the home of your Father and Mother." When rumors swept the neighborhood that Molloy himself might leave the parish, he announced at all masses that "I won't leave this parish until I'm carried to Mt. Olivet [a local cemetery] with six handles."[47]

Community organizers capitalized on these fears and persuaded Molloy to support the OSC. Given what one source called the "violence of his attitude toward negroes as expressed during the days of the Auburn-Highland Improvement Association," self-interest was clearly his primary motivation. Still, by the end of 1959, Molloy had personally donated $5,000 to the group and publicly denounced pamphlets implying the organization was a communist front. In October 1960, the OSC held its first convention, bringing together Catholic and Protestant, white and black, for the first time in neighborhood history. Within the year, the OSC began conservation and home-loan programs geared at encouraging young families to stay in the neighborhood. (Much of the area was already "redlined," with bankers unwilling to give long-term mortgages in the area on the self-fulfilling assumption that racial change, and a drop in property values, would occur before the mortgage came due.) Pledging to make the community a "more attractive place in which to live and rear families," the OSC began working to create what its Catholic admirers hoped would be "the most ambitious and the most serious attempt to cope with the hard problems of the racially changing neighborhood."[48]

Indeed by the end of 1961, community organizations with Catholic sponsorship were sprouting all over the city. In Southeast Chicago, priests played key roles in a new community group. One priest told the eight hundred delegates to the group's first convention that "It is essential that today the ugly name of segregationist be taken away from our organization" as he urged members to draw organizational boundaries without eliminating the one African-American neighborhood in the area. In Northwest Chicago, the city's largest community organization was formed in 1961. In an area containing 160,000 people and 55 percent Catholic, the signs of racial transition were less advanced than on the South Side, but the movement of African-Americans into the area had promoted fears of massive white flight. Again, Archbishop Meyer compelled the pastors of the twenty-two Catholic churches in the area to fund the ecumenical group, and IAF organizers began canvassing the neighborhood attempting to prevent the predictable cycle of resistance and abandonment.[49]

Most remarkably, a Catholic priest working in one of the city's most impoverished African-American ghettoes, Woodlawn, persuaded Alinsky to implement long-dormant plans for a community organization in an African-American community. The Reverend Martin Farrell had been working in African-American parishes for over twenty years by 1960. That spring, Farrell and another priest quietly pressured the archdiocesan school board into integrating a number of large, informally segregated Catholic high schools. Threatening to "go to the pulpits" if twenty African-American Catholic students were not admitted to each of six schools, Farrell recruited one hundred and twenty African-American Catholic eighth graders and convinced them to apply for admission. If principals at the various high schools balked—through special days of admission or requests for student "interviews"—Farrell appealed to Archbishop Meyer, who then forced the principals to acquiesce in the program.[50]

Farrell developed his interest in community organizations with similar zeal. Following a reading of Alinsky's *Reveille for Radicals,* he became convinced that only someone with Alinsky's hard-nosed trust in popular democracy could jolt city government into working for the African-American community. "It is quite evident to me," he wrote to Alinsky, "that all and any effort for neighborhood betterment must come from the people without any outside help." For three years Farrell pestered Alinsky, arguing that "You can't call yourself a great organizer until you organize Negroes." Finally, in 1960, Alinsky began constructing The Woodlawn Organization (TWO). One Presbyterian church in the area came up with $22,000 in seed money, and a foundation with close ties to Alinsky kicked in $69,000. In a remarkable signal of the archdiocesan commitment to community organization—and in a community no more than 5 percent Catholic—Farrell and Alinsky also persuaded Archbishop Meyer and Catholic Charities to give the project $150,000 over three years.[51]

The organization grew quickly, as residents rallied to campaigns against urban renewal and unfair merchant practices. University of Chicago officials eager to purchase sections of Woodlawn in order to finance campus expansion, as well as African-American politicians tied to the Daley machine, became the requisite enemies. The movement also latched onto the energy created by the civil rights movement just beginning in the South. (One memorable meeting in the St. Cyril's Catholic Church hall ended with freedom riders fresh from the South teaching mesmerized Chicago residents "We Shall Overcome.") Several weeks later, busloads of area residents, with Catholic priests and nuns in tow to intimidate overeager policemen, registered voters at City Hall. By the spring of 1962, in a marker of the organization's new-found influence, Mayor Daley and civil rights leader Ralph Abernathy, greeted by flower girls and altar boys from Father Farrell's parish, addressed a standing-room crowd of 1,200 gathered for TWO's first convention.[52]

II

Mayor Daley's appearance at the TWO convention suggests links between parish and political life. Since the late nineteenth century, priest and politician had created a well-worn path between parish rectory and ward office, using ties formed in one place to assist in problems encountered in another. Geographical boundaries, an all-male hierarchy, and organizational discipline were characteristic of both the Catholic parish system and the Democratic party wards. Unsurprisingly, members of the same parochial school class, or often the same family, rose to the top of each organization, creating sets of overlapping and subtle connections. One French sociologist described Irish-American priests and politicians as "parallel forces, neither one controlled by the other, but both working together for common ideals and interests."[53]

In Jersey City, New Jersey, (estimated to be 75 percent Catholic in the early 1940s), Mayor Frank Hague carefully cultivated contacts with local clergy. All officeholders, regardless of religious background, were encouraged to contribute to diocesan fund drives, the Democratic party rented parish halls for meetings, city contractors paved the occasional parochial school playground free of charge, and ward offices found employment for parishioners recommended by the local priest. Hague also established polling places across from local Catholic churches, hoping to catch the large crowds leaving the building on those Tuesdays that happened to be Holy Days of Obligation. In one memoir from the period, a Jersey City cleric casually used the expression "Thank God for friends in City Hall" after describing municipal assistance in the construction of a parochial high school.[54]

Similar connections existed across the urban North. "There is hardly a corner of the city," wrote one Chicago analyst, "that does not boast in its neighborhood lore of a monsignor who in days gone by was a political power in his own right." In Boston, Dorchester priests advised residents on the proper time to sell their homes and counseled government officials on programs within parish boundaries. One 1920s analysis of St. Stephen's parish, on Manhattan's Lower East Side, matter-of-factly noted that parish organizations "endeavo[r] to secure places for those unemployed. Sometimes this is only a matter of political influence. Employment is easily procured if the client belongs to the right political party."[55]

Even African-American Catholic parishes developed links to local political machines. In both New York and Chicago, outside observers commented on how frequently African-American Catholics landed the jobs doled out by white Catholic political chieftains. (Here the role of the clergy in connecting the two groups was essential, as was, perhaps, the sense among Catholic politicians that African-American Catholics were imbued with the same organizational loyalties.) An embittered African-American Protestant minister declared in 1940 that "The Negroes have joined the Roman Catholic churches in Chicago in large numbers largely because of the economic advantages. They help people into jobs through their employment agencies and conduct their own charity."[56]

These political and parish ties were an important backdrop for the development of postwar public housing, urban renewal, and highway programs. As the largest private landowner in the northern cities, Church officials helped link governmental agencies run by Catholic politicians with Catholic construction firms and building unions. And the Catholic dioceses themselves became important sources of construction contracts. The New York archdiocese, for example, poured $353 million into new buildings in a twenty-year span, with virtually all of the spending coming between 1947 and 1960.[57] "As one moves around the country," Monsignor John O'Grady concluded in 1950, "he cannot fail to note the extent to which priests are involved in discussions in regard to superhighways and new public utilities." These developments, he warned in another address, as well as in testimony before congressional committees, "will call for considerable shifts in our population. This of course in turn is bound to affect many of our parishes."[58]

Many of the reform-minded mayors sponsoring renewal projects during the 1950s—Daley in Chicago, John Collins in Boston, David Lawrence in Pittsburgh—were Catholics and facilitated access to local politicians. Indeed, Steven Avella demonstrates how Church officials in Chicago played pivotal roles in the complicated negotiations so needed for the success of urban building programs. Through lobbying of local officials, not only Cardinal Stritch but even individual parish priests, managed to alter the proposed routes of new Chicago expressways. Boston priests obtained positions on the Boston Housing Authority and later the Boston Redevelopment Agency. "Citizen seminars" at Jesuit Boston College served to introduce upwardly mobile Irish politicians to Yankee businessmen interested in downtown development.[59] One man, New York financial mogul Thomas Shanahan, served simultaneously as the chairman of Cardinal Spellman's Committee for the Laity (a fundraising group), vice chairman of the City Housing Authority and financial chairman of the Democratic party. (In 1957, Shanahan arranged for his daughter, a nun, to write her master's thesis on the career of Robert Moses.)[60]

Prior to and during the war, Catholic officials supported low-income housing programs—correctly assuming that a significant percentage of eligible tenants as well as laborers on the projects would be Catholics. Indeed, Bishop Stephen Donahue commanded New York priests to read an endorsement of one slum-clearance and public-housing measure in all pulpits the Sunday before the 1938 election. As one priest interested in urban issues later put it, "Let us remember that a large percentage of American Catholics live 'across the railroad tracks.' "[61]

After the war, the governmental focus shifted to projects designed to improve the housing of the city's poorest residents (increasingly, racial minorities) in conjunction with efforts to reverse the declines in population and business so worrisome to postwar city-planners. Evaluations by Church authorities were still generally supportive. Catholic spokesmen such as Monsignor John O'Grady referred to the "sacred obligation" to find decent housing for all citizens, and he lobbied Congress for federal housing programs. ''After all,'' O'Grady optimistically concluded, ''there should be no fundamental conflict between the interests of the Church and those who are concerned with housing and city planning.''[62]

Once the outline of postwar programs became more clear, however, Catholic skepticism increased. The actual implementation of an urban renewal program in the 1950s within the boundaries of New York's Holy Name parish left Bishop Donahue disenchanted. Embroiled in graft, the project resulted in the leveling of six square blocks on the Upper West Side, and then a delay of several years before private developers placed more expensive housing on the site. One woman from the parish wrote of the shame she felt when Grey Line tour buses cruised through the area identifying it as a ''slum,'' and of her fears for the children of Holy Name School, walking past the crumbling buildings each morning on their way to school. As Donahue noted in a letter to Robert Moses, city officials had forced three thousand parishioners to leave the area by 1956, leaving remaining parishioners with ''a large Church and large school at great expense.''[63] Another Holy Name priest complained to reporters about the inability of even the priests to get straight answers from city officials. ''Now the remainder of the neighborhood is deteriorating rapidly. . . . Family life can't survive when speculators move into an area.''[64]

Programs perceived as vital for business and civic interests—the building of freeways, the expansion of medical and university facilities—tore out large chunks of urban parishes.[65] Even as a limousine carrying Cardinal Spellman and Robert Moses glided through the Lincoln Tunnel on its opening day, for example, pastors on New York's West Side fretted over the lack of inexpensive family housing in the city. ''In the name of progress,'' wrote a St. Michael's parish historian, ''three tubes fed traffic into the neighborhood but sucked out the parochial life-blood, the faithful parishioners, until what was left was an emaciated ghost.''[66] In Milwaukee, city officials began clearing land for Interstate 94 in the early 1950s. To stem the movement away from his parish neighborhood of St. Rose of Lima—a name used interchangeably with the official Merrill Park—one parish priest initiated a community beautification and stabilization committee. Using parish funds and personnel, the committee immediately organized model-home shows, clean-up campaigns and parades through the parish in an attempt to halt racial transition and deterioration.[67]

The construction of a Lincoln Center campus for Fordham University demonstrated how the urban renewal process could take unanticipated twists and turns. To Fordham administrators, the steadily growing number of Catholic students wishing to take degrees in law, business, and education necessitated a more expansive Manhattan campus. Forced treks to Fordham's main campus in the Bronx would make the programs inaccessible. Robert Moses proposed that the Jesuits take ten acres of his Lincoln Square urban renewal project on the West Side, which also included two high schools, a repertory theater, a Philharmonic Hall, a new Metropolitan Opera House, and 4,000 units of luxury housing. When informed of this possibility, Cardinal Spellman urged the Jesuits to seize the opportunity, terming it "the greatest thing that has happened to Fordham since my predecessor Archbishop Hughes built the University."[68]

Moses resolutely quashed attempts to block Fordham's participation in the project on church and state grounds, just as he disdained protests coming from the 7,000 displaced families and 800 displaced businesses.[69] The dilemma for the archdiocese came from the fact that many of the displaced residents were Catholic, asked to move, in part, to facilitate the building of a professional school for a Catholic university. Indeed, one entire parish plant, St. Matthew's on West 68th Street, was destroyed.

Moses exploited these tensions ruthlessly. When Manhattan borough president Hulan Jack (an African-American Catholic) and a neighborhood priest expressed qualms about the project, Moses responded immediately. "It looks as if the support we counted on as to this and other Title I Slum Clearance projects in official and other circles is lacking," he informed Fordham president Laurence McGinley, S.J., and Cardinal Spellman in hand-delivered letters. "If this is the case, as it seems definitely to be, I shall have to drop out and leave the Slum Clearance to others. Tom Shanahan as Vice Chairman has the same opinion and will take the same action."[70]

Moses later told a colleague how he informed Cardinal Spellman that "this particular area was to be cleared for a new campus for Fordham University in the face of most bitter opposition and litigation and I asked His Eminence whether the Church lost a member when he moved into the next parish. He thought it was a pretty good joke and so do I." As Moses explained, "don't let's spend too much time with an individual pastor who thinks his jurisdiction and membership may be somewhat reduced. There must be adjustments in the churches to keep pace with the adjustments of the general population."[71]

Typically, government officials such as Moses defined African-American, largely non-Catholic areas as in need of renewal, but even these projects worked against the stability of heavily Catholic neighborhoods. City officials first used eminent domain to purchase and then destroy "slum housing." Since cautious private investors and government agencies tended to build only a relatively few low-priced housing units for the former residents, the rest of the displaced population moved throughout the city searching for any available housing stock. In Philadelphia, for example, the destruction of housing in lower north Philadelphia pushed African-Americans into the Catholic neighborhoods surrounding the Shibe Park baseball stadium. Human relations officials scurried throughout the area in the late 1940s and 1950s attempting to prevent violent incidents as one ''line'' after another was crossed.[72] Eventually, of course, as one Cleveland priest observed, such programs made ''it necessary that they [African-Americans] find housing and recreation in areas that previously were solely the locale of white Catholics.''[73]

A 1956 meeting of Chicago pastors moderated by Monsignor O'Grady reflected clerical frustration. ''How can [the] city acquire all this land for museums and campuses and not stop the blight?'' asked O'Grady, urging pastors to pressure city officials. ''I asked the Mayor's office for a plan of [the] next 10 years,'' commented one pastor, ''but there isn't such a thing.'' O'Grady concurred, noting that ''I have been clamoring for years to get a meeting with the highway people about this whole question of the use of City land, but we haven't even been able to get a meeting.'' ''[T]he real diabolical part,'' concluded one pastor of an African-American parish, ''is that all of this hatred will be thrown back upon the colored people that haven't one blessed thing to say about it.''[74]

The most searching critiques came from Alinsky's Chicago protégé, Monsignor John J. Egan.[75] The director of the Cardinal's Conservation Council, Egan spent the summer of 1957 as an intern on Alinsky's IAF staff and had become the primary conduit between Alinsky and the Chicago archdiocese. In a 1958 article, he described the sensation of being the only priest at a meeting of national planners, and yet knowing how crucial these issues were for ''Catholics, and especially for hard-pressed Bishops, who are trying to hold their dioceses together in an age of gigantic population upheaval.'' ''This business called urban renewal,'' he warned, ''demands some looking into.'' Privately, he argued that ''some light from the Catholic press has to be thrown on these questions so as to save many of our people and our institutions from the dictatorial powers which people like [Robert] Moses are arrogating to themselves.''[76]

These sentiments generated opposition. Boston's Monsignor Francis Lally, later chairman of the Boston Redevelopment Authority, urged Egan to ''take a positive attitude toward the possibilities for good represented in this concept.'' Egan first submitted his article to *America* magazine, which unsurprisingly sent the galley proofs to Robert Moses for comment. Moses's blistering response—''bitter, dogmatic, so biased, so factually wrong''—killed the article. (Moses also took care to send a terse one-sentence response to Father McGinley of Fordham: ''I can hardly express my disappointment.'') ''[T]he church,'' Moses argued, ''in all of its ramifications has never been overlooked, slighted or burdened with unannounced, unexpected and unconsidered problems.''[77]

Egan provoked an even more heated response by questioning the fifty-million-dollar Hyde Park urban renewal project in Chicago. Hemmed in by growing African-American ghettoes on three sides, Hyde Park residents had feared complete white flight since the end of the war. The one card area residents held was the location of the University of Chicago on its southern edge. ''It is not possible to operate and maintain a great university in a deteriorating or slum neighborhood,'' argued university chancellor Lawrence A. Kimpton, and representatives of the university in conjunction with city officials developed a plan to make the area a model for other communities contemplating urban renewal.[78]

Plans called for the destruction of several thousand dwelling units and the displacement of over 20,000 residents. Relocation efforts were scheduled over a five-year period, but with virtually no provision for low-income housing.[79] Given the liberal tenor of the community, proponents of the plan vociferously denied claims that race played a role in the project's planning. Middle-class African-Americans, the argument went, were more than welcome—the plan's only goal was to stop the spread of blight.

From the plan's inception, Catholic officials expressed skepticism. Congressman Barrett O'Hara of Chicago told a sympathetic O'Grady that he received ''letters, Monsignor, every day from people living in those [condemned] homes who don't know where to find housing within their means.''[80] The most cogent criticism of the proposal came from Egan as the representative of the Cardinal's Conservation Council. In testimony before the Chicago City Council and in a series of articles in the archdiocesan paper, Egan pleaded with governmental officials to look for metropolitan instead of neighborhood solutions. ''Remember,'' Egan told Chicago City Council members, ''you are not voting on a plan to build something; you are voting on a project to tear something down. . . . I must ask you what are we going to say to the rest of the city of Chicago, to the other neighborhoods? . . . [T]he anomaly you face is the city all slum but for one section, the one with those 'high standards.' ''[81]

Reaction to Egan's attack demonstrated the complexity of the issue. Supporters of the plan immediately accused the archdiocese of merely seeking to preserve its own investment in heavily Catholic areas of the city. ''There are very strong groups opposed to an integrated program in Chicago,'' declared the editor of the *Hyde Park Herald,* ''I am sorry to

Monsignor John Egan testifying in 1958 before the Chicago City Council on the Hyde Park urban renewal project. Archdiocese of Chicago Archives and Record Center.

say that the strongest of those groups are organized within the Catholic Church. The strongest resistance to integrated housing frequently comes from some of the Catholic parishes in Chicago.'' Catholic faculty at the University of Chicago registered their displeasure with the archdiocese, and liberals nervously feared Egan's statements ''would give any sensitive person grounds for believing that we were trying to keep Negroes from moving into other neighborhoods.'' Political insiders told Archbishop Meyer that ''Egan has disgraced us down there [City Hall], we used to have a good relationship with the Mayor.'' Conservative pastors found both Egan's association with the controversial Alinsky and public disputes distasteful. ''A little 'muscle' on the right person,'' charged Father John Gallery in a private letter to diocesan priests, ''has [always] gotten more results than headlines in the papers.''[82]

Unfortunately, ''a little muscle'' no longer sufficed. City officials pushed the renewal plan through when convinced that no single voice, particularly Egan's, represented the archdiocese.[83] By 1958, the many-sided dilemmas of racial integration and neighborhood change sliced through the Church structure. Dedicated priests and nuns continued to increase convert totals among the African-American population even as other clergy worked to prevent African-Americans from entering particular neighborhoods. Diocesan officials endorsed urban renewal and highway programs while co-religionists attacked a disregard for low- and middle-income families remaining in the city. The American bishops denounced segregation and theologians declared it a sin even as they watched Catholic parishioners fleeing racial change. A few Catholics took heart from the founding of the NCCIJ, while other Catholics castigated the easy liberalism of suburban outsiders.

What did remain clear by the end of the 1950s was the enormous scale of the problem. With larger families, and less likely, because of lower incomes and tighter neighborhood ties than other whites, to leave the northern cities, Catholics (and their church) emerged from the decade as the single largest group (and institution) in the northern cities. Even as the number of Catholics living in the central cities decreased, in other words, the Catholic share of the ''white'' population continued to climb.

Even the numbers in table 5.1 are conservative. A 1952 telephone survey, as opposed to estimates sent in by pastors, found 47.6 percent of New York City's population identifying themselves as Catholic, constituting 51.5 percent of the city's ''white'' population. A sample census of Boston's Charlestown area in 1950 found a stunning 90.3 percent of residents identifying themselves as Catholic. In some Boston census tracts the proportion of Catholics climbed to over 96 percent. Sophisticated

Table 5.1 Catholics in the Urban North

City	Catholics as % of non-African-American population, 1952	Catholics as % of non-African-American population, 1970–71
Baltimore	34	40
Boston (Suffolk County)	53	63
Brooklyn	26	44
Philadelphia	41	51
Queens, N.Y.	33	39

Note: Since estimates for the nation's African-American Catholic population are by diocese, I estimated that 90 percent of African-American Catholics lived in the relevant city. Also, the figures for Boston, are in fact, figures for Suffolk County, since data on religious membership exist only at the county level. Sources for this data: George Schuster, S.S.J., and Robert M. Kearns, S.S.J., *Statistical Profile of Black Catholics* (Washington, D.C.: Josephite Pastoral Center, 1976); Douglas W. Johnson, Paul R. Picard, Bernard Quinn, *Churches and Church Membership in the United States* (Washington, D.C.: Glenmary Research Center, 1974); *Churches and Church Membership in the United States* (New York: National Council of Churches, 1956); U.S. Bureau of the Census, *County and City Data Book, 1972* (Washington, D.C.: U.S. Government Printing Office, 1973).

studies of Chicago also placed the Catholic population at over 50 percent in large swaths of the city. In particular areas, the "white" Catholic total soared to over 70 percent.[84]

Scores of the nation's largest urban parishes continued to prosper, as yet untouched by the expanding ghetto. Not far from the troubled South Bronx, for example, Our Lady of Mercy near Fordham still held ten masses each Sunday for its over 8,000 members, many of whom lived in census tracts over 80 percent Catholic.[85] How the parishioners at Our Lady of Mercy and their contemporaries in parishes across the North would respond to neighborhood change would help shape the direction of America's racial crisis. "For the first time in our history we Catholics find ourselves fully involved in America's perennial problem," commented John McDermott, then a member of the Philadelphia Catholic Housing Council. He concluded: "[W]hile the proportion of whites in the northern cities has been declining, the proportion of Catholics has been increasing. In city after city in the North today, Catholics constitute far and away the largest single group. Thus it is that race relations in these cities to an increasing degree is a matter of white Catholic–Negro relations."[86]

SIX
WASHINGTON AND ROME

I

By the late 1950s, the formal stance of Catholic leaders on racial issues was well established. The 1958 statement by the American bishops and the excommunication of several segregationists in Louisiana received wide, favorable publicity from such disparate observers as Reinhold Niebuhr and psychologist Kenneth Clark. (Clark wrote privately of his gratitude for the "impressive and persistent contributions of the Catholic Church in America" in promoting equality.)[1]

Public awareness of these positions increased during the 1959 hearings of the U.S. Civil Rights Commission. Cardinal Spellman, speaking as a "Catholic American," attacked discrimination because "that is what the Church has always and must always believe and teach." In Chicago, a representative of the Chicago Catholic Interracial Council read an appeal by council president Sargent Shriver to recognize the "crucial role of housing in the attainment of civil rights." A few months later, the commission's acting chairman, Notre Dame president Theodore Hesburgh, expressed surprise that "so many Catholics can be so good in so many causes. . . . Yet in their social lives and social thoughts they are like a dinosaur—completely out of step with Christ and the mind of the Church on basic Catholic doctrine."[2]

As Father Hesburgh's comments suggest, Catholic leaders knew well that Catholic behavior often contradicted formal Church positions. The bulk of American Catholics lived in urban areas, often neighborhoods just becoming concerned about racial change.[3] In bishops, this knowledge produced caution. To liberals and community activists, it hinted at possibility. "Without question," argued one Catholic interracialist in 1960, "no other institution of our society could do more than the Church to solve this race and housing problem. For among the most powerful community institutions in the North, in city after city, are the Catholic parishes.''[4]

A burst of organizational energy across the country in the late 1950s and early 1960s reflected this optimism. Activities ranged from the Boston Catholic Interracial Council organizing parish chapters and public lectures to the inauguration of a study club at Buffalo's Bishop Turner High School, devoted to learning the ''Christian principles which are the foundation of good race relations.'' (Students fearful that their parents might disapprove informed them that they were simply attending Catholic Youth Organization [CYO] meetings.)[5]

The Philadelphia Catholic Housing Council sponsored a conference to discuss the integration of the fifty-one Catholic parishes with African-American members into the larger Catholic community, and Catholics in Warren, Ohio, started an integrated Trumbull County Catholic Council on Human Relations.[6] In Cleveland, the St. Augustine's Guild was formed as a result of a long battle to integrate the local Knights of Columbus. Eventually, Guild leaders ran a variety of educational programs and played an important role in establishing the Ohio Catholic Conference on Interracial Justice. (One priest in the group also persuaded the chancery office to eliminate the last strictly ''racial'' parishes). Also in Cleveland, Joseph Newman, an African-American Catholic, founded a lay group dedicated to fostering meetings between African-American and white Catholics in each other's homes. Despite what he termed Cleveland's ''sharp national boundaries in parochial and social life,'' Newman proudly noted the steadily increasing numbers of home visits.[7]

Catholic liberals became less hesitant to involve themselves in neighborhood disputes. In Boston, after threats were made to an African-American family moving into a formerly all-white housing project, Father John Lynch lectured parishioners on their responsibilities to welcome all newcomers to the neighborhood. ''Once again,'' he said, ''the entire parish and neighborhood have been put to shame by the vicious incident that occurred recently when another Negro family moved into the project. . . . We of all people should not deny their just rights to others when our forefathers suffered so much to gain theirs.''[8]

In Chicago, the Catholic Interracial Council worked to coordinate the move of African-American Walter Speedy and his family into the Visitation parish neighborhood after their home was scheduled for demolition in an urban renewal project. ''What we have all anticipated for a number of years,'' wrote one priest to the archdiocesan chancellor ''is about to happen. I am informed that a Negro family is moving into Visitation parish sometime this week.''[9]

Thirteen years earlier, several thousand neighborhood residents had protested against even the possibility of an African-American living in the neighborhood, and the parish church was viewed as the center of resistance to neighborhood change. (One estimate placed 85 percent of the neighborhood's white population as Catholic.)[10] Since three of the five children in the Speedy family attended Catholic schools, and the parents attended church with the children each week, Catholic liberals were particularly concerned about the reception of the family into the local community. (Only a few years previously, the pastor had embarrassed nuns in the parish high school by refusing the request of African-American parents to enroll their daughters.) The executive director of the Catholic Interracial Council emphasized that "in this heavily Catholic area, the parish can't help but be challenged and involved by a problem of this sort, but the religious background of this family particularly underlines this involvement."[11]

The process was painful. The Speedy family endured death threats, the discovery of some dynamite sticks on the front porch of their new home, and vandalism. With disconcerting specificity, hate mail warned the family against attending Visitation Church. The local priests offered the family little consolation. One cleric told members of the Catholic Interracial Council that "the parish could only guarantee their safety when they were *inside* the church" and predicted that "they would have a very hard time." At one point, Walter Speedy pleaded with Cardinal Meyer to use his influence to stop the harassment. "It was the Catholic Church that went on record," he argued, "that racial discrimination in our educational institutions must go." To his neighbors, in a letter to the editor of the local paper, Speedy announced that "We do not seek to socialize with the people in this neighborhood. We only want to be left in peace." He continued, "We purchased our house with our own money. We are not part of any plot or group. We had to move because the building in which we used to live is being torn down by an urban renewal project."[12] A columnist for the local African-American newspaper concluded that "the Visitation area symbolizes the reluctance of middle and working class white Catholics to admit Negroes to humanity, much less to brotherhood."[13]

For close to a year, Catholic Interracial Council members escorted the Speedy family into the church each week for Sunday Mass. Eventually, as other African-American families cautiously moved into the neighborhood, and neighborhood residents offered a few welcoming gestures, tensions eased. By the fall of 1962, the CIC could report that "Two of the Speedy children have enrolled and are attending Visitation school. The

Sisters were most gracious in receiving them and the experience of the children at school has been good.'' The following summer, priests and Catholic Interracial Council members flooded the streets of the neighborhood during a series of clashes between blacks and whites that resulted in 158 arrests and seven injured Chicago policemen. The pastor of Visitation parish led the clergy and also issued a statement urging ''no violence, no hatred, no revenge.'' Priests in area pulpits received statements to read, admonishing parishioners that ''When Catholics take part in racial conflict, they dishonor the Church and Christ who is its head. It disgraces our country around the world. It hurts the work zealous priests and sisters are trying to do in all the colored parishes in the city.''[14]

A few bishops also began to play more aggressive roles. To be sure, most Catholic leaders simply invoked the 1958 statement, rarely encouraging more substantive programs. (When the Vatican requested topic suggestions for the upcoming ecumenical council, only a handful of bishops even mentioned racial issues.)[15] Signals of change, however, came from a cohort of bishops appointed in the late 1950s and early 1960s. Baltimore's Lawrence Shehan moved into the city upon his appointment, deeming the suburban residence of his predecessor inappropriate. In the first months of his tenure, Shehan desegregated parochial schools and Catholic hospitals after receiving evidence of discriminatory practices. In a biting Lenten pastoral letter, Shehan also attacked those who failed to support an equal accommodations law then before the Baltimore City Council, including ''some Catholic legislators who represented districts heavily Catholic in population.''[16]

Albert Meyer, the new archbishop of Chicago, provided equally forceful leadership. Besides supporting Saul Alinsky's community organizations, Meyer quickly endorsed efforts by priests working on the city's South Side to desegregate Catholic high schools. ''Accepting a substantial number of Negro students,'' he conceded, ''may subject school authorities to criticism from former graduates and other persons. Be that as it may, we must do what is right without evasion, compromise or procrastination.''[17]

Meyer also called fifteen hundred diocesan clergy to a mandatory two-day 1960 conference on ''The Catholic Church and the Negro in the Archdiocese of Chicago.'' Speakers detailed the rapid expansion of the African-American ghetto and estimated that seventy-six of the city's parishes now contained African-American members. The question, as one priest put it, was ''how has the expansion of the Negro community affected the largest, most widespread institution in the city, the Catholic Church?'' The presentations emphasized the need for strong parish conversion programs, operating out of the school, if the archdiocese wished to avoid becoming a suburban institution. The archdiocese's one African-American priest, Rollins Lambert, concluded that conversion on a mass scale required the opening of all Catholic institutions to African-Americans, adding that "It is a tremendous scandal when he [the Negro] finds priests or Catholic organizations blocking his way [into new neighborhoods]." Meyer based his own address, "The Mantle of Leadership," on the 1958 bishops' statement. Noting recent papal efforts to build an African church, he called for a similar "missionary zeal" in Chicago's African-American community. He also ordered pastors to "remove from the Church on the local scene any possible taint of racial discrimination and racial segregation," as well as to deliver sermons on "this timely matter."[18]

The most impressive educational efforts occurred in Detroit, where laywoman Hope Brophy began a series of programs on racial issues under the auspices of the Archdiocesan Council of Catholic Women. Disgusted by racist homeowner groups operated by Catholics, and encouraged by Father John LaFarge to work within formal church structures if at all possible, Brophy and a few liberal clerics persuaded Archbishop Dearden to establish in 1960 an Archbishop's Commission on Human Relations (ACHR), the nation's first diocesan-funded group focusing on racial issues.[19] Dearden had arrived in Detroit only the previous year, with a nickname—"Iron John"—derived from his reputation as a theological conservative. In meetings shortly after his appointment, he startled Catholic interracialists with his ignorance about racial questions.[20]

The situation confronting the Detroit church, however, made Dearden a quick study. By 1961, sixty-three Detroit parishes had African-American members, and dozens more saw rapid neighborhood change on the horizon. Publicly, the archdiocese urged pastors to welcome new African-American parishioners and asked women's societies to provide information on the parish church to area newcomers.[21] Private reports to the archbishop painted a less optimistic picture. One priest admitted that the "Apostolate is hampered by reluctance of Slovak membership to accept wholesale invasion of their parish by the colored element." The pastor of Holy Redeemer observed that "At present Holy Redeemer parish is like a white island," while his counterpart at St. Catherine's sadly noted that "The change is continuing daily. We are reaching the point that will make it impossible to meet maintenance costs in a plant worth over two million dollars."[22]

Like his colleague, Chicago archbishop Meyer, Dearden organized a 1960 clergy conference on racial issues. At the conference, following private reports of the "absolute refusal and resistance of some pastors" to discuss integration and the willingness of Catholics to participate in segregationist homeowner groups, Dearden commanded priests to "show receptiveness to the Negro, meet them when they first move in and not after big groups come in." Another priest warned fellow clergy that only a concerted effort could bring "community attitudes in conformity with the teachings of our Church." One speaker compared Catholic commitments to the city with non-Catholic churches that "have sold out completely and gone suburban." "Time is fleeting here in Detroit," he warned, "if the pattern that has prevailed in Chicago would carry here, you might be in an integrated parish now and in a few brief years you might be serving the colored element almost entirely."[23]

Nothing better illustrates the dilemmas encountered by Catholic liberals than the education program organized by the Detroit ACHR. In 1961 and 1962, the ACHR sponsored a series of panel discussions in parishes throughout the archdiocese. Each panel included a priest, a city official or sociologist, and often a Catholic mother living in an integrated neighborhood. Programs began with the priest "outlining the Bishops' Statement of 1958, and by dealing to some extent with the morality and theology involved and with the responsibility of the individual Catholic." In the first year, panelists spoke to six thousand Catholic parishioners. A primary goal of the program was "to prepare our Catholic laity to take a more active part in property owners' associations and keep them on a rational and not emotional approach to their problem."[24]

Reports of the meetings provide a rare glimpse of parish life. Parishioners in "non-tension" areas, such as suburban Birmingham, tended to focus on "inter-faith relations" or discuss racial issues in a more abstract fashion. More heated conversations occurred in city parishes fearful of racial change. Occasionally, "foreseeing the possibility of some very unpleasant incidents," Catholic families or priests in the neighborhood asked ACHR staff for assistance. At the archdiocese's largest parish, for example, 5,000-family St. John Berchmans on Detroit's East Side, "there were no Negroes registered in the parish but a number of the priests were giving instructions to Negroes; it was apparent that a problem was approaching and the pastor began to be concerned about the matter."[25]

Occasionally, the meetings created a hopeful air. At St. Bartholomew, "the president of the local Homeowners' Association which had been known for years as a 'keep Negroes out' group, made a public admission of error, and at the last meeting was seen exchanging telephone numbers with a local Negro minister and the president of a Negro Improvement Association." Following another meeting, one observer commented on "the endless good will and fervent desire of simple people."[26]

More commonly, organizers acknowledged the difficulty in voluntary educational sessions of altering attitudes formed over a lifetime. One participant recalled the "extremely hostile" attitudes evident in many parishes toward discussion of open housing.[27] Another participant emphasized that attitudes on Detroit's East Side "had become so rigid that it was virtually impossible to get any sizeable number of parishes to call the congregation together in solemn conclave . . . and explain that race was very much a moral problem. Our contacts with parishes more often than not elicited such stock replies as 'We don't have the problem' or 'This is a very touchy subject, and we don't want to cause any trouble.' "[28]

In St. Francis De Sales parish, the ACHR ran an educational series "because of the continued panic in this area, and the failure to touch in depth this parish." Unfortunately, "this was not received with too much enthusiasm by either the pastor or his people, many of whom were in the older age brackets." At St. Bartholomew, three hundred people packed the first meeting, preventing the original plan for a "nucleus training program." An ACHR representative wrote that "This was our first experience with a Polish Catholic parish. Feelings were running so high that we developed the practice of chairing our own meetings." In addition, the "program was seriously hampered by the fact that we never saw a single parish priest in five meetings. These people look to their pastor for leadership."[29]

Another ACHR volunteer commented on the need for clergy education. "As it is," the volunteer noted, "we have situations where racial change is taking place less than a mile away, or where a priest will insist 'we have no problems' while their parishioners are busy out throwing rocks or attending Homeowner's meetings organized to threaten Negro citizens." One Detroit pastor in a changing neighborhood pondered this irony. "I see the appalling ignorance of the scornful indifference and stiff resistance to the social message of the Gospel. Maybe we are afraid to teach it. Further, on a parish level we have no lay leadership, nor any notion of it. We have not taught the layman his vital role in the Church."[30]

Of course, in a sense, laypeople had learned their roles quite well. A study of the city using data collected in the late 1950s emphasized the importance of institutional religion for Detroit's Catholics, 70 percent of whom attended services once a week.[31] For generations, these Catholics, as throughout the country, had absorbed a gospel linking neighborhood, family, and parish. This Catholic culture continued to shape the patterns of American urban life. Priests led processions through the neighborhood to mark feast days, and both parishioners and real estate advertisements referred to neighborhoods by the parish name. One sociologist concluded in 1962 that "the neighborhood with a heavy concentration of Catholics may exhibit a higher degree of neighboring and of community organization than will others of similar social rank and urbanism.''[32]

The universalist theology favored in Rome and by many American church officials coexisted uneasily with this traditional, and yet ''Catholic,'' culture. One scholar, referring to past practices, noted that the ''*sinfulness of enforced racial segregation was not recognized by ecclesiastical authorities* or *by the theologians of that time*.''[33] A pamphlet distributed by Catholic interracialists in the early 1960s also made the problem explicit. When asked ''If Segregation is wrong now, why wasn't it always wrong?,'' the pamphlet instructed interracialists to admit that ''Precisely to help the Negro, as she did to help the Irish and the Germans and the Slavs . . . the Church did make special provision for them'' but then to term these policies ''temporary expedient[s].''[34]

Clearly, the older, more literally parochial variant of the Catholic experience was fading. (''I have no parish to speak of,'' complained one pastor in 1962, ''I discourage them from fixing their property. It is like putting money down the drain. . . . The property is old and the people are old. The money needed is in the suburbs.'')[35] And yet Catholic liberals in the early 1960s knew that difficult struggles were yet to come. Heavily Catholic neighborhoods still dotted all of the northern cities, and these same white Catholics were, according to one survey, ''the most likely to be disturbed by Negroes moving into particular neighborhoods.''[36] NCCIJ director Mathew Ahmann complained that ''there have been very few imaginative attempts on the part of Catholic people or institutions to face the problems frankly and to work to solve them.''[37]

II

What sparked Catholic imaginations was the southern civil rights movement, especially the sit-ins and freedom rides of 1960 and 1961. While older Catholic interracialists expressed qualms over ''procedure'' and hinted at communist influences upon the freedom riders, younger liberals argued that the protesters were ''teaching our country and our Church a great lesson.''[38] As the turmoil in the South briefly subsided, the few Catholic veterans of the protests filtered North to generate more support. In Chicago, the Catholic Interracial Council chose to honor the four North Carolina students (one of whom was Catholic) who organized the first sit-ins. Catholic freedom rider Terry Sullivan also implored Catholic audiences to support the movement, arguing that ''our belief in Christ's Mystical Body [should] cause us to feel involved in what happens there.'' The Detroit Catholic Interracial Council urged one student from each Catholic college to join the freedom rides.[39]

From the liberal perspective, nothing demonstrated more conclusively the inadequacy of Catholic efforts on racial issues than the tentative response of the institutional church to the southern movement. One editorial entitled "The Peril of Pussyfooting" complained that "Our Interracial Councils have not only been unprepared for the protest movement. They are unprepared today in numbers and in nerves for the coming community spasms and readjustments that must take place if codes of racial separation are to be broken." Indeed, "If large segments of Catholic opinion repudiate freedom rides, sit-ins and picketing as crude provocation, what does this imply?"[40]

A 1961 conference sponsored by the National Catholic Conference for Interracial Justice on "The New Negro" is illuminating. In contrast to previous gatherings which emphasized the Catholic solution to America's racial problems, speakers directed their comments to the "general American community" and "not merely to a Catholic audience."[41] The most impassioned address came from an African-American Catholic student, Diane Nash. In 1959, Nash had traveled from Chicago, where she had attended a Catholic elementary school, to attend Fisk University in Nashville. Almost immediately, she began to participate in student sit-ins, and she eventually left Fisk to work for the Student Non-violent Coordinating Committee (SNCC). Nash challenged her audience to believe that civil rights work was "applied religion" and "the work of our Church." Following a moving description of the courage shown by freedom riders and her own recent stay in a southern jail, Nash demanded "directness" from Catholic pulpits. "If this is not an area in which the Church must work," she concluded, "what is?" The same argument, she added, "is true for the problems which exist in the North."[42]

Catholic liberals admired Nash and other protesters for not only their courage but their tactics. While a few Catholic activists had already been calling for "*Action-type programs* . . . even if it risk[s] some of the support [we have] won or would like to win," such voices had remained a distinct minority. Now liberals hoped that Diane Nash and her colleagues would not be "lonely within the Faith," and referred to the "outstanding example of Christian forbearance" displayed by the protesters.[43]

Following the example set in the South, then, a few Catholic interracialists moved from education to "direct action." In Chicago, the CIC played a key role in integrating area suburbs. Working with other religious and civic groups, council representatives attended local meetings, distributed fliers, and continually emphasized the right of African-Americans to live in any part of Chicago. In two instances, the entering African-American family was Catholic, suggesting the ways in which an upwardly mobile African-American Catholic community placed pressure upon Catholic institutions.

In Skokie, pastor Arthur Sauer publicly welcomed an African-American Catholic family after a brick crashed through their window, while reassuring his parishioners that "we do not face the same conditions that exist in Chicago, where there have been mass move-ins of Negro families." (Another Chicago priest, Father Daniel Mallette, wrote an impassioned article noting that on the same day as the Skokie incident the archdiocese welcomed an African cardinal, Laurien Rugambwa, to the city.) Less than a year later, in Niles, the council, together with the parish priest and local Protestant ministers, spoke to community groups about a new African-American family. Although a few local Catholics feared that the "move-in was a plot engineered by [the] CIC and the other two agencies," the Catholic pastor informed his congregation that "Religion and Americanism are sacred things" and that hostility to the incoming family negated both.[44]

The council also organized the "wade-in" of a city beach. Uneasy about "mixed" use of city beaches, and resentful of African-American dominance at other beaches along Lake Michigan, whites had informally prevented African-Americans from using Rainbow Beach off of 79th Street. Small groups of CORE and NAACP volunteers had attempted to integrate the beach that summer with little success. Finally, a group of ministers and priests escorted African-American youths onto the beach. One week later, the Catholic Interracial Council sponsored another "wade-in." One priest involved in the demonstration observed that "it was important to emphasize Catholic support in an area where many regular users of the beach were Catholics." Organizers commented on the "Great number of scapulars and miraculous medals" around the necks of the white teenagers, and complimented the local priests for helping to keep order.[45]

In August 1962, nine Catholic laypeople from Chicago traveled to Albany, Georgia, in support of the local campaign against segregation. Embarrassed that "Catholics have been conspicuously absent from the front lines of the struggle in the South" the group joined a largely Protestant contingent. (That the CIC could term it "the largest group of American Catholics that has participated in the non-violent movement" suggests how minuscule the Catholic presence in the civil rights movement had been until that point.) In Albany, the presence of the Catholic protesters had little effect. The pamphlet they wrote upon returning, however, "Why We Went to Albany," demonstrates how the southern experience opened protesters' eyes to discrimination within the Church. One African-

American Catholic student from Marquette wrote of his shame when an Albany priest refused to visit him in his jail cell. "If any one thing disturbed the serenity which settled over me during our imprisonment, it was this. I was at once shocked and hurt that a priest of my own faith could act so." Another participant was disappointed to find the views of a local cleric "similar to the general thinking of the southern whites."[46]

Cementing these tentative interfaith alliances were changing notions of both church and ministry. Throughout American religious circles, the civil rights movement—and continued resistance to religious pleas for tolerance and integration—provoked a rethinking of the traditional borders between religion and public life. Chicago's Monsignor John Egan informed an audience on Chicago's Southwest Side in 1963 that he doubted liberals who "say the problem facing us is one of 'education.' They often add that it is too bad the situation has come to this . . . because if they only had a little more time, they would be able to educate the people to accept Negroes. In passing I cannot refrain from observing that in all the years I have studied race relations I have yet to come across a community that declared it had enough time."[47]

More direct displays of religious commitment to racial equality were now necessary. Egan himself focused on community organization, but Catholic liberals more commonly invoked the themes of nonviolence and direct action already inserted by southern civil rights activists into the American religious vocabulary. The first issue of the Detroit ACHR newsletter, for example, admonished Catholics to obey the "positive obligation of direct action"; a *Commonweal* editorial similarly observed that in a crisis "action becomes the best form of preaching."[48] A New York interracialist infuriated Chicago's John McDermott by describing Catholic Interracial Councils as educational groups. "The Chicago CIC would never describe itself as an 'educational' organization in the terms of your editorial," McDermott retorted. "We believe that the time for an education-only approach in race relations is long past. Chicago CIC is a social action as well as an educational organization. We are committed to the direct action movement, have participated in it, and we strongly object to any statement which would seem to put the whole CIC movement under the banner of 'education'."[49]

Events on Chicago's Loyola University campus reflected this shift in emphasis. In the summer of 1963, a group of Loyola students, led by an African-American former seminarian, publicized the fact that African-Americans were not allowed into the Illinois Club for Catholic Women. (African-American students were also not allowed to use the club's swimming pool.) Located in a building on Loyola's downtown campus, the club performed charitable endeavors under the leadership of Julia Lewis, the widow of a local Catholic philanthropist. In support of the student protesters, the Catholic Interracial Council fired off a series of angry letters to the club and Loyola. "The un-Catholic, anti-Negro policies of the Illinois Club for Catholic Women," wrote John McDermott, "are a serious embarrassment not only to Loyola University as the club's landlord, but to the whole Catholic community."[50]

The initial demonstrations failed to sway Mrs. Lewis. In the club magazine, she defended her policies, noting that "We, as a private club, have every right to decide who shall be our members. . . . Probably I have dealt with more Negroes than the local rabble rousers and I know that the really sensible and sincere ones are not interested in associating with other than their own race." She added that "the trouble in Birmingham and elsewhere, even here, is not created by the Negroes themselves but a group of so-called heroic people."[51]

Mrs. Lewis's resolve, however, was shaken by the appearance of several nuns and a priest with picket signs outside the club doors. One year previously, 130 nuns from Milwaukee had participated in a weekend retreat to discuss racial issues. One well-received talk at the retreat emphasized the importance of nuns moving away from "Angelism" and becoming aware of "other people's problems." The following spring, several nuns from the same group listened to an African-American Catholic woman just returned from Albany, Georgia. When, she asked, would women religious join the picket lines? After learning of the Loyola situation, the nuns, six of whom were also involved in a tutoring program in the Cabrini-Green housing project, agreed to participate in the demonstrations. (One priest on the picket line joined because African-American basketball players from Loyola's 1963 national championship squad gave a free clinic at his inner-city parish and alerted him to club policies.) The story of perhaps the first demonstration by women religious in American church history generated wide media interest. Photographs of the nuns in full habit and holding signs condemning discriminatory policies appeared in papers across the country, startling both American Catholics generally and Mrs. Lewis in particular, who quickly announced a revision of club rules.[52]

Catholic interracialists supported the nuns' actions, but more conservative Catholics, including Mrs. Lewis's son, bitterly asked why nuns in "18th century garb, with 19th century rules, should suddenly vault into the 20th century." A few CIC board members also regretted the new proclivity for "Public demonstrations against a particular person or organization in protest of social injustices being practiced." Significantly, the nuns justified their decision using the new language of public witness. Sister Anthony Claret noted that "We did this not just because we were students at Loyola and teachers; we did this most of all because we are religious." A colleague, Sister Angelica, concurred. "[T]he unfavorable remarks we did receive are, I think, a serious indication of lack of understanding of just what is the Church."[53]

III

Of course understandings of the Church were rapidly changing. Even as Sister Angelica spoke, the American bishops were pondering the results of the first session of the Second Vatican Council, which had opened in the fall of 1962. Most of the American bishops had envisioned a brief, even perfunctory meeting. Only in the context of the council itself, particularly after the failure of attempts by the Roman Curia to channel the bishops away from controversial topics, did it become clear that both the pope and more activist bishops envisioned changes of breathtaking proportions.[54]

The most important change emerging from both the preparation and the initial session was a new interfaith spirit. For much of the twentieth century, Catholic officials at all levels had warned parishioners and clergy of the dangers inevitable in encounters with other faiths. Bishops and the apostolic delegate continued to set restrictions on Catholic participation in the National Conference of Christians and Jews, as well as forbidding attendance at meetings sponsored by the World Council of Churches. Parish priests cautioned against "mixed" dating, let alone marriage, and informed parishioners that entrance into a Protestant church or Jewish synagogue constituted serious sin. When combined with the extant Catholic ghetto of schools, social groups, and professional societies, these admonitions produced a culture unusually ignorant of other faith traditions.[55]

Catholic interracialists had long been an exception to this general rule. Catholic interracial councils had often welcomed non-Catholic members and generally emphasized the importance of united attacks on race problems. As early as 1942, John LaFarge had urged a "Long and careful and co-operative probing into the principles—ethical and religious—which we hold in some fashion in common with those not of our faith." Activist cleric Philip Berrigan more dramatically made the connection between a new vision of the Church and racial questions. "No current social problem tests the vocation of the Christian as fiercely as does segregation," Berrigan argued in 1961, adding that "in regard to segregation we must realize that the Church has now passed the rear-guard action of post-Reformation apologetics and has entered a new era with the world."[56]

Other Catholic experts in community organization urged cooperation on housing and neighborhood issues. One speaker at a 1953 conference warned priests that they "will soon discover that they must have the cooperation of the ministers and the rabbis." Monsignor John O'Grady urged Catholic charities officials to abandon "the traditional approach of the Church which has been to build up its own group and separate it from all other groups in the community."[57] "So many of us Catholics," argued a New York community organizer, "are now realizing that we have to work with all the people, not just those in our own parish. We have to get out of the [Catholic] ghetto and really move into the neighborhood."[58]

These efforts increasingly bore fruit. In 1960, Protestant, Jewish, and Catholic groups in Chicago combined for a conference on housing issues. In addition, Saul Alinsky's persistent emphasis on interreligious community organizations—and his equally strong support from the Catholic archdiocese—helped weaken denominational barriers on the city's Southwest Side. One Protestant minister who participated in the programs observed, "that was the first time that I had a sense of Protestant and Catholic clergy actually sitting down together and talking about issues."[59] In Detroit, groups working in changing neighborhoods routinely "contacted religious leaders of other denominations in the vicinity." Correspondingly, Protestant theologian Gibson Winter drew attention to the "suburban captivity" of the Protestant churches, and both Protestant and Jewish leaders "expressed their admiration for the Catholic refusal to desert a changing neighborhood even in the face of overwhelming financial crises for many parishes."[60]

Still, as late as the early 1960s, the walls separating the denominations remained firm. An account of an open housing petition drive in a Connecticut town over 60 percent Catholic noted that "Catholic priests and Protestant ministers barely knew each other, and there were no informal friendships among men of different religious persuasions." (Significantly, Protestant ministers viewed Catholic participation in the drive as vital since Catholics dominated the neighborhoods deemed most suitable for upwardly mobile African-Americans.) The organizers of the drive scornfully compared the enthusiasm displayed by Catholic pastors for the local Catholic school-funding drive with their apathy toward housing issues.[61] A Philadelphia Catholic interracialist wrote with disgust that "it is easier to obtain cooperation from a variety of non-Catholic sources for educational or leadership training work than it is to obtain cooperation from local Catholic leaders, even where an area is predominantly Catholic."[62]

Ministers and rabbis with lingering suspicions of "Catholic power" also expressed skepticism concerning Catholic motives, especially in new community organizations. The liberal *Christian Century* condemned "the Roman Catholic Church and the Democratic Party organization" for their efforts to build a community organization in New York's Chelsea area. To a group of Chicago opponents—one article was entitled "Church Supports Hate Group"—Catholic officials supported Saul Alinsky's community groups solely because "The archdiocese also has a stake in preventing relocation of racial minorities, which might endanger all-white Catholic parishes in the neighborhood."[63] An Episcopalian connected to the ecumenical Packard Manse center in Boston complained that "it is perfectly clear that the Roman hierarchy, happily related all up and down the line to our politicians, is not going to do much . . . about our screaming problems." One Protestant theologian recalled his perception of "all those Catholics on the other side."[64]

The new openness in Rome restructured these relationships. Even as bishops measured their words in order not to offend the non-Catholic observers clustered together near the front of St. Peter's Basilica, American Catholics began working with other religious leaders. Given the caution with which all religious figures approached controversial theological disputes, racial issues seemed a logical, even noncontroversial, avenue to joint action. In Detroit, Jewish and Protestant leaders welcomed Catholic overtures to form an open-occupancy conference to discuss racial and housing problems within the city. Leaders of the three faiths in San Francisco formed racial discussion groups, and in Jersey City clergy from all traditions met to "discuss the role of the Churches in the civil rights issue."[65]

The National Conference on Religion and Race marked the intersection of these trends. In June 1962, the National Catholic Conference for Interracial Justice released plans for a meeting of religious leaders to coincide with the one hundredth anniversary of the Emancipation Proclamation. (In fact, Catholic interracialists from Chicago had joined Martin Luther King in Albany the previous fall partially because "It seemed hypocritical to be staging a huge interreligious conference on race next year and be unable to act now with our Protestant neighbors in responding to a specific emergency.")[66] In remarks made before the conference, immediately following his return from Rome, Archbishop Meyer explicitly linked discussions on race to the Vatican Council. As one report noted, Meyer "pointed up the splendid atmosphere of genuine brotherhood under God that existed in those days—among the hundreds of Council Fathers, . . . men of all races and colors and ethnic and social backgrounds, and the brotherhood that shone forth among the Council Fathers and the observers from other faiths."[67]

In January 1963, six hundred and fifty-seven delegates, representing sixty-seven national religious organizations, journeyed to Chicago to discuss this ''common moral and social problem.'' The conference's ''astonishingly open spirit,'' according to John LaFarge, ''reflected to a notable degree the [Vatican] Council's attitude.''[68] Observers marveled at the sight of Catholic and Greek Orthodox bishops, rabbis, African-American and white ministers, all mingling in hotel corridors and commenting upon conference presentations. Cardinal Meyer emphasized that ''establishing a really integrated community is by its very nature a task for us all: not separately and alone, but jointly as well.'' Martin Luther King made the most dramatic appeal. Only weeks before beginning his participation in the Birmingham campaign, King scored the churches and synagogues for their inadequate response to the present crisis. ''In the midst of a nation rife with racial animosity,'' King maintained, ''the Church too often has been content to mouth pious irrelevancies and sanctimonious trivialities.'' He continued, ''Will we continue to bless a status quo that needs to be blasted and reassure a social order that needs to be reformed, or will we give ourselves unreservedly to God and His kingdom?''[69]

From that point, January 1963, following the first session of the Vatican Council and the National Conference on Religion and Race, Catholic interracialists could finally claim membership in the mainstream of the national civil rights movement. Understandably, John LaFarge, counting the twenty-five bishops at the Chicago conference, saw the meeting as both a ''turning point in our nation's history'' and a remarkable contrast to the embattled, courageous work of a few Catholic interracialists in previous decades. Catholic liberals could still bemoan the ''social distance between the Catholics and the white liberal-Unitarian-Quaker-Jewish civil righters'' but the trend toward ecumenical, aggressive action was now clear.[70]

In Baltimore, for example, a local priest active in the interracial movement received a phone call from the NCCIJ imploring him to promote Catholic participation in a Fourth of July protest at a segregated amusement park. After hurried consultations with the archbishop, a group of Catholic priests joined Protestant and Jewish clergy in a demonstration that led to their arrests by local police. (One priest recalled the stunned looks on the policemen's faces as they arrested their own clergy.) Following what the *New York Times* called the nation's ''first . . . [interreligious] direct concerted protest against discrimination,'' the Catholic Interracial Council then helped form a picket line outside the park as priests distributed copies of the archbishop's recent pastoral on racial justice. One African-American Catholic proclaimed that the protest ''has done more to change patterns and ways of thoughts in that city than years of concerted efforts by dedicated organizations.''[71]

More noticeable were the priests and laypeople streaming into the nation's capital for the August 1963 March on Washington. Catholic publications promoted the march throughout the summer, and representatives from the NCCIJ served on the planning group chaired by Bayard Rustin. Upon Rustin's request, Washington archbishop Patrick O'Boyle agreed to give the invocation at the gathering. (Together with attorney general Robert Kennedy, O'Boyle also pressured SNCC leader John Lewis into moderating the tone of what O'Boyle termed a ``quite incendiary'' address.) A few bishops cautioned clergy to avoid the march, but most endorsed the effort. One pastoral letter, read in all New York pulpits, called the march a ``major commitment to justice through charity'' and cities as small as Canton, Ohio, sent buses filled with Catholic Interracial Council members.[72]

The day itself reaffirmed liberal hopes. On the Mall, priests in Roman collars carried signs marked ``Archdiocese of Boston'' while young women held signs reading ``Manhattanville College of the Sacred Heart'' and ``National Federation of Catholic College Students.'' Several hundred of the marchers attended mass in a nearby church, where worshipers joyfully sang ``Faith of Our Fathers, Living Still'' as a complement to the music already heard that day. The sacrifice of the mass became truly meaningful, wrote one observer, ``by taking Him from the altar, and from the Communion rail and from the Church into the streets of Washington D.C.'' Two weeks after the march, a Chicago priest wrote that ``No one can say that the drive for human dignity and civil rights in the United States is simply a political effort. At this stage it is a deeply religious movement and the March on Washington was also a religious meeting.''[73]

Increasingly, the Catholic reference point on civil rights matters was less the parish community than the interfaith group of religious activists located in each metropolitan area. When comparing their efforts with those of mainline Protestant or Jewish groups, let alone when measuring themselves against African-Americans risking death in the South, Catholic liberals now evaluated their contributions as minimal. Father Robert Drinan, S.J., regretted ``passive acquiescence'' by Catholics on race relations, and told Catholic lawyers in Syracuse that ``This acquiescence is manifested by the fact that Catholics are not active in civil rights groups in proportion to the Catholic population.''[74] A *Commonweal* editorial commended the ``open spirit of the ecumenical council'' and ``large Catholic participation in the March on Washington'' but emphasized the ``free-

Members of Chicago's Catholic Interracial Council about to depart for the 1963 March on Washington. Archdiocese of Chicago Archives and Record Center.

floating, abstract official statements that are rarely followed up by official action."[75]

Catholic activism after the March on Washington stemmed from the desire to change this state of affairs. In the fall of 1963, Williamstown, North Carolina, police arrested several Protestant ministers from Massachusetts, but no Catholic priests, for protesting against segregated institutions. Back in Boston, St. John's Seminary professor Father Shawn Sheehan wept in shame that no Catholic priest had dared join the ministers. In response, Sheehan organized priests and ministers into a "penitential procession for racial justice" around the State House, Boston School Committee building, and City Hall. Many of the clergy carried signs quoting either Paul VI or John XXIII on racial discrimination. A representative of the priests declared that "In penance, the accusing finger is pointed only at self. We wish to silently bear witness and admit our guilt for all forms of segregation in the Boston area." Following the march, Harvard Divinity School professor Harvey Cox declared that "Relationships between Catholics and Protestants entered a wholly new phase." Significantly, Cox added that "The presence of Catholic clergy in impressive numbers, for example, could help the racial crisis from being complicated by religious divisions. This is always a danger since most the School Committee members are Catholics while most of the civil rights leaders are Protestants."[76]

Shortly thereafter, twenty-five priests from across the country met at Chicago's O'Hare Airport to discuss the need for increased Catholic activism on civil rights issues. Among the priests were the social action or interracial justice leaders of almost every large northern diocese. The meeting resulted from the reluctance of northern bishops to allow priests to travel South, into the territory of their possibly more conservative colleagues, to join civil rights demonstrations. "We have been wishing for a way," wrote one participant, "in which some of us could join Protestant and Jewish clergy in the kind of public witness they have been giving in the struggle for human freedom." The meeting also reflected embarrassment at the minimal Catholic presence in the southern movement. During the meeting one priest asked all present to answer the following question: "Was there one among us who had not felt embarrassment by the lack of Catholic participation in some of the non-violent protests of injustice? No hands were raised."[77]

Despite these concerns, the participation of priests and laypeople in civil rights protests was already becoming routine. One group of Catholic seminarians supporting the 1964 civil rights bill joined with Protestant and Jewish counterparts in a round-the-clock vigil on the steps of the Lincoln Memorial, all sitting next to a sign reading "Civil Rights is a Moral

Issue.'' At Georgetown, archbishops O'Boyle and Shehan spoke to an interfaith rally of 6,500, pledging their support for the bill. Priests and nuns organized letter-writing campaigns to persuade wavering congressional representatives.[78]

In the spring of 1964, North Carolina police arrested two Boston priests for attempting to integrate a local restaurant. Back in Lexington, Massachusetts, their pastor, Monsignor George Casey, excitedly congratulated them upon their arrest, saying ''You boys have a lot of spunk. You've brought us right back into the Ecumenical Movement.'' Later Casey defended the priests in print, arguing that ''They couldn't stand to see the place of danger wholly occupied by Protestants, Jews, and non-believers, with the Catholics home in safety, composing nice statements by way of compensation.'' Several priests and laypeople used their vacation to travel through Mississippi during the Freedom Summer campaign of 1964.[79]

Developments in Rome were equally hopeful. In April 1963, John XXIII's condemnation of racial discrimination in *Pacem in Terris* won applause from civil rights and religious leaders, with NAACP spokesman Clarence Mitchell calling it ''an unequivocal answer to those bigots and false prophets who like to justify segregation.''[80] Six months later, the headline ''U.S. Bishops At Rome Ask Clear Race Equality Stand'' topped the front page of the *New York Times.* Reporters detailed how Louisiana bishop Robert E. Tracy, speaking for the American episcopacy, had asked the Council to ''put clearer emphasis on the equality of everyone in the Church with no distinction on account of race.'' (Part of the problem was the absence of a Latin equivalent for ''race,'' an indication of how differently issues of discrimination looked from a Roman perspective.) The American bishops received support from South African archbishop Denis E. Hurley, who urged the Council to discuss racial issues. All nonwhite people, Hurley argued in one interview, are ''insulted by the racial sins of American, South African, and Rhodesian whites, because they are so deeply conscious of their identification with those who feel the lash of the white man's scorn in the United States or Southern Africa.''[81]

The next fall Paul VI met with Martin Luther King and, according to King, ''asked me to tell the American Negroes that he is committed to the cause of civil rights in the United States.'' (The pontiff discussed racial issues with King despite the frantic attempts of FBI director J. Edgar Hoover, using Cardinal Spellman as an intermediary, to scotch the

(*facing page*) Martin Luther King, Jr., and Ralph Abernathy after a 1964 audience with Pope Paul VI. UPI/Bettmann.

meeting.)[82] Paul VI also emphasized his interest in American race relations while meeting with President Johnson during the pope's 1965 visit to the United Nations. In Rome, Bishop Andrew Grutka of Gary denounced the scandal of "parishioners abandoning their neighborhoods when people of another color attemp[t] to settle in this same area." Washington's O'Boyle added that "our experience in the United States suggests that this is one area of social action which calls for close cooperation between Catholics, Protestants, Jews, and all men of good will."[83]

In retrospect, the warm afterglow from the initial burst of educational and community programs, the March on Washington, the lobbying for the civil rights bill, and the American statements at the Council signaled the conclusion of an era. Catholic officials and intellectuals at all levels now condemned segregation. The eighty-three-year-old John LaFarge, chatting with younger Catholic activists on the speaker platform at the March on Washington, provided a living reminder of the distance traveled by Catholic interracial pioneers, but his death three months later also marked the demise of a more cautious leadership style.

What Catholic activists could not foresee was the depth of the revolution wrought by the Second Vatican Council. On the bus trip back from the March on Washington, South Bend Catholics used "group discussions" to compose "resolutions for action on problems in the South Bend area."[84] Catholicism in South Bend, however, would soon become as difficult to define as the city's problems. Just as the frequency of civil rights protests in the North began to increase, a swirl of confusion engulfed the American Church. Formerly a fixed star in the mental universe of millions of American Catholics, the Church became another in a series of institutions called into question during the latter half of the decade.

SEVEN

CIVIL RIGHTS AND THE SECOND VATICAN COUNCIL

I

Catholic participation in the southern civil rights movement culminated at Selma in March 1965. Selma's tiny Catholic population was strictly segregated, with African-Americans and whites worshiping at separate parishes. (One attempt at integration of the city's "white" parish by a group of African-American Catholic teenagers met with fierce resistance.)[1] The white resident most active in the local civil rights movement was Father Maurice Ouellet, an Edmundite priest and pastor of St. Elizabeth's, the African-American Catholic church. A Catholic hospital and fundraising operation also sponsored by the Edmundites employed a sizeable number of African-Americans, providing good wages and unusual independence from Selma's local white establishment. These African-Americans, along with St. Elizabeth's parishioners, played important roles in the local civil rights movement.[2] Montgomery bishop Thomas Toolen, by contrast, remained comfortable with local traditions. When northern white Catholics began flooding into Selma after Martin Luther King's plea for assistance, Toolen attempted to halt the activists, maintaining that outsiders were "out of place in these demonstrations—their place is at home doing God's work."[3]

Ignoring Toolen's request, priests from fifty different dioceses, laypeople, and nuns flocked to Alabama. A Detroit priest shocked by the complacency of some of the local seminarians recalled that "my preaching and discussing about commitment, involvement, witnessing, etc. seem[ed] all of a sudden hollow, unless I take . . . this step now." With the support of Cardinal Ritter, St. Louis clergy and nuns chartered two planes directly to Selma. Twenty students and three faculty members from one small Catholic college resolved to make the trip, and along with clergy from Boston, New York, and Chicago arranged flights to Montgomery, rides into Selma, and places to stay. Many of the visitors gathered each night in the St. Elizabeth rectory, where one evening they saw Lyndon Johnson thunder "We shall overcome" on television.[4]

A Catholic participant proudly noted that many civil rights leaders "pointed out with happiness and gratitude that this was the first time that so many Catholic priests, acting with their bishops' permission, had joined them on the front lines of the movement."[5] Longtime activist Ralph Abernathy congratulated another priest on the Catholic turnout, jocularly adding that "the only ones they hate more than Negroes down here are Roman Catholics, especially Monsignors." Newspaper reporters and photographers emphasized the startling image of nuns in full habit striding down Dallas County roads. *Atlanta Constitution* editor Ralph McGill claimed that the women religious "inspired the committed and shamed the timorous," and on March 11 both the *New York Times* and the *Washington Post* carried front-page photos of nuns. The *Post* quoted one African-American nun, Sr. Mary Antona, as saying "I feel wonderful."[6]

She was not alone. Combining both "witness" and interfaith action, the demonstrations united the most dramatic changes in religious life during the previous four years. One commentator noted that "as the ecumenical movement has grow in significance, so has religious involvement in the civil-rights movement. One has fed on the other." Participants repeatedly used the language of the civil rights movement, one Chicago nun arguing that "[I]t is with the powerful prose of *presence* that a new dialog has begun." The joyous, ecumenical worship services were particularly moving for those participants still unaccustomed to interfaith ventures. One nun recalled her amazement at the rabbis, ministers, "well-dressed Harvard students," and Quakers listening to the "sober praying, singing and speech-making." (She also worried that Selma residents might see her habit and be reminded of "rabid anti-Catholic horror stories.") At one point, a number of the priests decided to say their divine office (a set of daily prayers) on the street. A participant described how barricades separated the protesters from the townspeople and police while supporters crowded around them. The experience, he concluded, "was for me better than any retreat I ever made in my life."[7]

Similar events took place across the country. While one Joliet, Illinois, Catholic school teacher and a student journeyed to Selma, priests and nuns from the city marched in sympathy through the business district. Over five hundred nuns, priests, and brothers joined 15,000 New Yorkers in a walk through Harlem. One Mother Superior declared that "It is right that we should suffer and show our suffering when—in Selma or anywhere—any of God's children are oppressed." One editorial concluded that "For a great many Catholics, the pictures of demonstrating clergymen and religious, flashed on TV screens or bannered across front pages, spoke more

Nuns marching in Harlem in sympathy with Selma protestors, 1965.
UPI/Bettmann.

clearly and directly than any conciliar decree could ever do about the effective presence of the Church in the world of today.'' A nun marching down Selma's Highway 80 made the same point more emphatically. ''We are the Church,'' she declared.[8]

II

Deciding just what constitutes the Church, of course, was not a simple task. Significantly, debates provoked by the civil rights movement echoed discussions occurring in Rome. Even as activist priest Daniel Berrigan asked at Selma ''What is the Church anyway? Is it where we came from, or is it here, being created by Negroes and their white acolytes?'' the world's bishops formulated answers to the same questions.[9] Several months after the Selma protests, three years of painstaking work in Rome concluded with a flurry of pageantry and Pope Paul VI blessing the 2,392 bishops assembled for the final session of the Second Vatican Council. Under unprecedented scrutiny from both the world's media and non-Catholic observers, the bishops had authorized a series of changes that would alter Catholic life more dramatically than any set of events since the Reformation.

The impact of the ''renewal'' upon American Catholicism was particularly profound. In contrast to their Western European counterparts, American Catholic leaders before the council could, and often did, point with pride to the continued faithfulness of most of their charges, numerous religious vocations, a vast educational effort, and financial generosity. These accomplishments, however, masked theological rigidity. With a few exceptions, American Catholic leaders and theologians toed a narrow line on doctrinal matters, especially in the wake of a Vatican attack on American ''liberals'' at the turn of the century. In the seminaries, professors glossed St. Thomas Aquinas for the answer to all questions. Church doctrines did not change—they were only illuminated. Catholics in the pews came to expect a persistent reiteration of established principles on moral and sexual matters, along with the occasional reference to a vaguely defined ''social doctrine.'' Indeed, part of the difficulty Catholic liberals faced in educating the faithful on social questions was their unquestioning assumption that papal categories developed in Europe—vocational guilds, confessional political parties—were equally relevant in America. With little dissent, then, parochial energies remained focused on the construction of a myriad of social, ecclesiastical, and educational institutions.[10]

This highly disciplined intellectual climate created bishops, priests,

A plenary meeting at the Second Vatican Council, 1963. UPI/Bettmann.

nuns, and laypeople unprepared for sweeping changes. "The faithful," as one theologian observed, "had been kept . . . carefully sheltered from any suggestion that certain issues were under discussion."[11] The most dramatic alterations involved ritual and discipline. The priest now faced the congregation during mass, the vernacular replaced Latin, folk guitars appeared on the altar, churches were constructed in the round.

Underlying these changes and, again, virtually unknown in America were new conceptions of Church and community. For American Catholics interested in racial issues, three themes proved central. First, the bishops radically reframed conceptions of the nature of the Church. In place of more traditional hierarchical definitions, the bishops repeatedly described the Church by using a biblical image, the "people of God." Instead of a building or structure, then, the Church must itself be a "kind of sacrament or sign of intimate union with God, and of the unity of all mankind." All Catholics—lay and religious—shared common tasks and neither should separate themselves from the Church or the world.[12]

Second, the bishops repeatedly emphasized that a truly Catholic Church placed its "concern . . . first of all on those who are especially lowly, poor and weak . . . Like Christ we would have pity on the multitude weighed down with hunger, misery, and lack of knowledge."[13] As part of this new focus, the bishops reevaluated the relationship of the Church with contemporary society, emphasizing the duty of all Catholics to improve, not simply reject, the modern world. The final conciliar document, *Gaudium et Spes,* pledged that the Church would be alert to the "signs of the times" and respond "in language intelligible to each generation." The first lines of the document, quoted fervently by American Catholic liberals during the rest of the decade, declared that "The joys and the hopes, the griefs and the anxieties of the men of this age, especially those who are poor or in any way afflicted, these too are the joys and hopes, the griefs and anxieties of the followers of Christ." As their most important task, the people of God needed to carry forth the work of Christ who "entered the world to give witness to the truth, to rescue and not to sit in judgment, to serve and not to be served."[14]

Finally, both the proceedings and the outcomes of the council suggested the formation of a global church. At the council itself, for the first time in the modern era, native bishops from Africa, Asia, and Latin America joined their European and American colleagues. The abandonment of Latin—a language obviously tied to Western Christendom—and the unprecedented tolerance of new liturgical forms implied a union of churches as well as promoting an exchange of ideas between North and South.

Indeed, one of the indirect thrusts of the council was to remind Catholics that the demographic future of the Church lay not in the industrial West but in the less developed world. Already by the early 1960s, in response to papal requests, hundreds of American priests and nuns had begun working in Latin America, where Catholics by 1960 outnumbered their counterparts in the northern part of the continent.[15]

The cumulative impact of these shifts in emphasis—from a hierarchical to a servant church, from an institution set apart from the world to one intimately concerned with modern life, from a Western to a more global vision—was enormous. In particular, a renewed focus on social justice—a later synod called "Action on behalf of justice . . . a constitutive dimension of the preaching of the Gospel"—reshaped the relationship between religious and civil authorities in much of the developing world. The most immediate effects were in Latin America, where a conciliar emphasis on a just distribution of resources became translated into a "preferential option" for the poor and, ultimately, a liberation theology.[16]

At precisely this moment, as Catholics reevaluated their roles in contemporary society, the national focus on racial issues and urban poverty provided a mechanism for engagement with the world. Particularly for younger priests and nuns, the council provoked an extraordinary reappraisal of religious identity. For many of those reading or discussing the documents, the semicloistered traditions so powerful in the American Catholic past suddenly seemed besides the point. "Relevance and involvement; effectiveness; responsibility—it is a striking thing," commented one priest-sociologist "that the challenge of the inner city to the Church sharply defines the crisis we face as Catholics in the world." Another priest commenting on civil rights demonstrations declared that "We are in the era of the Council and it is an era not of theoretical but of applied Christianity."[17]

Understandably, then, as American Catholics attempted to implement conciliar principles, they placed enormous stress on the problems surrounding them in the northern cities. Editors of the *National Catholic Reporter* pledged in their first issue to focus on the "events that really matter—the Vatican Council, for example, or the civil rights movement." One group of Boston church members asked Mathew Ahmann of the NCCIJ "what connection do you see between the racial issue in Boston and Ecumenicism?" Nothing, Ahmann replied, is "as intimate to the renewal of Christ's Church as the removal of racial discrimination in the Church and our relevant witness to the equality of all men in contemporary society."[18]

III

This crucial conjunction—of the Vatican Council and the civil rights movement—made and marked a fundamental realignment in American Catholic culture. Once placed in the crucible of racial and theological change, liberals began questioning a number of traditional parochial forms. Throughout one conference on "The Church and the Changing City," for example, academics and activists lamented the inadequacy of older parochial structures. While one participant argued that "the inner city is today the place where the Church is failing to be the Church," another declared that "We don't know the meaning of the Church in the inner city—we may not know it in the suburbs either." One Jesuit observed that the conference's real subject was "the *changing* Church in the changing city" and referred to "current internal efforts at clarifying its [the Church's] own identity."[19]

First on the list for renewed scrutiny were the connections between American Catholicism and the African-American community. One of the most remarkable achievements of the pre-conciliar era had been the development of an African-American Catholic community in the northern cities, using largely abandoned Euro-American Catholic schools and parishes as an institutional base. In turn, this African-American Catholic population supported attempts to integrate Catholic organizations and to develop race-education programs. In one Pennsylvania incident, white Catholic liberals were "reinforced by a young and competent group of Negro professional men, practically all veterans, mostly all married and with families—whose indignation spilled over into a demand that the 'neutral' position of pastors could no longer be acceptable." The Boston Catholic Interracial Council warned Cardinal Cushing that African-American Catholics faced a "crisis of faith" because of discrimination by Catholics in Boston. "It is hard enough to be denied justice," the letter concluded, "but when that denial comes from fellow Catholics, then it is almost unbearable."[20]

For the most part, however, African-American Catholics remained culturally conservative. To the disappointment of liberals, few African-American Catholics—clergy or laity—took leadership positions in the civil rights movement. (One white priest publicly wished for a "Catholic version of Martin Luther King.")[21] With little protest from African-American parishioners, parish energy still centered on the school, and priests working in the inner-city continued to hold traditional instruction classes and emphasize institution building. As late as the early 1960s, priests working in the inner-city still argued that "convert-making constituted the special mark of their apostolate." A Chicago priest noted with surprise that a 1964 meeting of inner-city pastors even considered "direct action protests on the civil rights question. This is an extraordinary and historical development, since the pastors by and large in the Negro areas have been completely quiet on the race question."[22]

Still, even this conservative community was not immune to change. By the middle of the decade, long-time Catholic observers in the inner city noted that traditional convert programs attracted less interest, and that young African-Americans often ignored the priests and nuns treated so respectfully by their parents. Potential converts, some analysts argued, were rejecting the Church because of the paucity of African-American leaders. The mass baptisms common in the 1940s and 1950s ended. Following one tiny ceremony, a Harlem Catholic regretted that it "looked more like a Baptist baptism" than the former grand rituals.[23]

Traditional norms for clergy and religious also shifted. In contrast to their white Protestant and Jewish counterparts, a sizeable cohort of Catholic priests and nuns possessed opportunities to both work and live in the heart of each northern city's African-American ghetto. The geographical parish system enabled these white Catholics to gain an insider's view of the hopes and tensions surfacing within the African-American community, as well as providing contacts with African-Americans eager to tutor whites in the intricacies of a social movement. The experience was frequently exhilarating. Priests who initially defined their ministry in terms of baptisms administered or marriages performed found themselves riding buses to Washington and marching at Selma; nuns eagerly volunteered for positions in inner-city Catholic schools. At Chicago's St. Columbanus, a bastion of resistance to racial change in the 1940s, African-American Catholics invited SNCC leader John Lewis and other activists for parish presentations. One African-American Catholic activist also described events in Mississippi during the summer of 1964, widening the horizons of St. Columbanus parishioners and clergy.[24]

Distinguishing themselves from their older mentors, a corps of priests, seminarians, and nuns rejected a "triumphalist" ethos. Where an older priest treated honorific titles such as "monsignor" with grave seriousness since he believed that African-American parishioners viewed such titles with respect, the younger priest scorned these titles as barriers to a more democratic community. The older priest emphasized the sacramental life of the Church; conversion programs in this view merely flowed from a desire to bring the sacraments to all people. For activist clergy, the very definition of a successful parish became its ability to address social justice issues. One conference on "The Catholic Church and the Negro," for example, concluded that "If the priest is not present [at protests], no amount of preaching will ever convince the Negro that the Catholic Church is his church.''[25] A young Chicago priest, later shot and seriously wounded while working in Lowndes County, Alabama, recalled walking the streets of his southside parish. ''Had I limited all of my attention to merely that 5% [that was Catholic],'' he wrote, ''I feel that I would have been doing our Church a dishonor.''[26]

Changing notions of the priesthood were most evident among seminarians. In the context of the civil rights and anti-war movements, many students found the efforts of elderly instructors to parse the *Summa Theologica* irrelevant. The contrast between the contemplative, isolated aura of the seminaries and the conciliar thrust toward identification with the impoverished also rankled. In New York, angry rumblings from Jesuit seminarians were sharpened by daily journeys into the city to take classes at Fordham or St. John's, and the mandatory return each evening to the bucolic surroundings of a suburban seminary complex—complete with gilded altars and a lush, marble entrance—named Shrub Oak.[27]

Rules limiting seminarian involvement beyond the classroom were inevitably modified. A group of Boston seminarians studying the conciliar documents resolved to spend part of their summer vacation in Roxbury instead of relaxing at an archdiocesan retreat house. Operating out of a parish loft, they spoke with community activists and knocked on neighborhood doors. The next summer, in response to a plea from Cardinal Cushing for new ideas in inner-city Boston, the seminarians spent a month in Roxbury, and upon ordination, three of the priests received assignments to work in the area.[28] Inspired by a ''sense of mission,'' roughly one-third of the 1965 ordination class in Chicago volunteered for work in inner-city parishes, gathering monthly to compare notes.[29] Monsignor Geno Baroni noted that in Washington, D.C., ''the civil rights movement attracts many sems and they became vital partners with local SNCC and CORE groups,'' adding that these connections were immeasurably strengthened by developments in Rome. ''The results are hard to measure,'' he concluded, ''but there is a 'new breed' emerging that is knowledgeable and concerned about the witness of the Church in the modern world.''[30]

Events in New York demonstrate how rapidly the civil rights movement and the Vatican Council reshaped traditional parochial structures. On July 11, 1963, New York's Cardinal Spellman traveled to Harlem to dedicate a public housing project—the Cornelius Drew Homes—named after a longtime Harlem Catholic pastor.[31] In charge of St. Charles Borromeo parish for fifteen years, Drew had epitomized the older Catholic pastoral style. Upon his arrival, he promptly planned a new one-million-dollar community center; the fund drive began with an initial $10,000 check from

Cardinal Spellman, who emphasized the center's usefulness in combatting communism. Along with his curates and the nuns in the school, Drew supervised a plethora of organizations, ranging from convert classes that ushered 300 new members into the Church each year, to parish societies and youth boxing clubs. School tuition was one dollar a month, but Drew quietly waived it for families unable to pay this fee. The style was meticulously Roman. Drew even prohibited the previous pastor's practice of singing hymns in English at high mass and ordered the parish choir to return to the Latin. Despite such authoritarian gestures, however, Drew, like many of his colleagues in inner-city areas across the North was genuinely popular with both parishioners and non-Catholic leaders in Harlem. Local celebrities participated in parish programs, and Drew received warm praise for his efforts on housing and employment issues. (Indeed, precisely because priests like Drew had made such a permanent commitment to African-American neighborhoods, public opinion polls showed them to be among the most trusted whites in the African-American community.) Understandably, Cardinal Spellman in his dedication speech emphasized how Drew symbolized "the devoted and dedicated priests, brothers, sisters, and laity who for more than fifty years have cherished the privilege to live, to work, and to die among their friends and neighbors in Harlem."[32]

Even as Cardinal Spellman spoke, however, support for this parochial style was crumbling. That same summer, Father Philip Murnion, fresh out of the archdiocesan seminary, was assigned to a neighboring Harlem parish. At the seminary, Murnion had been one of the few seminarians to attend talks by Kenneth Clark, Whitney Young, and Roy Wilkins on civil rights issues. Upon his arrival in Harlem, Murnion noted the declining numbers of converts and the apathy of many young parishioners to traditional programs. By contrast, he saw large, excited crowds listening to Malcolm X at 114th and Lenox, only a few blocks from the parish hall. Murnion quietly put together a summer program for youth, as well as encouraging the use of gospel music during a few liturgies. (Ironically, longtime parishioners disliked what they called "other people's music." Only the students loved it.) Murnion also urged Harlem's Catholic parishes to move beyond narrowly spiritual concerns. Why, Murnion asked, are few Harlem Catholic parishes involved in riot mediation and civil rights protests? ("In the matter of the school boycott," Murnion observed, "the Church only belatedly took action by opening the doors of a few individual institutions.") Could the Catholic Church even speak to African-Americans when its white members resisted or fled integrated neighborhoods? Explicitly referring to John XXIII's request for an overhaul of church structures, Murnion pleaded with fellow priests to become involved with other local churches and institutions. A colleague at Harlem's Resurrection parish made a similar argument. "[I]n the past," he argued, "the work of the Harlem parishes was almost exclusively the work of conversion. Now the Church in Harlem is beginning to see the need to become involved in the social problems."[33]

Clergy in Brooklyn took similar positions. By the mid-1960s, one older Brooklyn priest noted, younger clergy refused to view conversion programs as the foundation for a genuine Christian community. (Indeed, at one 1966 clergy conference priests debated the proposition "Should the parish be regarded as a social agency and the priest as a social worker?") Catholic priests and sisters became the main impetus behind Christians United for Social Action and began organizing residents and sponsoring rent strikes in the Brownsville ghetto. The author of a memorial volume for Our Lady of Victory parish declared that the parish "is part of the Bedford-Stuyvesant community and must be involved in community programs or be isolated from reality and have no justification for its existence."[34]

Equally dramatic changes occurred in communities of religious women. No group in the American church remained more wedded to traditional practices. Mother superiors and, less directly, priests and the hierarchy monitored all aspects of daily life. Convent leaders rationed television and radio privileges, handled financial transactions, occasionally inspected mail, and carefully circumscribed contacts within the parish, let alone with the secular world. Daily life moved to medieval rhythms—each day began at dawn with mass or community prayer, meals were taken in common, and lights were turned out early in the evening.[35]

Two-thirds of the nation's 181,000 nuns worked as teachers. During the 1940s and 1950s, increasing numbers of sisters, spurred by state certification requirements and encouraged by Vatican officials, had received advanced degrees from Catholic universities. American sisters also formed organizations bringing together women religious from previously isolated religious orders in order to discuss common concerns. By the 1960s, these women were the best-educated nuns in the world. Still, many Catholic leaders and parishioners treated them with a patronizing mixture of awe and scorn. As one 1965 columnist put it, "By canon law and long-standing traditions, nuns are treated like the children most of them spend their lives with."[36]

Throughout the North, however, women religious did become involved in tutoring and education programs related to prejudice and discrimination. In Chicago, sisters encouraged Catholic Interracial Council educational programs in the schools during the 1950s, and a few sisters wrote theses on racial issues. In the early 1960s, sisters from several West Side convents began "discussions about the problems of the area in which they were living"—discussions which became the genesis of a diocesan-approved Urban Apostolate of the Sisters in which nuns tutored disadvantaged students and visited parishioners' homes. "These nuns now must realize," one priest commented, "that the great foreign mission apostolate of the Church is in their own backyard."[37] One nun expressed the belief that "Sisters have more responsibility than simply to save the saved. They must see themselves as daughters of the Church; not merely as teachers of third grades."[38]

The process was similar in Boston and New York. In December 1964, several groups of nuns attended a seminar on "The Church and Civil Rights" at a Boston high school. The wife of the city's leading African-American activist, Joyce King, led the nuns in a discussion of "What the Religious Teacher Can Do about Civil Rights." The day concluded with a rendition of "We Shall Overcome." The next summer, several small groups of sisters involved in independent programs formed a committee in the local Catholic Interracial Council. (This despite Cardinal Cushing's admonition that a "nun's place is in the classroom—not marching in street demonstrations.") New York Sisters of Charity embarked upon a series of monthly meetings, scripture readings, and retreats directed toward understanding the role of the sister in the modern world. In 1964, order superiors asked for two hundred volunteers to spend the summer in New York City, "learning to see Christ in the poor" and "ministering to the sick, the poor, the deprived, the culturally handicapped, the socially segregated." The following summer, members of the order looking for summer projects in the city asked sponsors to avoid "very Catholic missionary kinds of things."[39]

These efforts merged with the excitement of the Second Vatican Council. More than any other group of Catholics, women religious took seriously papal commands to reexamine their work. Conciliar pleas to become engaged with the world's problems particularly resonated with women religious frustrated by rules distancing them from modern life. Both the convent and the traditional habit became metaphors for isolation. (One book on the subject was entitled *Vows But No Walls*). Speakers at conferences for nuns contemplating the renewal of their orders suggested serving inner-city populations by "tearing down our convents and renting out apartments" since "the conventual form of theology was no longer adequate to meet the needs and answer the problems confronting the modern world." One nun maintained that "the habit has become a kind of 'skin coloring' so that prejudices and stereotypes have to be broken down before

Philadelphia nuns working in an inner-city recreation program, 1970.
Urban Archives, Temple University.

the person can gain the response due to a person and not to 'one of the good sisters.' "[40]

As with the clergy, the launching pad for engagement with the world was in the inner city. By 1967, Sister M. Charles Borromeo Muckenhirn could argue that "it is the city near or in which a convent exists which is the church, the people of God in that area." She added that "Although many were shocked at the events in Selma, there has evolved more clearly the awareness that the Christian, most of all the religious, belongs with those who have no power in this world."[41] By 1966, 800 Chicago nuns were involved in the Urban Apostolate of the Sisters. Two leaders explained to diocesan officials that they were "studying the Council documents and questioning how they can best put the vision of the Council Fathers into effect." That same year, three teams of five nuns each, all with Ph.D.'s, toured the country as representatives of the NCCIJ, giving race-relations workshops in Catholic parishes to a total audience of over 25,000.[42]

Concern over urban issues even led to questions about the foundation of American Catholic communities—the geographically defined parish. Theologian Father Richard McBrien argued that "This strong tie [of the church] with the residential system has produced an individualistic, pietistic morality which relates only to the private spheres of one's life." He continued, "The neighborhood community is not to be the sole nor even the primary locus of Church activity and mission." Why, sociologist Sister Marie Augusta Neal asked, should the "geographically-based parish be the only or main link with the people of God?" A Notre Dame professor declared that "To continue to talk about the parish as a community is a little far-fetched, since the very way in which it is defined—by territorial divisions such as major streets—has nothing at all to do with the concept of community."[43]

IV

Viewed through the prism of racial and urban issues, then, two distinctly "Catholic" but also quite different peoples emerged in the mid-1960s. On the one hand, theological and social liberals called into question the traditional roles of clergy and women religious as well as parish forms. These tensions had been evident in subtle ways since World War II, but what distinguished events in the 1960s was the conviction that the theology of the Vatican Council necessitated a less structured, more inclusive institution and the adoption of the tactics of the civil rights movement. Sister Marie Neal's commentary on one Boston parish reflects this spirit. She noted that 65 percent of the people within the parish boundaries had identified themselves as Catholic, but "only three percent were identified with any parish action program. Three percent!" She concluded that parishes "indifferent to or unaware of the plight of the inner city" risked becoming "the pious legitimation of social evil that, in this case, reinforces class superiority."[44]

For many Catholics, these criticisms came as if in another language. Large numbers of American Catholics continued to labor, socialize, and worship in patterns barely distinguishable from those developed by their immigrant ancestors. Communicants still answered with a parish name when asked "where are you from?," pastors emphasized the importance of Catholic education, and nuns oversaw the schools. Especially to Catholics remaining in the northern cities, discussions of "social evil" paled besides the fear of plummeting property values and the abandonment of traditional communities. Catholic parishes had once sanctified particular neighborhoods, fostering a particular sense of community throughout the urban North. Now, representatives of the very institution that instilled a sacramental template onto the idea of "neighborhood" seemed willing to sacrifice these same neighborhoods on the altar of integration.

In areas where the Church itself had named the local community and helped define the meaning of its parishioner's lives—through nationality, the parish school, the spires of the church building towering over the neighborhood skyline—the anger by the mid-1960s was palpable. A Cleveland woman addressed her bishop in 1964. "These people," she wrote, "are concerned about their homes, their children, and the parishes that they worked so hard for. The Catholics have been more than willing to sacrifice for all the building programs you have thought necessary, but when these are at stake, after all their efforts, you are going to see the resentment." She then listed relatives robbed or mugged in the neighborhood, concluding that "My folks and my husband's are in the St. Thomas area. 18 to 20 years ago, it was possible to live with the first ones that came around, and the priest encouraged people to stay. The ones that moved were the smart ones. Now those that are left live in constant fear for the safety of their families."[45]

One man wrote in disgust to the Chicago Catholic Interracial Council after reading about their support for demonstrations in Albany, Georgia. "I was raised in Lawndale, attended Blessed Sacrament School and Church. We had a mixed population of Irish, German, a few Italians all American born. People of Bohemian extraction and those of Polish background lived about 1 mile east of us." He continued, "Father O'Brien had built a new Church some time ago. We were all very happy about that. It took a long time to get the money together. Along came the influx of negroes. We decided we would not move to get away from them. They kept swarming in like the black plague.'' A woman, also from Chicago, added that ''They have also taken over all our parishes on the south side, first St. Columbanus, St. Dorothy's, St. Laurence's, St. Clotilde's, St. Carthage's, now they are getting St. Leo's and the poor old white folks, who built these parishes from scratch had to give up everything and the parishioners are scattered all over. . . . I hope I am as good a Catholic as you. 'Human Dignity'—my eye.'' A New York Catholic informed the Brooklyn diocesan paper that ''In the past history when the Jewish, Italian, Irish, Chinese etc., came to America, they too were segregated, but they did not cry 'we want integration' or go marching. . . . ''[46]

Reactions to the participation of priests and nuns in the Selma marches suggest the depths of this division. To Catholic liberals, the demonstrations at Selma were a triumph. ''We made the Church visible,'' proclaimed one St. Louis nun, ''illustrating that the twentieth century church has a social conscience.'' One observer commented that ''From the point of view of the ghetto we Catholics inhabit, Selma was a demonstration of the whole Church as community. . . . This time the Catholic presence amounted to something.''[47]

Liberals also connected the Selma demonstrations to conditions in the North. One Minnesota seminarian who traveled to Selma argued that his experience in inner-city ghettos had convinced him that ''the race problem was clearly all of a piece.'' In the Cleveland diocesan paper, correspondents angrily contrasted the courage of Martin Luther King with the silence of ''frequent communicants and leaders of church groups.'' Another letter writer praised the local bishop for allowing two priests to join the Selma demonstrations, arguing that in conjunction with the Vatican Council ''the Diocese of Cleveland has concretely acknowledged the relevance of religion to contemporary social problems.''[48]

That other Catholics drew different conclusions is evident from the same diocesan newspaper. One woman professed herself ''sick at heart'' from viewing pictures of marching nuns in the newspaper. Instead of public protests, she argued, the nuns ''should be down on their knees . . . praying.'' She continued, ''I was taught that the habit is blessed to help in prayer and obedience, not public demonstrations. I feel the superiors of these nuns should be punished just as we mothers should be punished for any neglect towards our children.'' Another correspondent urged priests ''to preach Christ's views on the Mystical Body . . . as did St. Paul. To join in with Communists and college brats is hardly a way to appeal to rational thinking.'' Along the same lines, another Cleveland reader maintained that coverage of events in Selma allowed ''communist leaders . . . to divide the Catholic population.'' Indeed, ''[s]egregation, birth control and civil rights are issues that are held up especially to draw the expression of different Catholic spokesmen in order to destroy unity.''[49]

Catholics in Chicago reacted with equal vehemence to the arrest of five nuns and seven priests in a protest against de facto school segregation. (The daily papers splashed photos of the nuns stepping into police paddy wagons across the front page.) A statement issued by the priests argued that ''it is an appropriate and religious act for priests and nuns'' because of the ''moral and religious issues'' involved. One African-American priest maintained that the arrests of the nuns proved that ''Chicago is worse than Selma'' since ''for the first time in the history of this country, Catholic nuns were arrested.'' African-American activist Dick Gregory applauded the demonstrators. ''Social, financial, and political control [of Chicago],'' Gregory told reporters, ''is in Catholic hands. When the priests and nuns get out there and march, it makes the whole city stand up and take notice. I'll say that if we ever do break the back of the Daley machine than we can all say a special halleluia to those priests and nuns who march with us here.''[50]

The official response of the archdiocese was to urge Catholics to obey the law. A few priests and laypeople also urged the demonstrators to avoid giving the impression that a ''Catholic'' position existed on such a complicated issue.[51] More blunt objections came from other Catholics. ''Certainly,'' argued one letter-writer, ''there are more effective ways of working for racial justice than associating with a group of unwashed beat-niks in breaking the laws of the city.'' A Mrs. Dorgan protested ''this show of nuns and priests marching for this and that. . . . We can accept the changes in the liturgy or theology, but these peacemakers stirring up trouble because of their conscience, God forbid. I have all I can do to keep my faith in God the ways things are going in this world. What has happened to the beautiful Catholic Church's unity, togetherness, same belief[?]''[52]

Mrs. Dorgan's plaintive plea for unity correctly identified the central development of the era. Scholars have recently begun to recognize the waning of denominational cohesion during the 1960s as a harbinger of a cultural shift that would shape American politics and society through the 1980s. Alliances on issues as diverse as abortion, school busing, and censorship, in other words, became predicated less on denominational affiliation than on perceptions of authority, historical change, and social justice.[53] In no place was this transition more evident than in American

Catholicism, once renowned for its organizational strength and cultural cohesiveness but increasingly torn by dissent. By the mid-1960s, the shared political and theological concerns that drew Protestant, Jewish, and Catholic liberals together in the civil rights movement, also pushed other Catholics apart from their putative religious leaders. Essentially, two moral languages—an older, highly structured communalism and a new attempt to build a "community without walls"—challenged each other for religious and cultural recognition.[54]

Before the 1960s, these tensions rarely surfaced outside ecclesiastical circles. The decision of civil rights activists to direct their attention to discrimination in the North, however, and the simultaneous tumult resulting from the Vatican Council, tore at the seams knitting together what theologians now called the People of God. When visiting with Paul VI, Martin Luther King informed him that the Church could help African-Americans in the urban North since "in these counties the Catholic Church is very strong and a reaffirmation of [the Church's] position [on civil rights] would mean much."[55] "[I]t is the Roman Catholic community which has played a large role in the development of the big Northern urban areas," agreed NCCIJ secretary Mathew Ahmann, "and which frequently makes the difference if problems are to be adequately met and solved."[56] The most vivid tests of these claims would soon occur, as Catholic liberals brought the message of Selma and the Vatican Council to the neighborhoods of the urban North.

RACIAL JUSTICE AND THE PEOPLE OF GOD

I

New approaches to racial and urban questions were evident in Philadelphia by the mid-1960s. For a generation, the city's Catholic liberals had been attempting to prod a series of church officials into active involvement with racial issues. (One archdiocesan chancellor in the 1950s demonstrated the tenacity of older Catholic traditions when he told interracialists that "communists are behind these changes" and observed that Philadelphia's Italians, New York's Puerto Ricans, and Boston's Irish "never needed any organization to work for them.")[1] In the late summer of 1963, an African-American family moved into a previously all-white section of Folcroft, just outside the city. Residents immediately vandalized their home, and a crowd of 1,000 protested their presence. Quickly, one group of Catholics agreed to repair the damaged home, while a number of priests and laypeople "worked mightily" but unsuccessfully to persuade the local pastor to calm any parishioners who might be involved in the protest. "Herculean," but futile, "efforts were made to move [Archbishop] Krol."[2]

Within the week, the archdiocesan newspaper condemned the violence, noting that Folcroft was "reportedly 34% Catholic" and urging those causing the disturbance to "make their peace with God." Five years earlier, Catholic liberals would have settled, happily, for this statement. In this instance, however, they pressed for more substantive programs. On September 9, 1963, after failing to get an appointment, a group of white and African-American Catholics surprised the archbishop at his residence. As one interracialist, Dennis Clark, noted in his diary, "They encountered him, well attended. He dodged [the protesters]. . . . The [archdiocesan] Chancellor reprimanded them." Clark added that "the point is that the inaction of those recognized as Catholic religious leaders on social matters and things of civic import is notorious, scandalous, unpardonable . . . a source of cynicism and alienation."[3]

During the next few months, Catholic liberals repeatedly visited the area, forming groups of neighborhood residents to discuss racial issues. In January, the Catholic intergroup relations council scheduled a meeting in the local parish hall to discuss racial issues. (One previous conclave had collapsed when the local pastor prohibited a meeting on such a controversial topic and called local police.) At the start of the meeting, the local pastor declared to several hundred parishioners that "These men are invading our parish and forcing it on me. I have pleaded with the chancery, 'you are hurting my people' but these people have forced their hands." He continued, "Hear what they have to say. If you don't agree with it, just quietly stand up and walk out." Five minutes later, as members of the Catholic intergroup relations council read the 1958 bishops' statement to the departing crowd, an audience of twelve remained in the hall.[4]

Stung by this latest rebuff, Catholic activists increased the tempo of their protests, hinting that African-American Catholics were threatening to leave the Church over archdiocesan inaction. Interracialists also connected problems in Philadelphia to the ecumenical movement. Dennis Clark argued that "numerous requests for official Catholic participation or representation in work for interracial justice have been made over a period of years by various civic and religious groups. These requests have been almost uniformly ignored or refused." From the interracialist perspective, "opposition to current forms of public demonstration on behalf of interracial justice" reeked of scandal, and, Clark argued in a national magazine, proved that "loyalty to Catholicism is no guarantee of common understanding."[5]

In March, Clark and his colleagues publicly reiterated their belief that "it is not possible to work in any effective formal relationship or collaboration with official Catholic leadership in Philadelphia on social action matters."[6] Finally, in a private meeting with diocesan officials, Archbishop Krol commanded one of his seminary professors to draft a proposal. Nine months after the Folcroft incident, the archdiocese announced the formation of an Archdiocesan Commission on Human Relations. The commission, Krol announced to the press, "will direct both the ecumenical movement in the Archdiocese and the apostolate for racial harmony" since the problems were "interrelated." That June, in the first joint religious ceremony within memory, representatives of several faiths prayed for racial harmony in the Cathedral of Saints Peter and Paul.[7]

II

In Philadelphia, protests worked. By placing pressure on parishioners, priests, and the archdiocese, Catholic liberals forcefully integrated Catholic institutions and developed substantive human relations programs. Divisions between the two ''Catholic'' worlds, however, were frequently less malleable. Again, at one level, the changes were remarkable. In an event unimaginable even a decade previously, Martin Luther King spoke to a large crowd at Boston's archdiocesan seminary in 1965. (Women from the Catholic Interracial Council attending King's address on the Boston Common arranged for African-American members to hold white babies, and white members to hold African-American babies.) Soon after King's visit, Father Robert Drinan, S.J., the dean of the Boston College law school, urged civil rights activists to step up the tempo of demonstrations.[8]

The next year, the Boston archdiocese formed a Commission on Human Relations to coordinate inner-city and racial education programs. Increasing numbers of priests (including a group of Benedictine monks from New Hampshire) volunteered for inner-city positions, and in 1966 four sisters began full-time work in a nondenominational school and community center in Roxbury. Women religious advocating the Roxbury program argued that it would ''eliminat[e] the barriers which separate Negro and white, the inner city and suburb, Catholic and non-Catholic, the sacred and the secular. Such endeavors seem logical outcomes of Vatican II and of the Constitution on the Church and the Decree on Christian Education.''[9]

In 1965, the city's Catholic Interracial Council marched in the annual St. Patrick's day parade in support of the local African-American community. (The previous year, teenagers in Dorchester and South Boston had pitched tomatoes, cans, and bottles at an NAACP float bearing the message ''From the fight for Irish freedom to the fight for U.S. equality.'') One hundred and fifty priests and nuns, including a few veterans of the Selma marches the previous week, sang verses of ''We Shall Overcome'' as well as ''When Irish Eyes are Smiling.''[10]

But these developments sketch only a partial portrait. A few white youths harassed the St. Patrick's Day marchers, for example, and police had to separate one of the more aggressive youths from a priest willing to take him on.[11] Members of the Boston Catholic Interracial Council also acknowledged the vast distance between episcopal statements and the actual process of ''bringing 'outsiders' '' to speak to parish groups in South Boston and Charlestown.[12]

Even more revealing are records from meetings of members of the archdiocesan Commission on Human Relations with Catholic school faculties in racially tense Dorchester. The response was generally unenthusiastic. One nun working in a local Catholic high school admitted that ''Part of the reason why we are not aware of the problem is because we never hear the news, never read the newspaper and never get out.'' (Intriguingly, discussion also turned to why African-Americans first moved into Jewish neighborhoods.) Beneath the deferential forms—when a priest from the commission opened the meeting with a "good afternoon," the teachers responded with a unanimous "good afternoon, father"—existed extraordinary tension. "All I have to do," noted one nun, "is go home and make a comment at all about this and there will be an explosion in the convent."[13]

At one meeting, tempers flared as commission members kept returning to their outlines on prejudice and racism while teachers condemned African-Americans who, in their view, moved North simply to swell the welfare rolls. "The final presentation of the specific problems facing Dorchester (the presentation included a colored map of the ghetto and the surrounding ethnic population densities) was cut short," according to the minutes, "by a barrage of heated comments." Many of the teachers also advocated a change in topic to discussion of the "white poor." Ultimately, the commission members deemed it "inadvisable to break into small groups as planned" since "[g]eneral discussion was so heated." Following the meeting one mother superior informed her charges that the sessions were a "waste of time—all this talk—it's not affecting us at this time." One pastor encouraged a young nun to "go home and say your prayers and forget about all this nonsense."[14]

Similar divisions became evident in the long dispute over urban renewal plans in the heavily Catholic, working-class Charlestown area. On the one hand, the Catholic clergy, an Alinsky-trained organizer hired by the Boston Redevelopment Agency, and a few prominent citizens favored renewal plans for the perennially depressed area. While the community organizer worked to unite neighborhood groups, Cardinal Cushing's top advisor, Monsignor Francis Lally, took up residence in a Charlestown rectory and directed the Church's support through a series of editorials in the archdiocesan and local papers. Lally emphasized that urban renewal could help "the poorest of our citizens" improve their lot by eliminating "huge areas of blight." Conceding that urban renewal had destroyed the also heavily Catholic West End, the pastors of the three Charlestown churches promised that "The aim is to save homes, not destroy them. The Charlestown renewal program is for our benefit, not the outsiders." Without action, the pastors warned, "your home and your parish will vanish." One pastor used the phrase "resurrection of Charlestown."[15]

Many Charlestown residents were less sanguine about the prospect of neighborhood change. The most prominent fear was that renewal simply meant the replacement of Charlestown's working-class residents with a more affluent population. "We want people back," announced one local politician, "not a professional man, his secretary and a dog." A town

A Boston woman protests a proposed urban renewal program, 1965.
Boston Globe Photos.

meeting to discuss the matter in 1963 ended in a raucous shouting match. (The archdiocesan newspaper the next week described renewal opponents as "expert[s] in noise making, advice-giving and simple long-winded orations.") Two years later, proponents of renewal called for another hearing. In support, Cardinal Cushing issued a statement endorsing "the most hopeful promise of rehabilitation of that beloved part of our great city." Various local priests spoke at the meeting, and one priest, according to opponents, rammed the plan through with a sudden vote.[16]

Again, the response from some town residents reflected a sense of betrayal. One parishioner rented a huge billboard overlooking the town's expressway with the simple message, "Father Flaherty is a Judas." Other opponents of renewal muttered about "thirty pieces of silver" and a flyer posted around the area showed five guards—recognizable as caricatures of the local clergy—protecting officials of the Boston Redevelopment Authority. One resident complained that "in our town you don't even ask how you are any more, you ask about urban renewal. We used to get a rest on Sundays, but it has gotten even now we get it off our altars."[17]

Tensions within the Catholic community were even more evident in areas fearful of immediate racial transition. During the 1960s, the African-American population jumped from 22.9 percent of the population to 32.7 percent in Chicago; in Cleveland from 28.6 percent to 38.3 percent; and in Detroit from 28.9 percent to 43.7 percent. Cities with relatively small African-American populations saw equally sizeable gains—in Boston the African-American population rose from 9.1 percent to 16.3 percent, in Milwaukee from 8.4 percent to 14.7 percent, and in Buffalo from 13.3 percent to 20.4 percent. The pockets of integration that did exist in the northern cities were usually ephemeral—areas midway through a complete turnover in population. In fact, African-Americans became significantly more segregated from whites in the northern cities during the 1960s.[18]

When combined with a growing African-American population, this segregation placed enormous pressure on heavily Catholic neighborhoods. National survey data indicated that Catholics were as likely as the average citizen—indeed Irish Catholics were more likely—to support open housing measures and school integration. Catholic parishes were also more likely to be integrated than Protestant congregations.[19]

The plethora of racial incidents also suggested, however, that abstract commitments to equality weakened in the local arena, especially when residents perceived their entire community as at risk. In New York City, white Catholics were the group least likely to view integration as desirable, and most likely to discount African-American claims of discrimination. Surveys of Catholics also found little enthusiasm for integrated congregations as an ideal.[20] Since many Catholics were themselves veterans of racial transition, heavily Catholic neighborhoods contained populations determined to avoid a repetition of previous patterns. "We are certain," wrote one Chicago priest, appalled that a colleague had refused to speak with African-American Catholics on the issue of integration, "that you are aware of the great fears which exist in your community area on the subject of race, fears which have their root in the fact that many families have moved there to escape integration."[21]

In this context, a few community organizations, often inspired by Saul Alinsky's efforts, struggled to create stable, integrated neighborhoods. To their credit, these organizations did bring together previously separated groups into common organizations, helping to slow the pace of change. In Detroit, eighty ministers from various faiths, led by three Catholic pastors, "realized the need for community involvement in solving the problem of resegregation" and formed the Northwest Community Organization.[22] "St. Leo parish," commented one analyst in a 1965 report on Chicago's Organization for a Southwest Community, "changed from all-white to practically all Negro in six years, whereas in some neighboring parishes where no community organization existed the total change occurred in two years or less."[23]

Still, many local pastors and residents continued to view even these efforts as misguided. In Chicago, one pastor refused to participate in the program, arguing that nothing but purchasing all available homes in the neighborhood and making "sure it is sold only to the right people" could "stop the current pattern of massive racial change." A Polish pastor struck out on his own and began planning a low-income housing project in conjunction with city aldermen, hoping that he would be the "final arbiter for exactly which families will be permitted to live in the new residential district." (A frustrated member of the archdiocesan Office of Urban Affairs noted that "as soon as it becomes publicly known that he will accept only white—mainly Polish—residents of the middle-class . . . this could be an explosive issue for the papers and the civil rights groups.") One South Side Irish pastor concluded that "In this area O.S.C. spells not so much integration as failure."[24]

Even more important, organizers struggled against metropolitan problems—limited low-income housing, a segregated real estate market, African-American poverty—beyond neighborhood control. Particularly devastating was the problem of crime. By the mid-1960s, the combination of a soaring city crime rate (especially for violent crimes) and images of urban rioting flashed on the nation's television screens prompted city residents throughout the country to place "crime" or "law and order" as the top priority in public opinion polls. In one poll of Boston homeowners, Catholics were particularly likely to list crime as the most significant urban issue. Michigan Catholics also emphasized recent riots and demonstrations when asked to identify problems facing the state.[25]

In New York City, the issue came into sharp focus during an impassioned 1966 campaign over whether to fund a civilian review board for the city's police. (Debate on the issue overshadowed a simultaneous gubernatorial race.) One day before the election, a group of two hundred priests, nuns, and laypeople took out a large advertisement in the *New York Daily News* and eight-six priests held a televised press conference in Bedford-Stuyvesant in support of the board. (An anonymous advertisement from "Roman Catholic Priests Against the Civilian Review Board," with a thug manhandling a policeman, appeared in the same issue of the *Daily News*.)[26] Following the vote, researchers discovered that a startling 83 percent of Catholics within the city—as opposed to 55 percent of the city's Jews and 63 percent of the city's total population—had opposed the board. (Fifty-four percent of the Catholics also claimed to have a relative on the city's police force.) "In virtually no instance," concluded one report, "was the Catholic-Jewish difference obliterated when other factors (national origin, education, or class) were held constant."[27]

In large part, of course, concerns over crime merged with already present racial stereotypes.[28] Many heavily Catholic urban areas contained the bulk of the urban firemen and policemen, groups whose contacts with African-Americans reenforced standard prejudices.[29] Placed in this perspective, a single incident could eradicate years of patient community organizing. Eileen McMahon details how on Chicago's Southwest Side, the murder of a seventeen-year-old white St. Sabina's parishioner by an African-American youth after a yelling match in the church parking lot—with a priest from the rectory rushing out to give the boy last rites—shattered the remaining hopes of the local community organization. "Instantly, instantly, everybody knew that name—Frank Kelly," recalled one priest, "The horror that spread! The fear it engendered! Up to that time people had been figuring they were going to buck it out." That year, one thousand families left the parish, and the next year an equal number dropped off parish rolls. Soon, the parish neighborhood was almost entirely African-American.[30]

III

Linking the various Catholic cultures proved increasingly difficult. During a painful racial transition which included door-to-door calls by real estate salesmen, a Cleveland priest noticed hostility to any sermons on the theme "loving your neighbor"; bitter parishioners assumed the topic would be a cover for a lecture on racial tolerance. In South Bend, Indiana, policemen registered their disgust with an African-American neighborhood group sponsored by local priests and seminarians. Speeches on black power in the St. Augustine parish meeting hall, they argued, precipitated the riots that swept through the city during the summer of 1967. "[T]hese guys don't know enough about the outside world," complained one white officer who had grown up in St. Augustine's parish, "They live in their own world . . . these priests-to-be are not familiar with these problems." In response, Father Daniel Peil noted that police officers habitually overreacted to even the vaguest rumor. What else, he argued, could explain the arrival of six squad cars packed with heavily armed officers at his rectory one evening, all in response to a false rumor that the parish was helping to arm African-American youths? Peil also publicly criticized priests who "traditionally obtained paved streets and political jobs and all sorts of things from City Hall for 'their people' but who have not noticeably been campaigning for civil rights legislation."[31]

Participation by priests and nuns in civil rights events and marches, as well as attempts to calm residents of heavily Catholic neighborhoods, provided the most vivid evidence of two separate worlds. Even as Chicago priests asked residents to return to their homes on hot summer evenings, then, they heard resentful catcalls. One parishioner of Visitation argued that "The church has no business in this" while another predicted that "Monsignor Wolfe will find pop bottle caps in the collection on Sunday." Why, other residents asked, were priests who should be leading "the fight to save the community" lecturing them?[32]

In Cleveland's Murray Hill area, twelve priests attempted to calm angry crowds protesting the integration of the local public school through forced busing. The crowd's response was to throw garbage on the street and jeer the clerics. "Mind your own business, Father," yelled a few bystanders, while others started a mocking chant, "Pray for us, Father." One priest from the local parish tolled his church bells for a prayer service, but only fifty people chose to attend.[33] Also in Cleveland, rumors swept St. Leo's parish that a curate planned to purchase a home and give it to African-Americans in order to integrate the parish neighborhood. The priest denied the charges, although he admitted to advocating open-housing laws after a trip "down South" where he "noticed the poverty and misery of the Negro and the way he was treated." A St. Leo parishioner generated controversy by mailing copies of the bishops' most recent statement on race to a few parishioners. "Living in a spiritual ghetto," he wrote, "one can not help to become convinced that if the Church does not start forgetting about numbers and physical plant it will destroy itself.''[34]

Members of Philadelphia's Archdiocesan Commission on Human Relations spent several tense evenings moving through crowds in the city's Kensington area after an African-American family moved beyond an informal neighborhood boundary line. (The commission organized a neighborhood tension subcommittee with priests willing to respond as city and Church officials alerted them to potentially violent situations.) Following the initial disturbances, all children in the local parochial school took home letters to their parents asking them to ensure quiet streets. Since parochial schools enrolled roughly half of the area's students, priests from both the parishes and Northeast Catholic High School attempted to spot pupils and order them away from the scene upon the threat of disciplinary action. Generally, the priests were treated with respect. At one point, however, a local priest spoke to the crowd from a bullhorn, ''asking the parents to take their little children home.'' He added that ''I hope all the Catholics in the crowd will go home.'' In response, many in the crowd booed and jeered.[35]

Philadelphia priests also mediated a dispute between African-American students at Bok Vocational School and neighborhood residents on the city's south side. African-American students bused into the white ethnic area understandably feared for their safety, particularly after one white student was stabbed in a confrontation. Along with police officers, a team of eleven priests flooded the neighborhood, visiting homes and patrolling the streets. One pastor issued a moving pastoral letter to what he termed this ''presumably Catholic'' neighborhood. A group of priests found themselves in the middle of the street at one point, with angry neighborhood residents on one side and frightened African-American students on the other. Residents cried ''get back to the altar, Father'' and ''draft dodger.''[36]

A particularly heated set of confrontations took place in response to civil rights marches during the summer of 1966 in Chicago. Beginning in the early 1960s, liberal Catholics in Chicago had become increasingly involved in the city's civil rights coalition, with numerous priests, nuns, and laypeople participating in demonstrations against the city's recalcitrant school superintendent and marching through Mayor Daley's Bridgeport

(*facing page*) Philadelphia priests attempting to calm a 1966 crowd agitated by a possible ''move-in'' of an African-American family.
Urban Archives, Temple University.

neighborhood. In addition, the city's active Catholic Interracial Council played an important role in the Coordinating Council of Community Organizations (CCCO) that directed local civil rights efforts. Along with their CCCO colleagues, the few Chicago Catholics in contact with Martin Luther King and the Southern Christian Leadership Conference had long urged King to forcefully attack the city's segregated neighborhoods and inadequate schools. (One of the key reasons the SCLC chose Chicago as the location for its first northern campaign was their perception of strong support from the city's religious community.) The previous fall, King had challenged listeners at a Catholic Interracial Council dinner to "dare to live in racially integrated neighborhoods."[37]

In July 1965, King addressed the city's Catholic Interracial Council again, preceded by the executive secretary of the council and one of the nuns arrested that summer in protests against school segregation. (Before the meeting, groups of nuns led the audience in freedom songs.) King heaped praise upon the nuns' efforts, and asked that the weight of the Catholic Church in Chicago be placed behind the movement. He also speculated that, as in Boston, the Catholic response might determine the success or failure of his upcoming campaign.[38] That fall, several priests and CIC members attended an SCLC/CCCO-sponsored retreat to discuss the Chicago campaign's direction, and the following January King met to discuss these issues with recently installed Archbishop John Cody. (Following the meeting, King reminded reporters of the Catholic presence at Selma and urged "activ[e] involvement" in the Chicago campaign by priests, nuns and laypeople.) The first highly publicized incident of the campaign—Martin and Coretta King occupying an unheated slum tenement housing five families on the city's West Side—occurred when a priest from St. Agatha's parish informed King of the families' plight.[39]

During that same period, however, questions about the relationship of Chicago church authorities to racial issues had begun to surface. Upon his arrival in Chicago, liberals had hailed Archbishop Cody as a decisive leader on racial issues, noting his willingness to excommunicate segregationists during his tenure as acting archbishop of New Orleans. The African-American newspaper, the *Chicago Defender,* praised his appointment because "For the first time in its history, this diocese has a hard-fighting prelate who will not give quarter to the enemy. . . . [His predecessors] were captive of a hard core of religionists who saw nothing sinful or un-Christ like in the separation of the races either at the communion rail or in the church pews."[40]

While liberal on racial issues, however, Cody still wielded authority in extraordinarily blunt fashion, disappointing Catholics eager for a less dictatorial style in the aftermath of the Second Vatican Council. (In New Orleans, Cody had prohibited clergy from writing letters to newspapers on any civil rights issue without his personal permission.) Upon arriving in Chicago, he removed two activist priests from diocesan advisory positions and placed them in parishes. Convinced by FBI reports that King possessed communist sympathies, Cody, like Boston's Cardinal Cushing, kept a cautious distance from the civil rights leader. (Bureau agents briefed Cody just after the archbishop's first meeting with King, detailing the ''communist backgrounds'' of some King advisors and King's own ''hypocritical behavior.'') The city's one independent alderman, Leon Despres, worried that Cody might halt progress made in the past decade. ''The church,'' Despres argued, ''could set the rights movement here back many years, or push it forward depending on what it does.''[41]

The most important test came in the summer of 1966. After the tension-filled completion of James Meredith's Mississippi march, King returned to Chicago to address a rally at Soldier Field, the kick-off for what organizers called the ''action'' phase of the campaign. Although disappointed by a low turnout, SCLC members were pleased by the strong support given to the movement in a statement read by a representative of Archbishop Cody and then reiterated in mandatory sermons in the over four hundred and fifty archdiocesan parishes. Urging Catholics to accept African-Americans into local unions and to support fair housing laws—the decision to buy or sell a home, Cody emphasized, is ''a moral decision''—the archbishop linked the civil rights movement to the Second Vatican Council as part of a ''gigantic social revolution.''[42]

That week, King struggled to control a series of disturbances sparked by police turning off a fire hydrant in the stifling July heat. (Both Archbishop Cody and Mayor Daley appeared on television and promised to open public and Catholic recreational facilities in the area. Cody asked priests and nuns in the area to walk through the West Side in religious garb.) Under pressure to demonstrate to supporters that the Chicago Freedom Movement could produce concrete results, King then authorized the first in what became a series of marches into the city's heavily Catholic, all-white far Northwest and Southwest Sides. Focusing on real estate practices, the marchers demanded the opportunity for African-Americans to live anywhere in the city. By sparking what they knew could be violent confrontations, march organizers hoped to pressure Mayor Daley and real estate officials into substantive negotiations.[43]

Inevitably, these marches illuminated the chasms separating various groups within the Catholic Church. The first marches and prayer vigils occurred in Gage Park—where precisely two out of 28,244 residents were

African-American in what one observer called the "most heavily Catholic neighborhood" in the city. (In one 1952 sample of a nearby area, 62 percent of residents identified themselves as Catholic.) Because of its location to the west of the African-American ghetto, the area had become home to many of the roughly 400,000 Catholics displaced by the expanding ghetto in the 1940s and 1950s.[44]

One representative of Archbishop Cody warned Catholics connected to the Freedom Movement that a "danger of violent response" existed in any attempt to march into Southwest Side neighborhoods without lengthy preparation. John McDermott of the Catholic Interracial Council agreed with this analysis and urged better communication with local priests, but he also informed colleagues organizing the campaign that he supported the marches. Jesse Jackson urged McDermott and other Catholics to join the proposed march in Gage Park "instead of sitting here and discussing what we do." King expressed skepticism that Cody could go "all the way down the line" since "He has taken a lot of heat for his Soldier Field statement."[45]

Ultimately, large groups of priests, nuns, and Catholic laypeople joined the marches. While a few marches produced only verbal confrontations, huge mobs of whites tossed stones and bricks at the marchers during several demonstrations. Initial police efforts to halt the violence were tentative, and one observer noted that "it was obvious that some officers were torn between their duty and their identity with their friends and neighbors in the crowd." The crowds screamed with such unrelenting fury that King declared he had "never seen anything so hostile and so hateful."[46]

Following the first set of marches, Archbishop Cody again acknowledged the legitimacy of the marchers' goals, while asking for a moratorium aimed at avoiding violent outbreaks. The editor of the archdiocesan newspaper echoed Cody's request, arguing that the marches required "the services of a good portion of our police department—leaving other areas with less than normal protection." Catholic liberals and other Freedom Committee leaders rejected these pleas. The Catholic Interracial Council urged "Catholic parishes and schools in closed communities" to "teach the relevance of the gospel to the achievement of a just and open community." Freedom Committee spokesman Al Raby appealed to the archbishop "to use his great influence not to stop legal and peaceful protests against injustice but rather to stop the racist practices in the real estate profession." Activist James Bevel was more blunt. "When there's trouble," Bevel sneered, "Daley sticks up his liberal bishop to say, 'You've gone far enough'. Well we've got news for the man. If the bishop doesn't have the courage to speak up for Christ, let me join the devil."[47]

Eventually, CCCO leaders, Mayor Daley, and a group of prominent Chicago residents (including Cody) negotiated an open housing agreement that ended the formal protests. During the negotiations, Andrew Young concluded that Cody, while more sensitive than the mayor, still operated through "political relationships rather than theological." Still, Cody did immediately authorize an impressive educational program for archdiocesan priests and parishioners immediately following the housing agreements. (A letter introducing the program acknowledged that the archbishop was "[s]timulated in some respects by the unbelievable racial hatred which has been released here in Chicago over the past few weeks.") When full details of the program were released several months later, Al Raby described the Chicago Freedom movement as "overwhelmed by the comprehensive nature of the archbishop's program. . . . Had such a dialogue begun ten years ago we might have easily have avoided many of the serious problems of the last ten years."[48]

Many Catholic liberals emerged from the protests with their commitment to a more cosmopolitan and ecumenical vision of community affirmed. Only Martin Luther King's example, wrote one Chicago nun, enabled her to learn what "Christianity is all about." When she returned to the African-American section of the city after one perilous march, the hugging and crying made her realize "what a community was in a way I have never known at even the most beautiful of Catholic Masses." A Catholic instructor at a local inner-city community college asked Archbishop Cody whether "any member of the Church in Chicago has thought about the wisdom of fostering in Chicago our past and present 'tribal' social system. I can understand how much of a consolation it must have been for a Pole to live in a neighborhood with other Poles, but is this sort of thing needed any longer?"[49]

Many of the city's liberal Catholics, however, continued to lick their wounds. The fury of the crowds, and their particular hostility to priests and nuns, had stunned Catholic activists. "The scores of nuns and priests who marched," noted John McDermott, "became a special target. Spectators yelled 'You're not a real priest,' 'Hey Father, are you sleeping with her?' " During a July 31 march, a rock struck one suburban nun working in a summer volunteer program, Sister M. Angelica, O.S.F. When she fell, the crowd cheered. ("For the first in the history of this city," the archdiocesan newspaper thundered, "a nun was attacked on the streets of Chicago in a public demonstration. And the attack came from a mob of howling Catholics.'') A number of priests from local parishes moved through the crowds, attempting to persuade area residents to leave the scene, but they learned that ''the sight of a Roman collar incited [the crowd] to greater violence and nastier epithets.'' Only police intervention saved an African-American priest after he was pulled from his car by a gang of toughs during one protest. One discouraged priest wrote of seeing a high school senior in his religion class toss rocks at the marchers, one of whom was an African-American fellow student. Another Catholic high school student recalled seeing ''people I went to church with, screaming 'Nigger!' and throwing rocks and dirt at King—these nice people I knew all my life. I couldn't believe it.''[50]

Unsurprisingly, liberals repeatedly referred to their inability to pierce barriers in local communities and, at times, even their own families. ''For [those still in the neighborhood],'' noted one commentator, ''the civil rights issue may . . . represent the possible loss of a home, the transformation of a familiar neighborhood into a ghetto—a threat to family, community and, not least of all, to the Church itself.'' John McDermott noted that ''they consider themselves good Catholics, yet utterly reject integration. And they are particularly bitter toward priests, bishops and organizations who tell them they are in conflict with their religion. 'Since when?' they retort.'' Unfortunately, one Chicago priest concluded, ''In no sense do they [liberals] represent the people. They speak another language and, it now seems very clear, practice a different religion.''[51]

More conservative Catholics also professed unease. At one parish meeting on the Northwest Side, members of the Chicago Freedom Committee attempted to explain their position, only for the meeting to disperse after a priest-moderator proved unable to control the crowd. Even as liberals picketed Cody's residence in opposition to his call for a halt to the marching, audiences in heavily Catholic areas of the city listened to speakers denounce the archbishop for ''prostituting his religious power.''[52] During the marches, observers spotted signs denouncing ''Archbishop Cody and his commie coons.'' Rumors swept one parish that Cody had commandeered parish collections for his own, presumably nefarious, use. Priests were forced to deny the allegations in the parish bulletin. The area, concluded one cleric, ''has acquired a large number of people who moved from what's called 'changing neighborhoods'. . . and they're fearful the same thing that happened there is going to happen all over again here.''[53]

Divisions widened in the aftermath of the protests. In early 1967, the Cicero leader of an anti-integrationist group urged five hundred local residents to resist the archbishop. ''They try and force integration on us and we'll rebel,'' announced John Pellegrini, ''Cody wasn't elected by us. He doesn't have the right to take away our rights. I suggest you refrain from contributing to your parish. Don't support your defeat.'' Pellegrini added that ''This new breed of priests, in my opinion, haven't studied any theology or any scripture; they studied law books and even these they misrepresent.'' At this point, the name most viciously booed in local meetings, according to one source, was ''not Mayor Daley, nor Senator [Paul] Douglas, but John Patrick Cody.''[54]

That January, Monsignor Edward Burke, the former chancellor of the archdiocese, a longtime supporter of Saul Alinsky and now the pastor of St. Bartholomew's parish on the Northwest Side, poured out his frustration in a parish bulletin article entitled ''Let's Do Some Thinking for a Change.'' ''When we fight for the rights of the Negro,'' Burke maintained, ''we cannot overlook the rights of the white person. He has been forced to support, unaided, himself and his family. If he owns property, he purchased it by the sweat of his brow and is a true Christian when he asks that his possessions be not disturbed.'' Burke readily conceded that ''the Negro has God-given rights—equal to the rights of the white'' but he also argued that ''My reasons for opposing integration are based on the conviction that proponents of civil rights possess a superficial viewpoint of what integration really is.'' Integration, in Burke's view, did not equal ''physical proximity.'' Why do ''Catholic newspapers who continuously urge the whites to love the minorities, never fight the landlord, who [is] responsible for the slum[?]'' If Catholic liberals found his remarks unpalatable, Burke warned, he would simply resign his post.[55]

Catholic interracialists quickly chastised Burke. Monsignor Daniel Cantwell noted that Burke's comments would only ''entrench a very important area—the Northwest Side—in its white ghetto mentality.'' John McDermott observed that ''It's another sorry example of how the church can sometimes be more white than Christian.''[56]

A few of Burke's supporters also expressed their position. ''Why should we give money to priests that are buying property to GIVE to the colored?'' asked one Berwyn correspondent. ''Look at what happened to Blessed Sacrament . . . or Our Lady of Sorrows, did the colored keep them up? I love Father Burke for his stand.'' A Chicago resident recommended to Archbishop Cody that ''the church preach the neighborhood policy in housing and schooling.'' One woman responded that ''I'm all for Monsignor Burke and I'm sure I can speak for all of Cicero and Berwyn. I have three cousins that are priests and two that are nuns and thank God none of them think like you do. How can anyone have respect for a priest when he makes a monkey of himself by marching down the street with a bunch of nitwits or sits on a curb to demonstrate[?] I tried living with them when they moved into our neighborhood in Chicago. . . . I worked all my life and still have a mortgage for what I lost. . . . Now I live in Cicero and Monsignor Burke knows exactly what he is talking about when he says Negroes must learn to accept obligations, *including respect for property.*''[57]

IV

The close overlay between religion and politics in the northern cities ensured that the engagement of American Catholics with racial issues would have political implications. Mayors, like bishops, found themselves pincered between liberals concerned about racial integration, a more assertive African-American population, and a shrinking, but increasingly fearful body of working and middle-class whites. Indeed, city officials were often products—not simply representatives—of neighborhood and parish systems pressured by the dynamics of racial transition. (In Milwaukee, for example, almost 70 percent of the city's elected and appointed officials were Catholic.)[58]

In this context, the traditionally cozy relationship between the Church and city government—with the neighborhood alderman in close touch with the local monsignors and the mayor appropriately deferential to the local bishop—no longer sufficed. Perhaps the most vivid example occurred in Albany, where the political machine directed by Dan O'Connell and Mayor Erastus Corning dominated Albany's public life. Albany was heavily Catholic—as evidenced by the publicity given each year to the custom of priests giving communion to police department employees as a group, as well as the occasional gift of free paving, snowplowing, and sewer work for local parishes and schools.[59]

In the mid-1960s, a group of liberal Catholics joined an interfaith task force aimed at addressing poverty issues in Albany's South End district. One member, Father Bonaventure O'Brien, urged Albany residents to join new antipoverty organizations and denounced the mayor's office for not pursuing available federal funds. On election day, O'Brien encouraged neighborhood residents to serve as poll-watchers. ``His presence at a polling place on election morning,'' noted one reporter, ``stirred heated exchanges with politicians who wanted to know what the Catholic Church was doing in politics. Father Bonaventure replied that Pope Paul went to the United Nations didn't he?''[60]

O'Brien quickly found himself requested not to comment on public issues and, eventually, to leave Albany. Albany's bishop announced that ``the reasons for such decisions must be left to God, and the conscience of the Bishop,'' but most observers assumed that civic and diocesan leaders had tired of the priest's accusations. (One conservative priest praised a new traffic light as ''the latest proof of the solicitude of the administration for its citizens'' and attacked those who ''snipe at the administration on the slightest pretext.'') In response, liberal Catholics led a series of teach-ins, protests, and fasts protesting the bishop's decision. Four hundred Siena College students attended a prayer service in support of O'Brien, and members of the Catholic Interracial Council publicly argued that Church officials were contradicting their own teachings. (Many of the protesters compared O'Brien to another Franciscan troublemaker, Francis of Assisi). O'Brien himself asked whether he should follow the ''Constitutions on the Church and the Liturgy [from Vatican II] to their conclusion? Or should I follow my given superior?'' Ultimately, under pressure to demonstrate that its decisions did not depend upon mayoral approval, the diocese funded its own settlement house, as well as calling for governmental participation in all available Federal anti-poverty programs.[61]

In Baltimore, Cardinal Lawrence Shehan spoke at a tense 1966 City Council meeting in favor of an open housing ordinance. Along with some scattered applause, Shehan's address was interrupted by a hail of boos. Of the thirteen Catholics on the twenty-one member council, only one member supported the ordinance. A local conservative attacked Shehan for pursuing a ''double career as a priest and politician and he is lousy at both of them.'' That fall, the editor of the archdiocesan newspaper denounced the gubernatorial campaign of Catholic George Mahoney (Mahoney's none too subtle slogan was ''Your Home is Your Castle.'') The election of Mahoney, the editor argued ''would mean a victory for forces which are hostile to everything honorable which being Catholic, American, Democrat, and Irish should mean.''[62] In New York, two liberal priests joined a protest outside the home of a Queens politician. According to one newspaper account, the priests saw their work ''as a deliberate attempt to shock middle-class Catholics into a better realization of the problems of the slums.''[63]

Faith also merged with politics in Chicago. Much of Richard J. Daley's life had revolved around the Church—as an altar boy, a student at De La Salle High School and De Paul University, a regular at daily mass, the father of a nun, and an honored guest at an endless series of parish dinners. During the 1950s, Mayor Daley had kept the city's Catholic Interracial Council at a polite distance, attending the occasional banquet while ignoring requests for action in moments of crisis.[64] Ironically, Daley spent much of the following decade jousting with Sargent Shriver—former head of the Catholic Interracial Council and by the mid-1960s director of the

Office of Economic Opportunity—over control of federal monies streaming into the city.

At times, nerves were rubbed closer to home. According to one account, Daley threatened to pull his son out of a Catholic high school when a religion instructor assigned a book lauding Saul Alinsky's Catholic-backed efforts to challenge the local political machine. ("We thought this was a religion class," Daley told the priest teaching the course.) Activists connected to a Catholic settlement house tried to move an African-American family onto Mayor Daley's block in Bridgeport; a Jesuit seminarian protested inadequate playground facilities in the West Side ghetto by holding a Bridgeport "play-in." At one point, Daley pulled aside Monsignor John Egan and warned him to be "careful of your association with some of those people in community organizing. They're not your kind of people." In another instance he lashed back at the persistent questioning of a local nun. "Look sister," he fumed, "you and I come from the same background . . . houses as old as on the west side, but the people took care of them, worked hard, kept the neighborhood clean."[65]

Election results reflected the tensions tearing at the various Catholic communities. Due to their large numbers, Democratic loyalty, and strategic location, Catholics had propped up the New Deal electoral coalition in the North since its emergence in the 1930s. (Essentially, Catholics supplied one-third of the total Democratic vote despite numbering only 25 percent of the national population.)[66] In the 1950s, these ties had begun to fray, with strongly anticommunist Catholics increasingly likely to vote for a candidate like Eisenhower, although Catholic Democrats at both the local and national levels still dwarfed the number of Catholic Republicans. In 1960, religion proved the central issue of the campaign, and the extraordinary Catholic support for coreligionist John Kennedy added up to almost half the votes cast for the Democratic candidate.[67]

The many issues surrounding "race" upset traditional political calculations. In retrospect, the emergence of "race" as a partisan political issue in the 1960s—with conservative Republicans staking out positions on crime, welfare, busing, and affirmative action distinct from the positions of liberal Republicans as well as northern Democrats—proved to be a central political development of the postwar era. Not only did white southerners desert the Democratic party en masse in national elections, but middle- and working-class whites in the North also exhibited a new political independence.[68]

As early as 1964, use of the term "backlash" was routine. "Already," one commentator concluded, "there are in the cities disquieting signs that various nationality groups—heavily Catholic, let it be said—are turning sharply against the Democratic party on the race issue.'' Another analyst interviewed 200 Catholics in New England. ''The Negro is an obsessive preoccupation of the ethnic Catholics with whom I spoke, the prejudice is real.'' He added that ''The point is often made . . . that state power was never used preferentially for the Italians or the Irish [and that] they cannot quite understand why the Negro should want to move into alien neighborhoods, why the Negro should not prefer, in spite of housing conditions, to stay within the ghetto.''[69]

In the fall of 1964, the startling success of George Wallace in a few presidential primaries further unnerved Catholic liberals. Despite vocal hostility from liberal Catholic groups, and the refusal of students at a local Catholic college to applaud any of Wallace's statements during one tense speaking engagement, he received over 30 percent of the vote on Milwaukee's South Side, and made a stronger than expected showing in Gary, Indiana. In the presidential election, most Catholics supported President Johnson, but Senator Barry Goldwater's best totals in New York State came from heavily Catholic Staten Island, which, significantly, had also been the county most supportive of Franklin Roosevelt's 1936 reelection bid.[70]

The political resonance of themes connected to race became even more clear following a celebrated address by William Buckley to over five thousand New York City Catholic policemen attending a 1965 Holy Name Society communion breakfast. Buckley first denigrated media coverage of events at Selma, arguing that an overemphasis on police brutality overshadowed more restrained police efforts. He also linked these problems to the situation in New York City. Why, he asked, does the death of a civil rights worker in the South merit far more coverage in the *New York Times* than the death of a city policeman? Buckley concluded that ''we live in a world in which order and values are disintegrating.''[71]

Buckley's remarks drew a spirited rebuttal from more liberal groups, and his own mayoral campaign existed only as a protest vehicle, but the connections many Catholics drew between race and crime exacerbated existing tensions within the Democratic party. In Boston, voters elected conservative Louise Day Hicks to chair the city's school committee, and Hicks almost became Boston's mayor using the slogan ''You know where I stand.'' Following one Hicks victory in a school committee election, Boston's African-American Catholics informed Cardinal Cushing that ''the white voters of Boston, overwhelmingly Catholic, seem to have locked the door of the ghetto and thrown away the key.'' NAACP leader

Thomas Atkins derided Catholic leaders for their inability to influence Hicks and city officials. The Church, he concluded, "has failed to be relevant to the central moral issue of our age."[72]

Precisely the areas of Chicago into which Martin Luther King had led civil rights marches during the summer of 1966 abandoned liberal Democrat Paul Douglas that fall in his campaign for reelection to the Senate. Most of these voters still remained Democrats—party loyalty, in other words, proved stronger than division on a particular set of issues—but they also demonstrated an increased willingness to abandon particular Democratic candidates, especially beyond the local level.[73] Voter opinion on the crucial issue in Chicago's local races—open housing—was so evident that candidates often simply disavowed national party platform planks. One candidate for 13th ward alderman asserted in his campaign literature that he was "in full accord with the Rt. Rev. Monsignor Edward M. Burke, pastor of St. Bartholomew Roman Catholic Church, who stated recently that the rights of whites must not be overlooked in a fight for the rights of minorities." Another candidate—a graduate of a local parochial high school and then DePaul University—asked "Why should these people risk their homes, their schools and their church for some ideological experiment that has never worked?"[74]

V

Answers to the question were not self-evident. In retrospect, these neighborhood conflicts demonstrated some of the distinctive qualities of northern race relations. At one level, of course, as in the South, conflicts took place along racial lines. Convinced that the entrance of African-Americans into particular neighborhoods would inevitably lead to the "ruin" of neighborhoods and schools, some white residents fiercely resisted all African-American newcomers. As suggested by the inability of heavily Catholic neighborhoods to welcome African-American Catholics, skin color mattered more than income or culture. Each northern city still contained a handful of white neighborhoods marked by strong ethnic identities, but the heterogeneity of these areas belied the notion that resistance to African-Americans was solely an attempt to preserve traditional ethnic communities.

Still, in contrast to the southern focus on voting rights or segregation in public areas, notions of housing, "turf," and community did emerge as the primary northern racial battlegrounds. And as Martin Luther King in Chicago and other activists throughout the North discovered, geographical and literally parochial definitions of community—originally fostered, in large part, by the Catholic Church—proved less amenable to change than southern discriminatory laws. SCLC worker Dorothy Tillman recalled her amazement at the hostility she encountered in the North. ''And the sad thing about it,'' Tillman noted, ''was that most of these neighborhoods we went to was like first or second generation Americans. I mean they had not been here as long as we had been here. . . . Most of them were fleeing oppression. Down South you were black or white. You wasn't Irish or Polish or all of this.''[75]

In short, neighborhoods still existed whose functional identity—for the majority, if not each resident—derived from religious structures. Given past experience, community residents feared that the dynamics of American racial transition would forecast a rapid collapse of traditional institutions following more than a token integration. By the mid-1960s, these neighborhoods had also become more solidly working-class, a consequence of the steady movement of young, upwardly mobile white Catholic families to suburban areas. Criticism from more affluent, suburban coreligionists, as well as outsiders generally, was especially unwelcome, since, in Nathan Glazer's words, ''it also served to ascribe, or impute, a good deal of wrongdoing to working-class Catholics who weren't especially conscious of wrongdoing at all.'' Why, the question was often asked, did activists focus on open housing within the city as opposed to the suburbs, or not concentrate on eradicating poverty within the African-American ghettos?[76]

All of these issues came into sharp focus in Milwaukee—site of the most sustained Catholic encounter with racial issues. Well over 40 percent of Milwaukee's ''white'' population was Catholic, and parochial schools enrolled one-third of the city's students. Into the mid-1960s, officials in the Milwaukee archdiocese had moved cautiously on interracial and civil rights programs. One discouraged Milwaukee priest wrote to the NCCIJ office in 1963 that ''We would love to join you in the March on Washington but the Chancery office has said that it is very 'rash and imprudent' and has forbidden all priests [from] participating.'' In the spring of 1965, Catholic conservatives cancelled the appearance of John Howard Griffin, the author of *Black Like Me,* an account of a white man who colored his skin and traveled through the South, before a local Catholic youth convention. (Eventually, after liberals protested, Griffin spoke at a North Side theater.)[77]

Equally significant was that fact that one of the main speakers at the student conference did address the topic ''Race and the Renewal of the Church.'' The next month, a group of Milwaukee Catholics, including several priests and two nuns, traveled to Selma to join the protests organized by the Southern Christian Leadership Conference. One of the priests, Father James Groppi, found the experience particularly moving. "That's something I've always wanted to do," he later told one reporter, "sit down in the heart of the segregationist south with a group of priests and ministers in an ecumenical protest."[78]

Groppi himself spent his seminary vacations in the inner city, and in 1963 had been appointed as a curate to St. Boniface parish in the heart of the city's small African-American ghetto. This central-city neighborhood had witnessed steady white flight during the 1950s, with Poles and Germans the last to leave the area. From a peak of over 800 students in 1945, virtually all of whom were white, school enrollment at St. Boniface had fallen to 255 in 1965, 91 percent of whom were African-American. Upon his return from Selma, Groppi and his colleagues threw themselves into community work with renewed fervor. That summer, the NAACP youth group asked Groppi to be their chapter advisor and he joined a local school-integration organization.[79]

In the fall of 1965, twenty-four Catholic priests and a number of sisters working in Milwaukee's inner city decided to participate in the upcoming Milwaukee school boycott by teaching classes and allowing the local parochial school buildings to be used as "Freedom Schools." The pastor of St. Boniface, Father Eugene F. Bleidorn, later explained his reasoning. "[T]he time was now to prove that the Catholic Church was not a white Church which happened to be located within a Negro ghetto. For future work in the apostolate to the community and identification with the community, my conscience dictated that we become a part of the Boycott and the Freedom School Movement." One poll of his parishioners, he continued, revealed 83 percent in favor of parish support for the boycott.[80]

With Milwaukee's Archbishop Cousins away in Rome for the final session of the Vatican Council, chancery officials attempted to handle the situation. Following consultations with the local school-board president and the district attorney, auxiliary bishop Roman Atkielski commanded priests not to use parochial property during the boycott. Initially, the inner-city priests simply viewed the issue as one of freedom of conscience. In what Father Bleidorn called a "historic" statement, the priests contrasted mere legal opinions with the "force of Christ's words."[81] A day later, after supporters notified Bleidorn that coverage on national television and in the newspapers was emphasizing conflict within the Church and not the boycott, he retreated slightly. Priests and nuns, he declared, would work in the freedom schools without using church property. That night, on television, Bishop Atkielski warned of "ecclesiastical consequences,"

Father James Groppi leading freedom songs at a 1965 Milwaukee public school boycott. State Historical Society of Wisconsin, WHi (×3) 37641.

even as a group of fifty protesters opposed to archdiocesan restrictions marched in front of Atkielski's parish rectory.[82]

On the opening day of the boycott, twelve Catholic priests and twenty-five nuns were highly visible in the operation of the freedom schools. Instead of using the St. Boniface school building, Father Groppi led a group of 500 African-American students to a neighboring Baptist church for a two-hour lesson on African-American history and civil rights songs.[83] The next day, the two Milwaukee dailies carried an advertisement entitled "An Open Letter To Our People," paid for by the inner-city priests. By not using parochial facilities, the priests argued, "our church fails to give its full Christian witness here. On the other hand, should we disobey the bishop we feel that at this time in our Church many would not be able to understand our actions, and so would suffer some spiritual harm because they are not yet ready to receive the full impact of Vatican II." Father Bleidorn began his next Sunday's homily by reading from James, 2:12–17: "My brothers, what good is there in anyone's professing his faith if he does nothing in practice?"[84]

Local reaction to these events was mixed. To the dismay of some parents, a group of Marquette High School students already participating in a discussion group focusing on racial issues applauded Groppi's actions. (In the fall of 1964, the group had resolved that "Discussion . . . was fruitless if divorced from actual experience.")[85] Many of the letters to Groppi and his colleagues also expressed support, with a typical comment being one woman's declaration that "I have felt for some years that the church has not been as active as it should be in the field of civil rights." She added that "Sixteen years of parochial school education taught me that Christ was always on the side of the underdog and gave not a second thought to civil disobedience or its consequences." Archbishop Cousins refused to discipline the priests upon his return from Rome, even after Groppi chained himself to a school construction site to protest segregated education. While terming Groppi's behavior a "little immature," Cousins also told a reporter that "He [Groppi] has a lot of guts and is doing things maybe others of us don't have the courage to do."[86]

Groppi's opponents were no less impassioned. Letters of opposition poured into the chancery office and filled the editorial columns of local papers. A group of thirteen "Catholic Laymen, workers in an industrial complex," mailed a painstakingly composed missive to the archbishop, Groppi, and Bleidorn. "It is regrettable," the letter began, "that Catholic clergy should encourage contempt for the law by urging children and young people to aid and abet this school boycott. During their formative years, children should be taught discipline in accord with the fourth commandment. Rebellion is a primary tenet of communism.'' Other letter writers denied that they were prejudiced, while attacking the tactics of the protesters. ''[I]f our priests and nuns must demonstrate,'' wrote one woman, ''why not demonstrate for the Catholic parents who have not only supported our Public school system but struggled at great sacrifice to build and maintain the Catholic Church, the Catholic school [and] the Catholic convent.'' Another correspondent added that ''When problems like this arise among the lay people and priests I am ashamed to tell my fellow workers I am a Catholic.''[87]

The stakes of these conflicts escalated over the next two years. In 1966, Groppi led African-American protesters into a white suburb to protest the membership of local officials in a segregated fraternal club. (One of Groppi's more conservative colleagues noted in his parish bulletin that ''Father Groppi and the dear sisters have gone far beyond their rights, and have definitely abandoned all Christian principles.'') Only the intervention of the police, and eventually the National Guard, prevented violent confrontations. Groppi also changed the name of the NAACP youth council to the ''Commandos,'' and members began wearing dark glasses, army fatigues, and black berets. At St. Boniface, priests placed banners in both Swahili and English around the church, and Groppi took to wearing black on the altar—not in mourning, he emphasized, but in celebration. Reporters commented on the incongruity of a white priest leading parishioners in chants of ''Black Power'' during mass.[88]

In 1967, the local civil rights movement switched its attention to housing. Although one of the most segregated cities in the nation, Milwaukee had not passed a local open-occupancy law—in part because Mayor Henry Maier refused to support a law for the city unless the surrounding suburbs passed similar measures. That summer, after three deaths in ghetto riots, Maier placed a curfew on city residents. Immediately following the lifting of the curfew, Groppi announced a new series of marches, beginning as always in St. Boniface's parish, but this time crossing the Menominee Valley's 16th Street bridge into the city's heavily Polish South Side.

The marchers displayed tremendous courage. Although to white observers Groppi's NAACP commando bodyguard looked especially threatening, the unarmed commandos, under Groppi's direction, prevented marchers from lashing out at their assailants. Groppi often stepped directly into crowds of whites to verbally confront hostile observers. On the first night, a crowd of around 5,000 whites started the chant ''Eee-yi-eee-yi-eee-yi-oh, Father Groppi's got to go.'' The next night, crowds of whites chased the marchers back across the bridge, and by the end of the week police were using tear gas to prevent whites from attacking the protesters.[89]

Profiled on national television, Groppi immediately became one of the most recognized figures in the civil rights movement. Dick Gregory and other activists flew to Milwaukee to participate in marches, and Martin Luther King sent a congratulatory telegram praising Groppi's ability to "be militant and powerful without destroying life or property." King also praised Groppi as an inspiration for those activists seeking "a middle ground between riots and sentimental and timid supplications for justice."[90]

As in Chicago, the marches displayed a deeply fractured Church. Despite enormous pressure, Archbishop Cousins refused to discipline Groppi, and chastised Catholics who threatened to withhold money from the Church. (The city's four television and nine radio stations carried a statement by Cousins on the subject after the summer's riots.) Predictably, a small countergroup of Catholics formed to protest Groppi's actions, and marched several times to the archbishop's residence. One priest affiliated with the group complained that "the whites had to fight not only the black man, but the Catholic church and clergymen who supported open housing as well." White Catholic youths occasionally found themselves screaming at nuns who had instructed them in elementary school, and police were particularly bitter about arresting priests for civil disobedience. (One policeman informed a priest as he dragged him into the paddy wagon that "the only thing I want to do is to slip you faster into hell.") Eventually, Groppi led marches into the city's South Side for two hundred consecutive evenings, stopping shortly after the city council finally passed an open housing ordinance. "Marching," he announced, "is not only a protest, it is a prayer."[91]

For Catholic liberals, the marches were an epiphany. According to one Boston priest who flew to Milwaukee to join the marches, Groppi's actions "unmasked the hypocrisy of the lily-white Catholic community. . . . As Milwaukee is the prophetic voice of our nation today, so St. Boniface is the prophetic voice of the Church—not only the Catholic Church but religion itself." A Chicago priest observed that at a local settlement house "the big topic of conversation was who was going to march in Milwaukee this weekend. . . . I was talking to one sister who had been teaching for forty years . . . and who had marched with you and the group for 22 miles a week ago Sunday." A Boston CIC member added that "The Milwaukee situation is a crucial issue for another reason. It's a confrontation between the basically Protestant Negro community and the heavily Catholic white community."[92]

From across the country came encouragement. Religion instructors and students from one Providence, Rhode Island, parish expressed their "support of your objectives'' and a fourteen-year-old African-American Catholic student from Pittsburgh wrote that ''When you began your work you made me very proud to be a Catholic.'' A Boston priest observed that ''White Catholics as a community . . . are not escaping black disenchantment and disdain but Father Groppi, sharing opprobrium and imprisonment with his black parishioners, is keeping us in there. He and a few like him, are keeping a corner in the heart of the American black man for Catholicism.'' ''We are Catholics,'' noted one married couple, ''the second generation of Irish emigrants—white, of course. It frightens us to witness the attitude of some of our close friends—even relatives—in regard to this problem. It is a stigma on our Faith and Irish inheritance.'' A Watertown, Massachusetts, couple complained that ''We are ashamed and mortified to listen to the comments of our Catholic friends (Irish and Italian and all other nationalities alike). Their expressions and voices change when they discuss justice and brotherhood where black people are concerned.''[93]

Groppi's supporters linked his marches with Vatican calls to reinvigorate the Church. A Milwaukee couple argued that ''After Vatican II there is really no question about the suitability of a cleric or religious 'meddling in politics' when moral issues are involved.'' Another admirer added that ''We should by now be 'sick and tired' of our self-serving comfortable, mediocre, bigoted so-called Christian Community,'' while a group of Jesuit seminarians insisted that ''The sooner we stop trying to impose an artificial ecclesial system on our people and freely join hands with Christ/our brothers in the (so-called) secular city—the more honest will we be, and possibly, the more successful.''[94]

Other correspondents noted an increased sense of belonging to a global church. ''A few years ago, this check would have gone into the Bishop's collection,'' wrote one couple, ''but after spending three years in South America with the Papal Volunteers, our views have been changed. We want our money to go where the need, and action is. God bless you.'' From Peru, an American seminarian on retreat noted that ''the retreat master is pretty good, but the thing that is a constant pre-occupation and source of searching and inspiration, is you, and your work.'' A Catholic missionary in Tanzania offered his support: ''The world is very small these days and things that happen in some hick town in the states can effect our message here in the back country of Tanzania. God knows enough things have happened in the last few years to make us ashamed of our background and also to make people here suspicious of a white American's presence in Africa.''[95]

Many of the thousands of letters Groppi received objecting to his activities displayed a virulent, obvious racism, with anonymous correspondents repeatedly using the word ''nigger'' and threatening Groppi's life. An equal numbers of letters, however, expressed a more nuanced anguish. ''We feel that you as a learned priest,'' wrote Mrs. A. C. Caruso from New York City, ''belong with church matters and not state. You are not a mother, a father, a home owner, nor have the responsibilities of the life of your own flesh and blood to rear in these rioting and innocent blood shedding days.'' Fred Gordon of Jersey City wrote that ''You're supposed to have a divine vocation directed by heaven and yet, when the chips are down, you succumb to the thinking of a bunch of radical low-lifes!'' (Gordon added that ''This is the first time I've ever written to a Radical or even tried to express myself against something I abhor.'') From Philadelphia came the complaint that ''It is not fair to shove people of another race down our throats. The Archbishops, Bishops, priests, the president, Robert Kennedy and other people advocating open housing sit in their single homes and tell us who to live with. We are as good as you are and when all the people I have mentioned live in the middle of row houses with people of other races I will too.''[96]

Letters from Milwaukee area residents were particularly bitter. At least, commented Mrs. Elenore R. Haubert, those citizens you attack as racist have remained in the city. ''You should know,'' she continued, ''the all-pervasive attitude of long standing which I've lived with all my life, that this area is something in the nature of 'other side of the tracks'—meaning literate or intellectual inferiority, but truly meaning only less material affluence.'' The mother of a ''German-Polish Catholic Family'' asked ''Father whats [sic] so good about the south side that the Negroes want to move over there?. . . . My husband had a hard life and he didn't finish school because he wanted to help his parents get a little ahead.'' A local monsignor concluded that African-Americans were no ''poorer than the whites were in the depression of the thirties.'' Mrs. J. N. Hipp recalled the days when St. Boniface sheltered a German congregation. ''I can remember the first Mass that was read in the new church Even then it was a sacrifice. . . . They [the whites] have been chased out of their homes, by negroes and now they and you are desgrating [sic] the church they built.''[97]

Just as liberals connected civil rights marches to the need for an attack on traditional parochial structures, Groppi's opponents linked his actions to what they perceived as the decay of contemporary Catholicism. A Mrs. Dugan—she signed her note ''an old scrub lady''—complained that ''Instead of making converts you are making enemies for our Religion just like some of these others [sic] Priests who leave the Church and condemn it then . . ." Mrs. John C. Fonuke commented on a news broadcast that emphasized the exuberant services in Groppi's parish. "[W]hen I see people screaming, clapping hands and acting far from natural in God's home, so near the tabernacle, I wonder how long it will be before the Church is a three ring circus." Mae Baxtis wrote that "I am a very strong Catholic, I should say, *I was,* but demonstrations like yours, sicken me, in fact most of the 2000 Fathers in Rome this past four years have made me heartsick to see what 2000 men could do to a 2000 year old Church."[98]

In short, "race" helped mark the cultural changes sweeping through American Catholicism and American society. Between 1964 and 1967, two distinctly Catholic visions of church, community, and authority clashed in the streets, parishes, and Catholic schools of the northern cities. Catholic liberals questioned traditional parochial structures while becoming active participants in local civil rights coalitions. Their activism illuminated not only the enduring strength of Catholic racism but the distance now separating various parts of the Catholic community. One Catholic activist involved in a racial education program confided that she had "rediscovered how the local parish women felt; just as she had felt in the years before she had moved up to what is now considered a higher social echelon."[99]

These divisions extended into the symbolically charged arena of ritual. A remarkable facet of the preconciliar twentieth-century Church had been the range and depth of popular religious practices—the parades through the parish on feast days, the endless, well-attended rosary crusades, the 70,000 Catholics who might pack into a single church during one week for a novena, the traveling statues of Mary.[100]

The most important project of Catholic liberals in the 1940s and 1950s had been to refocus attention on the celebration of the mass, and away from these less structured, individualistic acts of piety and petition. To liberals, the parishioner saying the rosary during mass, or the statues of saints drawing attention away from the altar, destroyed the communal purpose of the liturgy, as well as obscuring corporate possibilities for social change, especially on racial issues. By emphasizing the "collective implications of Catholic worship," argued John LaFarge, "the parish becomes a laboratory of interracial justice."[101]

Many interracial activists, including LaFarge, played important roles in the burgeoning liturgical movement. Chicago's Monsignor Daniel Cantwell, for example, celebrated the then avante-garde "dialogue Mass" (with the priest facing the congregation and the worshippers allowed to respond at certain points of the liturgy) at retreats in rural Wisconsin for Catholics interested in racial issues. In South Bend, Indiana, the pastor of the city's one African-American parish also helped usher in liturgical reforms—requesting more "policing" of liturgically "non-complian[t]" pastors even as he issued a plea for greater parochial school integration.[102]

In 1963, the annual liturgical conference was held in Philadelphia, just before the March on Washington. Hundreds of participants placed small "March on Washington" buttons next to their conference badges, and made plans to stop in the capital on the way home from the conference. Listeners continually interrupted Benedictine Godfrey Diekmann's keynote address with ovations. "[W]e trumpet the blasphemous triumph of Satan," Diekmann thundered, "if we eat of the Bread and drink the cup, and refuse to accept the Negro as our daily table guest."[103]

The Vatican Council ratified and extended these notions. Virtually no mention was made of the pious practices so evident in Catholic life, and there was relatively little discussion of the ecumenically suspect Mary.[104] Again, liturgical enthusiasts linked these changes to racial issues. "No Christian who understands the liturgical renewal," wrote one Pittsburgh priest in 1964, "can be indifferent to the interracial situation in our country today." Another interracial activist, Philadelphia's Dennis Clark, called for a "new piety for the new age." Clark urged reformers to be sympathetic to older parishioners' concerns, but he added that in his view "traditional pietistic practices seem odd and rather occult."[105]

Diary entries from a 1965 sojourn in Harlem by a Jesuit seminarian, Joseph Roccasalvo, demonstrate these connections. For Roccasalvo, the experience of knocking on area doors and visiting with Harlem residents was powerful. "Why this stark contrast between my abundance and their penury?" he asked. At the conclusion of the project, the seminarians attended a mass said in the traditional manner. After working in Harlem, the experience seemed to Roccasalvo "a meaningless pantomime, a kind of 'closet drama' with no participation except inert presence. . . . Thank God for the liturgical changes which so readily admit the people to God's sacrifice."[106]

Only in this context do the fervent pleas of those who begged Father Groppi to conduct a more dignified liturgy—or stop protesting and "pray the rosary"—make sense. In contrast to Father Roccasalvo's disgust with the Latin mass, more conservative Catholics registered displeasure with, as one disgruntled cleric put it, "liturgists, men who promote race relations and various kindred subjects."[107]

From this vantage point, long-standing parochial forms—including the various Catholic institutions, the cohesiveness of the neighborhoods, the distinctions between religious and laypeople, and traditional religious rituals—were under siege. In Chicago, a bitter fight broke out between parents and clergy when parents discovered that a new religion textbook for third graders contained the sentence "Dr. Martin Luther King is like Jesus." Debates on the matter often lurched into attacks on the withdrawal of vigil lights from the church building or the lack of emphasis in Catholic schools on praying the rosary.[108]

In Philadelphia, a rock concert in the parish hall sponsored by a local community organization degenerated into a small riot, and the largely African-American crowd ran through the parish neighborhood fleeing the fighting. As parish priests attempted to restore order, longtime parishioners stood on their porches and yelled "Thanks, Monsignor, for bringing them into our neighborhood." At roughly the same time, without consulting the parishioners, the priests arranged for architect Robert Venturi to renovate the church in the spirit of Vatican II. The most dramatic innovation was a neon cathode tube encircling the altar area. Some parishioners (termed "pussyfooters" and "Kafka-esque" by architectural critics) refused to hold their weddings in the church and eventually persuaded the priests to place the tube in storage.[109]

Common to all of these episodes was a novel juxtaposition of Church and culture. Encouraged by developments within the Church to more fully engage American society, priests, nuns, and laypeople immersed themselves in an America itself in tremendous flux. In no other era in American history would traditional notions of "authority" come under such scrutiny, in institutions ranging from political parties to the military and the family. This scrutiny inevitably extended to perhaps the nation's most authoritative institution, the Catholic Church. By the late 1960s, Catholic liberals, inspired by both Selma and the Second Vatican Council, had begun to apply a language of "freedom" and "rights" to what they perceived as oppressive Church structures, even as more conservative coreligionists recoiled from the theological and social implications of this new parochial world. In this way, the crisis within Catholicism both mirrored and shaped the more general reorientation of American society.[110]

NINE
CATHOLIC FREEDOM STRUGGLE

I

Efforts by Catholics to address racial questions continued in the aftermath of the neighborhood confrontations of the mid-1960s. During this period of civil rights agitation, Pedro Arrupe, S.J., the leader of the world's Jesuits, asked "what is the role of the Society of Jesus in her service to the spirit of the Second Vatican Council?" Following an apology for past Jesuit blindness on the issues of slavery and segregation, Arrupe attacked the "insulation of far too many Jesuits from the actual living conditions of the poor, and hence of most Negroes." The American provinces, he warned, must reassess "ministries, manpower, and other resources" as well as establish "separate Jesuit residences in a poor Negro section of one or more of the major cities in each province."[1]

The Association of Boston Urban Priests—a group of 100 activist clergymen—issued its own "direct and public criticism of the stance of the Church in Boston towards the city." In language evocative of the Vatican Council, the priests pleaded for the Church to become "part of the world" as opposed to a "sanctuary from it." Boston's faithful needed to revamp often "homogeneous" parochial schools as well as reform a police force "predominantly Catholic in composition." All Catholics, the clergymen argued, must move from "parochialism [to] universal concern" and choose "servanthood" and "prophetic risk" over "triumphalism" and "security."[2]

The most imaginative efforts to link the Second Vatican Council with American racial concerns occurred in Detroit, where Archbishop John Dearden had become convinced by the early 1960s that race relations were the central moral issue of the era. (After the 1963 Birmingham, Alabama, protests, Dearden commanded all pastors to read from the pulpit a letter on race "in its entirety with no change or alteration whatsoever.")[3] Immediately following the end of the Vatican Council, Dearden inaugurated one of the nation's largest racial educational programs, "Project Commitment." Roughly forty members of each parish in particular regions of the archdiocese were invited to a series of eight evening sessions on race relations. According to archdiocesan sources, attendance was remarkably good, and within a year 230 parishes and 5,000 parish "leaders" were participating in the program.[4]

The program aimed to "create in each parish a committed core of Catholics informed and active in human relations, particularly in racial matters, who will work within their own parishes and in the community." One official promised that "After people from an all white parish on the extremities of the city spend eight evenings in dialogue with other Catholics from changing or resegregated areas they will feel profoundly different." Dearden himself explicitly connected the program to the Vatican Council, informing 1,500 Catholics in the Gesu parish church that the Pastoral Constitution on the Church in the Modern World "begins by calling attention to the fact that we, the People of God, are intimately bound up with all the concerns of those among whom we live." The titles of the individual parish sessions—"Post-Conciliar Catholicism and Race Relations," "The People of God"—reflected the importance of new theological conceptions.[5]

Project Commitment took its place among a group of archdiocesan initiatives. By 1967, the budget of the Archdiocesan Opportunity Program—partially funded by the federal Office of Economic Opportunity—topped three million dollars. The various programs focused on impoverished children, particularly the fifteen thousand pupils enrolled in parochial schools located in areas targeted by the war on poverty. All archdiocesan contracts began to include equal opportunity clauses, and Dearden encouraged donations of funds and materials to inner-city parishes. Parishes must help each other, argued one archdiocesan official, just as "The bishops at the council learned that they must not only be concerned with their own dioceses." Seminarians stayed for a weekend with African-American families as part of their training (African-American seminarians stayed with Jewish families), and archdiocesan officials reminded priests of the obligation to be "in the forefront of the rights movement all the time."[6]

Some of Detroit's white Catholics resented this activity. Occasionally, pastors refused to participate in Project Commitment discussion sections, content to sit in the back of the room and observe the proceedings with obvious distaste. At one parish, the decision to bring groups of African-American Catholics into the homes of participants sparked a fierce debate, with some parishioners wondering whether program organizers intended to move African-Americans into the neighborhood. Father Charles Coughlin, still brooding after almost thirty years of enforced silence at Little Flower Parish in Royal Oak, admitted that he opposed the programs.

Members of Detroit's Archdiocesan Commission on Human Relations singing "We Shall Overcome" at a 1965 conference on "Human Dignity in Detroit." Archives of the Archdiocese of Detroit.

"This open housing," Coughlin argued, "is forced. I don't believe in force. . . . What irks me in Detroit as in other big cities is the emphasis on housing and race and not one bit about crime on the streets."[7]

Detroit's extraordinary civil disturbances during the summer of 1967—leaving forty-three dead and roughly $80 million in property damage—placed these issues in a new perspective. Because several of the city's parishes were located in riot-torn areas, many priests, nuns, and lay volunteers viewed events from close range. Terrified residents poured into church buildings during the fighting, and the priests, like other residents, became accustomed to diving onto the floor upon hearing the sounds of gunfire. ("I found myself asking," one cleric later wrote, "what the Church in our time will do for the poor.") Priests from St. Agnes parish attempted to move among the crowds, but African-Americans responded with a command for "Whitey . . . [to] get off the streets." At one point, a stunned Dearden drove alone from his downtown office through the area of the most severe rioting to view events at St. Agnes and offer his assistance. At the enormous Catholic seminary just west of the most severe rioting, students and faculty watched the smoke hovering over the city, obscuring the skyline. During one lull in the fighting, seminarians noticed that the face, hands, and feet of the statue of Christ on the seminary grounds had been painted black.[8]

To liberals, the riot signaled the need for a redoubling of efforts. In an impassioned address to the National Catholic Council for Interracial Justice shortly after the riot, Dearden argued that "While we have been doing something, it is clear that we must do more. . . . It is precisely this racial crisis, with its manifold economic and sociological overtones, to which we must react in the Church." He added that since "[t]he Negro-white confrontation in American cities is in great part a Negro-Catholic confrontation" a concerted effort by Church officials on race and poverty issues was "demanded by [the Church's] role as 'the sign of salvation.' "[9]

The next year Dearden ushered in a new series of programs geared at eradicating racism and assisting inner-city residents. Following the 1968 release of the Kerner Report describing two separate American societies—one black and one white—Dearden called an emergency meeting of the archdiocesan clergy and ordered a series of mandatory sermons on racial issues. "We have to stand up and admit we are white racists," announced

(*facing page*) Statue of Jesus on the campus of Sacred Heart Seminary in Detroit. The hands, feet, and face were painted black during the 1967 riot. Copyright 1987, Daniel M. Rosen.

SACRED HEART OF JESUS
THY KINGDOM COME

one priest, ''then maybe we can get somewhere.''[10] One seminary professor began a clothing and food program entitled ''Summer Hope.'' An auxiliary bishop warmly welcomed the Poor People's March to Detroit, and then, along with Father Groppi, joined the marchers at Cobo Hall. Most important, Dearden announced that over a million dollars from the archdiocesan development fund would be donated to promising inner-city programs. A board composed primarily of inner-city residents would make funding decisions.[11]

Some white Catholic parishioners obviously viewed the riots from a different angle. ''My husband and I have worked on Project Commitment since last September,'' wrote one woman, ''After speaking with people that have attended at least one of the three series our parish held, I feel we have lost almost all we have worked for as far as neighborhood integration goes, as well as communication, or person-to-person friendships.'' A few parish meetings on ''racism'' turned into shouting matches. Frank Tully wrote to the *Michigan Catholic,* the archdiocesan newspaper, to voice his disapproval of ''so-called leaders of Project Commitment quite willing to trample on the freedom of the white community in order to highlight the freedom of the black community.'' Another correspondent refused ''to support a system that discriminates against older clergy and clergy of ethnic backgrounds and groups of ethnic peoples like the Poles, the Italians etc., while at the same time telling us not to discriminate against the good Negro people.''[12]

Reaction was particularly hostile, in part, because as the archdiocese expanded inner-city programs, it had begun to close large numbers of parochial schools. No single parish school's expenses, Dearden ordered, could equal more than 40 percent of the parish budget, since in light of the Vatican Council, parochial education was only one among many obligations of a parish community. Calling for a ''kind of Christian heroism,'' Dearden added that ''We can't wind up with a fringe of relatively substantial schools with everyone else neglected.''[13] Archdiocesan lawyers and supporters in the state legislature did attempt to win public funding for some school expenses, but voters rejected a referendum supporting financial assistance. (In the midst of discussion on the issue, one archdiocesan spokesman even questioned the morality of appealing for public funds for Catholic schools, since the schools, in his view, failed to produce racially tolerant citizens.) An editorial in the archdiocesan newspaper chastised those who ''object to the [archdiocesan development] fund being used for the poor when we have apparently lost our campaign for state aid to private schools. . . . In a sense the very fact that most of those we will help in the ghetto areas are not Catholic is an indictment of our past.''[14] By 1973, 137 of the archdiocese's 360 schools had shut their doors. In 1971 alone, parochial school closings forced 18,000 students to search for new schools.[15]

Again, these developments sparked bitter dissent. One priest reported that some parishioners on the city's East Side now viewed the cardinal as the enemy. ''In closing the schools he is launching the ultimate attack on their values and the system which maintained them.''[16] ''Our good old cardinal at the church here,'' complained one woman to pollster Samuel Lubell, ''he says to hell with whites. He talks about the 'affluent suburbs'. We're not so affluent. He asks us to give and give, and it all goes to the colored in the city.'' A parish usher maintained that ''They knew we needed that money for our schools. They gave it to the colored just so we couldn't use it.'' In one largely Polish parish, the parishioners grilled a newly appointed pastor who had previously worked in an African-American parish. ''And what do you think of the archdiocese taxing the Poles to give money to the Blacks?'' asked one parishioner at an open meeting. An elderly woman explained that ''Why, at first we thought he was an 'agent' of Cardinal Dearden. We were sure he was going to say we should all move, and turn over our neighborhood to Black folks.'' Rumors swept through the parish that the new pastor hoped to replace the statues of ''white'' saints with ''black'' ones.[17]

II

Such tensions were no longer novel. During the 1960s, hundreds of incidents, including the participation of clergy and women religious in civil rights marches through heavily Catholic neighborhoods, had widened the cultural divide between two ''Catholic'' peoples. Inspired by both the civil rights movement and the Second Vatican Council, liberal Catholics attacked not only racism but what they perceived as outdated organizations, rituals, and structures. Other Catholics, especially those living in neighborhoods threatened by racial transition, now perceived fellow communicants as threats to a painfully established local moral order. During the Detroit riot, a white policeman spied priests attempting to calm bystanders. ''Hey, Father,'' he asked scornfully, ''why aren't you out demonstrating?''[18]

What marked events in the late 1960s was the insertion of a new vocabulary, developed in the context of the racial crisis, into discussions of authority within the Church itself. Before the late 1960s, no group had been more reliant upon (or grateful for) the teaching role of the hierarchy than Catholic interracialists. Catholic doctrine on segregation or discrimination, as interracialists perceived it, created no problems. Only the application of that doctrine—by timid bishops, recalcitrant laypeople, ignorant priests—presented difficulties.[19] A 1963 Broadway play, *The Deputy,* ignited a series of heated editorial exchanges by accusing Pius XII of abandoning Europe's Jews during World War II. Catholic liberals worried that a similar drama might be written about the American Church and racial issues. "Should the Church sail along with the complacency of its white majority," asked one priest, "making theoretical statements on race-equality and waiting to see what happens next?"[20]

The logics of the civil rights movement and the Second Vatican Council, however, were anti-authoritarian. The key term was "freedom." The primary American contribution to the Vatican Council had been in the area of church-state relations, where the American Jesuit John Courtney Murray, silenced by Roman authorities during the 1950s, provided the intellectual ammunition for a recasting of Catholic conceptions of religious liberty. Instead of the traditional notion that Catholicism should eventually become an established church—error, in other words, had no rights—Murray persuaded the bishops that the American experiment of religious freedom provided a worthy model. Inevitably, however, as Murray pointed out in his last published essay, formal acceptance of religious freedom within society would lead to questions concerning freedom within the Church.[21]

At the same time, Catholics involved in what became termed the "freedom movement" (with its "freedom songs," "freedom rides," and "freedom schools") discovered a language applicable to other relationships—especially those between bishop and parish priest, pastor and curate, clerics and laypeople, an all-male clergy and Catholic women. Prior to the 1960s, Catholic authority was venerated; half-hearted implementation of papal teaching was scorned. By the end of the decade, precisely those individuals most visible as symbols of authority—priests, the hierarchy, and the pope—received the sharpest criticism. What began as attempts to ensure the right of African-Americans to use public facilities and vote, developed into an assault on inequality within the Church itself.[22]

The most obvious tensions were between clergy and laypeople. For two generations, a few liberals had been attempting to establish a more prominent role for laypeople within a remarkably clerical church.[23] Many clergy, however, continued to view lay initiatives as ill-advised. (Cleveland's Bishop John Krol reminded lay activists in 1957 of "a tendency" to emphasize "social adjustment and good fellowship" more than "good relations between man and God.")[24] Even the growing numbers of highly educated, active laypeople found it difficult to emerge from the clerical thickets. One speaker at a dinner honoring founding members of the New York Catholic Interracial Council complained that "There is very little awareness that the Catholic layman represents a vast unused potential for constructive social actions."[25]

These tensions were particularly evident in Chicago. While progressive on racial matters, Cardinal Meyer instinctively chose to work with other priests, and not laypeople, on archdiocesan issues. Indeed, his career as a theologian, seminary professor, and administrator had allowed him little contact with laypeople outside the most formal settings.[26] By contrast, the lay head of the local Catholic Interracial Council, John McDermott, persistently emphasized the need to foster lay leadership within the Church. When the Chicago archdiocese contemplated setting up its own race relations program in 1961, McDermott protested. "Only a few [priests]," he complained, "seem to understand that how this job is done is of crucial importance to the lay apostolate as well as to interracial justice and charity." McDermott also argued that the National Catholic Welfare Conference (run by the hierarchy) worked on racial issues only because of "their fear about the growing power and influence of the lay-controlled National Catholic Council for Interracial Justice." He congratulated Philadelphia colleague Dennis Clark for publicly criticizing the Philadelphia archdiocese. "We are not going to get anywhere in social action in the Church," McDermott warned, "until laymen get up off their knees before the clergy." In 1964, he concluded that "we are not just passive followers in the Church, but citizens with some rights."[27]

The Vatican Council pushed these developments forward. In place of older notions of the "lay apostolate" that presumed direction by the clergy, the bishops shifted discussion to the independent role of the Catholic layperson in society. The primary image used to describe the Church during the council—the people of God—deemphasized hierarchy. The trajectory of one activist Catholic couple was perhaps typical: first, participation in a Friendship House interracial and ecumenical weekend. Then a close reading of various conciliar documents, carefully marking "everything about laity, everything about women." Finally, a growing sense that "the laity is *obligated* to speak out on issues in which they have expertise." Within the parishes, laypeople began to read the scriptures during the liturgy and play substantive roles in the growing number of parish and even diocesan councils.[28]

The contrast between soaring expectations and actual change, however, was often stark. Priests accustomed to unquestioned authority and responsibility often found extensive consultation with parishioners a trial, and newly formed parish councils struggled to balance loyalty to the pastor with the need for an independent voice. Bishops only hesitantly moved toward more inclusive diocesan structures.[29]

By the late 1960s, a few frustrated laypeople took the language developed both in Rome and Selma and directed it toward local church leaders. "Since the Vatican Council" commented longtime activist Ed Marciniak, "the improved position of the Christian proletariat can be compared to the evolving status of the Negro in American society." Both African-Americans and laypeople, he added, were segregated from real power. An entire chapter of one jeremiad advocating a larger role for laypeople compared "the situation of the American Negro" to "the situation of the Catholic layman in his efforts to establish a viable relationship with the Catholic clergy." Paul Twine, newly appointed leader of the Chicago Catholic Interracial Council, made the comparison even more explicit when protesting a $40 million fund drive announced by the archbishop without lay consultation. "I sometimes think the place of the layman in the Catholic Church," Twine argued, "is similar to the place of the Negro in American society. He is treated as a powerless nobody and he must struggle and shout to be treated like somebody."[30]

The path of the relationship between Catholic women and the hierarchy followed a parallel arc. By the middle of the twentieth century, Catholic men and women were firmly ensconced in a vast array of schools and parochial organizations segregated by sex. Excepting the nuns in control of many Catholic schools and hospitals, however, women were absent from the corridors of ecclesiastical power. Hope Brophy, for example, developed a series of educational programs on racial issues under the auspices of a Detroit Catholic women's group in the late 1950s. When the archdiocese agreed to sponsor and expand the program, Brophy was politely asked not to attend the kick-off luncheon since her presence might make the assembled clergy uncomfortable.[31]

Statements by John XXIII before the council and by a few bishops during the sessions hinted at dramatic changes, but the extremely limited role of women during the sessions confirmed the fears of activists. Along with women involved in the civil rights movement, some Catholic women began comparing racist ideologies to assumptions about fixed gender roles.[32] Theologian Mary Daly, for example, compared the theological habit of attributing certain "characteristics" to women with "fallacies in the racists' arguments."[33]

American nuns encountered these new currents at a particularly opportune time, since the Vatican command for a mandatory "renewal" of all religious orders coincided with the rush to participate in the civil rights movement. Even as several Sisters of Loretto received enormous media attention for marching at Selma, for example, the leader of the order, Sister Mary Luke Tobin, journeyed to Rome to serve as the first woman "auditor" for discussions on the future of the world's nuns.[34] One analyst commented that "The nuns at Selma were not a prelude to more demonstrations but serving notice to the entire church that they must not be counted out when the Church is giving out first-class citizenship within the People of God."[35]

Those nuns most active in racial and social issues also pressed for changes in community life. As one activist put it, "the Sisters who have been exposed to Catholic Action groups, the Sister Formation Movement, or the inner-city apostolate are the ones most anxious for adaptation and renewal."[36] Sister M. Berchmans Shea argued that both social protest and religious renewal reflected the "nun's quest for identity as a person rather than in terms of function or role." Mercy Sister Maureen Fielder recalled discussing civil rights issues with other nuns, and then remembering that she was not allowed to attend high school basketball games without permission. Pittsburgh policemen might stop traffic and allow Fielder to drive through a red light, but when she proposed a march in support of Selma protesters, her mother superior answered "We can do more by staying home and praying" and "We don't go out late in the evening, sister." As one of the speakers preceding Martin Luther King at one Chicago rally, Sister Mary William noted that "It is hard to march in Chicago because on every street corner . . . is a nice, big policeman who has either a daughter or aunt or cousin who is a nun." She concluded, "Why are Sisters like Negroes? Because they, too, are segregated, ghettoed, and because leadership is not encouraged in their ranks."[37]

Similar logic restructured relationships between priests and the hierarchy. Far more than laypeople, and even more than nuns partially sheltered within religious orders, diocesan priests depended upon the goodwill of their pastor and local bishop. Conflicts between pastors and bishops accustomed to obedience, and clerics and seminarians imbued with new notions of "rights," were endemic by the late 1960s. One Boston interracialist endorsed a 1965 protest by local seminarians. "[T]he same 'unfreedom'," he concluded, "which imprisons Negroes in the Roxbury ghetto creates a repressive and regressive atmosphere in the seminary."[38]

The most celebrated example occurred in Los Angeles, where Cardinal Francis McIntyre championed an ultramontane vision of Catholic life. (McIntyre's critics never tired of pointing out that before joining the seminary he had been a successful Wall Street investor.) In one sense, McIntyre's accomplishments were noteworthy. In his first five years in Los

Angeles, he supervised the establishment of 26 new parishes and the building of 83 schools for the one thousand Catholics moving into the archdiocese each week during the postwar boom. For parishes located in impoverished areas, McIntyre arranged for generous archdiocesan subsidies. Less successfully, McIntyre attempted to manage the growing archdiocese with an iron hand. Censors plucked "liberal" Catholic publications from seminary libraries, chancery officials frowned upon ecumenical discussions, and controversial theologians found archdiocesan buildings impossible to reserve for public addresses. At the Vatican Council, McIntyre pleaded for the retention of the Latin Mass since "active participation is frequently a distraction."[39]

This conservatism extended to racial issues. To the chagrin of other bishops, McIntyre habitually argued that, in Los Angeles, "Kind lives with kind—Irish with Irish, Poles with Poles, Mexicans with Mexicans, Negroes with Negroes." Based upon his experience in New York City, McIntyre privately maintained that segregated churches were simply "natural grouping[s]"—a viewpoint roughly corresponding to Catholic orthodoxy thirty years previously but now viewed as utterly untenable. Chancery officials quashed attempts to form a Catholic Interracial Council chapter and McIntyre routinely praised conservative local politicians. Archdiocesan officials also commanded local Christian Family Movement (CFM) chapters to halt attempts to discuss racial issues; the replacement topic was the innocuous "Happier Families in CFM."[40]

In 1963, a group of liberals demanded "the leadership of our clergy and hierarchy in the struggle for interracial justice" by holding a four-day sit-in at archdiocesan offices. McIntyre responded with the extraordinary statement that "The Negro is better treated in Los Angeles than anywhere in the United States." Similar utterances inspired William DuBay, a young priest, to announce to the press that he had written Paul VI and asked for McIntyre's removal due to "gross malfeasance." McIntyre reacted with typical obtuseness—first transferring DuBay away from his predominantly African-American Los Angeles parish and later suspending him from the priesthood, actions which provoked a flood of unfavorable news stories and some support for DuBay from fellow clerics. "Vatican Justice is like Alabama Justice," chanted DuBay's supporters at numerous rallies, and DuBay himself asked "Why should Catholics be deprived of their civil rights just because they are Catholics?"[41]

Significantly, DuBay spent the final years of the decade attempting to form a national priests' union. "The reform of the Church," DuBay concluded "rests not upon achieving consensus but upon the right to dissent." Priests, in other words, joining the path trod by activist laypeople and nuns, moved from concern with civil rights for African-Americans to discussions of rights within the Church. Cardinal McIntyre was widely regarded as an embarrassment, but less recalcitrant bishops also confronted local priests' associations and caucuses, developments unthinkable only a few years previously. At New York's St. John's University, a priest leading a bitter faculty strike emphasized how "The Catholic Church should have been in the forefront [of civil rights]" even as he made an impassioned plea for academic freedom. In Chicago, one spark for formation of the Association of Chicago Priests was Archbishop Cody's unilateral decision to replace activist priests involved in racial and urban affairs with more conservative colleagues.[42]

Attacks on clerical authority culminated in an issue unrelated to racial matters. In the summer of 1968, after much agonizing, Paul VI overruled a papal commission appointed to review the subject and reiterated traditional opposition to artificial methods of contraception. One component of Paul VI's decision, of course, was the desire to avoid a weakening of Church authority by reversing a traditional position. The ironic result was to weaken Church authority on a variety of issues connected to sexuality. While hardly the only cause, the encyclical played a role in the sharp decline in weekly mass attendance and formal identification with the Church, a decline not halted for almost a decade. An immediate firestorm of criticism also swept through the American Church structure: theologians signed formal dissents, priests worried about what position they should take within the confessional, and the majority of laypeople, according to public opinion surveys, rejected the papal position.[43]

These events underscored a central dilemma for Catholic interracialists. Appeals to authority tended to undercut lay initiative and strengthen a hierarchical order, but traditional authorities offered a more appealing perspective on racial questions. What then, would a position on racial issues that reflected the views of the faithful be? Two New York priests pondered this question during the dispute over African-American activist James Forman's demand for 500 million dollars in reparations from American churches. Most parishioners clearly viewed Forman's demands as unacceptable, but many priests wished to at least discuss the matter. "In these days of active participation by the laity in decisions concerning the allocation of Church resources," the priests commented, "the leaders of the Church face a real problem."[44]

Causing added anguish was the realization that some of the most progressive bishops on racial issues were also the most authoritarian. Washington archbishop Patrick O'Boyle, for example, a longtime supporter of racial integration and the civil rights movement, became embroiled in an attempt to discipline priests who dissented from Church teaching on birth control. San Antonio archbishop Robert Lucey, famed for his work with the Mexican-American and African-American communities, and a consistent advocate of the United Farm Workers, provoked fifty-one archdiocesan priests into calling for his immediate dismissal. The high-handed tactics of Chicago's Archbishop Cody, despite consistent financial support for Catholic work in the inner city, sparked a decade-long series of agonizing disputes.[45] Studies completed in the early 1970s documented vast differences in how priests and bishops perceived questions of authority—providing a partial explanation for the rapid increase in priests leaving the ministry.[46]

III

Just as a logic of freedom and rights encouraged individual Catholics to rethink Church structures, the more pluralistic theology emerging from the Second Vatican Council led to questions concerning Church ritual and form. Eventually, this questioning extended to the notion of interracialism, which since its first articulation by John LaFarge in the early 1930s had come to dominate formal Catholic thinking on racial matters. In contrast to the Church of the late nineteenth and early twentieth centuries, advocates of interracialism had argued that the separation of Catholics into various ''racial'' groups destroyed the unity necessary for Christian community. Segregation of African-Americans was particularly pernicious because of its comprehensive nature, but the separation of European ''racial'' groups was also seen as an unfortunate barrier to a unified communion. Integration, then, became a theological necessity. Archbishop Joseph Ritter of St. Louis commanded his priests in 1953 to ''never permit references to the different races or nationalities in his parish. 'Our people' should always be on his lips as the outward expression of his interior pastoral charity.''[47]

A minority position before World War II, this interracial ideology had become orthodoxy by the early 1960s. Bishops quietly phased out European national parishes as well as strictly ''racial'' parishes once used to segregate African-Americans. ''With all the agitation at the present time for desegregation,'' wrote one Toledo priest in 1962, ''one thinks twice before establishing a new center for the colored. . . . we think that the trend for the Negro to identify himself with an existing parish will keep on growing.'' Another priest predicted that ''As the integration of the Negro takes place, the racial Church will disappear. Such seems a good guess, just as the national Church is gradually moving by itself away from its former isolation.''[48]

Unity assumed uniformity. The achievement of the Church, in the standard view, was its ability to unify various cultures across the globe into one set of liturgical rituals and practices. (A common boast of Catholics was of their ability to attend mass anywhere in the world and feel immediately "at home"). Language and ritual unified the Catholic community, marking a theological reality distinct from any particular or local expression.[49] Attempts to wean European national groups away from specifically immigrant practices, as well as the rigidly Romanist style adopted in African-American parishes, reflected this pervasive confidence in "integration" broadly understood.

Two developments altered this situation. First, the Second Vatican Council, while reaffirming the long battle of interracialists against discrimination, altered the terms of the discussion. No longer would the mark of the universal Church be its uniformity across time and culture. The most obvious change was the openness to a variety of liturgical forms, as well as the mandate to use the vernacular in all liturgies. Crucially, conciliar documents emphasized the links between evangelization, properly understood, and a sensitivity to local situations. According to *Gaudium et Spes,* the Church is "not bound exclusively and indissolubly to any race or nation, to any one particular way of life, or to any customary pattern of living." In effect, the bishops self-consciously strove to create a pluralistic institution, one more deeply rooted in local cultures. Following the Council, regional and national associations of theologians, bishops, and church organizations developed with astonishing rapidity.[50]

Second, in America, the ramifications of these reforms fused with the Black Power movement. As in the broader African-American community, a handful of African-American Catholics began to question universalist claims. Older interracialists termed standard liturgical rituals "Catholic"; dissenters now viewed them as "white." From the dissenting perspective, the depictions of Irish saints in the stained-glass windows of many inner-city churches, the decorous music, and most of all, the paucity of African-American priests, nuns, and bishops, belied the notion of a neutral institution. African-American Catholics accustomed to the message that the Church subsumed all cultures now learned of a "black theology" and a "black spirituality." One priest noted that only integration "in the true sense of the word" could serve as the basis of the church's work. "Black Power," he continued, "does not necessarily mean the end of the Catholic Church's work in the Black Community, but it may just be the beginning of real effective work."[51]

These arguments did not win unanimous approval. Many African-American Catholics—just like Catholics more generally—moved away from the Latin mass and traditional hymns with reluctance. A Chicago priest recalled the tension in his parish when a few parishioners attempted to replace the sign of peace handshake during the mass (itself a not entirely welcome product of the Council) with a "soul" grip. The African-American head of the Catholic Interracial Council in Detroit scornfully attributed the city's 1967 violence to " 'Black Power' advocates."[52]

Still, a significant number of African-American Catholics began to reflect upon their experience in an overwhelmingly white institution. "The Church has tried to convert blacks to whiteness," charged Sister M. Shawn Copeland of Detroit, "Consequently, mental and physical harm has been caused to black people."[53] "Everybody was pretending that there was no such thing as black or white" recalled Chicago priest George Clements in describing what he called "the gospel of integration" preached during his seminary days. No one dared mention "anything about blackness." An African-American New York priest, Lawrence Lucas, recalled Malcolm X stopping him while Lucas walked across Harlem one day during the late 1950s. "Are you out of your God damned mind?" asked Malcolm when Lucas confessed his goal of becoming a Roman Catholic priest. Lucas himself penned a polemical attack on racism in the Church in 1970 which included an endorsement of the just-war theory as a justification for African-American self-defense. He concluded that "The history of the American Catholic Church shows that at the height of each group's ethnic consciousness, such mixings [of ethnic groups] often proved disastrous."[54]

A cluster of strictly African-American Catholic groups reflecting this vision formed in the late 1960s. The most publicized events took place in Detroit. Prior to a 1968 national conference of clergy interested in the "interracial apostolate," an African-American priest sent a letter to the nation's approximately 170 African-American clergy and asked them to attend a special caucus during the conference. At this meeting, held only two weeks after Martin Luther King's assassination, sixty African-American priests, including the nation's lone African-American bishop, discussed their situation. After a full day of wrenching debate, and four votes, the priests passed a resolution terming the church "primarily a white racist institution" and demanding African-American control of inner-city programs. The Church in the inner-city, the priests warned, is "rapidly dying" and black clergy felt an "utter frustration" because of their inability to "provide a ministry for black people in the Church."[55]

Equally important, African-Americans attempted to place Catholicism in a new context. The hope of the interracialists, that racial and ethnic differences would give way to an overarching religious identity, faded admidst the fear that the Church would become wholly irrelevant to African-American concerns. Statues of African saints and African representations of Christ and Mary began appearing in more churches, and a few pastors hired artists to paint murals in the church building depicting African-American heroes. African-Americans in one national Catholic youth group formed a separate caucus. The phrase "black first, Catholic second" entered the lexicon. "We have given up the myth," one African-American priest told a meeting of the nation's bishops in 1969, "that we could be white, that if we did all things in proper and approved fashion, we could be acceptable to the white people."[56]

This assertion of an African-American Catholic identity naturally led to demands for "community" control of monies and institutional resources—a pattern evocative of confrontations between Polish pastors and Irish bishops in the early twentieth century and, of course, one replicated throughout American society in the late 1960s. "The present attitude of the Black community," announced the African-American priests in their 1968 statement, "has developed to one of Black people controlling and making decisions for the people in the Black community."[57]

Frequently, liberal whites and African-Americans both used the language of Black Power, a marker of how for some whites inner-city parish work was a radicalizing experience. In Newark, twenty priests working in inner-city parishes (nineteen of whom were white) accused the Newark archdiocese of racism. (They entitled their statement a "Declaration of Brotherhood to Our People.") "We have survived riots, burnings, beatings [and] shootings," the priests announced in a statement. "We cannot survive as Christians unopposing [sic] the contrived archdiocesan apathy." Other members of the group added that the archdiocesan high school system was "nothing less than the largest white racially segregated private school system in the state of New Jersey." Father Lawrence Lucas addressed a rally of 1,000 in support of the priests. "What Negroes need," Lucas declared, "is the right to determine certain aspects of their own lives." The protesting priests followed a similar line of attack when the archbishop announced the appointment of two priests to inner-city advisory positions. "This process of selection," retorted the priests, "is one more example of the failure of those in authority to consult the community in forming any official policy."[58]

In Chicago, parish appointments became a central issue. The position of pastor at St. Dorothy's parish became vacant in 1968. Parishioners in the predominantly African-American parish expressed support for the assistant pastor, African-American priest George Clements. Instead, Archbishop Cody named Rollins Lambert, the longest-serving African-

American priest in the archdiocese, to the position. The rejection of the outspoken Clements and the selection of the more moderate Lambert became viewed by some observers as part of an attempt by Cody to silence militant African-American voices. Four of the city's African-American priests, including Clements, threatened to quit their posts if the appointment was not rescinded. Jesse Jackson, with Clements sitting behind him, informed a meeting of 450 archdiocesan priests that "It is wrong for George Clements to be ridiculed, to have no opportunity to grow in this diocese." At one meeting in St. Martin's gym, protesters termed Lambert an Uncle Tom and irrelevant "to the problems of the black community." Also present at the meeting was Fred Hampton, leader of the Chicago Black Panthers. (The Panthers had entered the gym and ordered everyone not to leave.) Hampton's booming voice drowned out the attempts of Catholic Interracial Council leader John Hatch to forge a compromise on the issue.[59]

Lambert himself held a press conference to denounce his "political" appointment. He also termed Cody an "unconscious racist." Two days later, Clements, Lambert, and African-American priests from across the country celebrated a "Black Unity" mass using African-American dance and music. The featured speaker was Jesse Jackson, and Black Panther members stood along the sanctuary walls.[60] The next month, the African-American priests again attacked what they termed the assumption that "the black community should become in every respect like a middle-class Irish or German neighborhood and parish." After several months of protest and negotiation, Cody finally agreed to appoint Clements as pastor of Holy Angels parish, also on the city's South Side. (Privately, Cody expressed his belief that "Father Clements is a black priest, blatant, arrogant and one of little knowledge. . . . It is indeed regrettable because priests, Irish, Polish, Italians, Germans, who were beginning to be well-disposed toward the black cause have been 'turned off' completely.")[61]

Catholic liberals observed these developments with some uneasiness. One nun wrote her colleagues in 1969 to urge support for the National Black Sister's conference. "The last two days," she added, "are open to white sisters. Positive attitudes about the need for such a conference are important." Mathew Ahmann of the NCCIJ noted that when "Catholic Negro laymen recently organized a Council of Negro Laymen in Cleveland, many looked askance. The group replied that the Poles had their priests and their Church, the Irish had theirs, and the Italians theirs. . . . Not a very wide vision, perhaps; but who is to say that it is not a realistic one?" The NCCIJ, Ahmann later added, despite favoring integration, recognized "how this ideology has been used as [a] chimera which in fact may have blocked the thrust of the Negro community to self-determination.''[62]

Ahmann's remarks suggest the fragility of the interracialist vision by the late 1960s. The primary difficulty remained white intransigence. As Ahmann and other activists had discovered, many residents of white, heavily Catholic neighborhoods were still quite willing to resist attempts to integrate particular areas of the northern cities. The novel aspect of the situation was that liberal white Catholics now questioned their ability to work in African-American areas. Cleveland's Commission on Catholic Community Action acknowledged that ''we can no longer delude ourselves into thinking that we are helping people regardless of their race or color. . . . It is time now for the Church and the Commission to manifest and give evidence of the fact that the only way to help anyone is by having a special regard for that person's race and color.'' A New York priest concluded that ''The time for whites who consider themselves missionaries or volunteers . . . has ended.''[63]

In short, could whites foster a Catholic community in inner-city areas without inhibiting, by their very presence, the development of an African-American Catholic church? Did the cultural forms upon which Catholicism was carried limit the possibility of religious community across ethnic barriers? White liberals who attended one church in Detroit's African-American ghetto were quietly asked to leave by the parish priests, since the priests feared that the parishioners viewed the white presence as ecclesiastical tourism. In their 1968 statement, the African-American priests demanded more ''black-thinking white priests'' willing to support and learn from the African-American community—a role, the priests added, ''that white priests in the black community have not been accustomed to playing and are not psychologically prepared to play.''[64]

Both of these developments—the crisis in authority and the collapse of interracialism—came into focus during the operation of Project Bridge in Cleveland. In 1966, a severe riot swept through the city's Hough area, shocking civic leaders. The following year, tensions between the white, heavily Catholic population and African-Americans flared during the mayoral campaign that ended with the election of African-American Carl Stokes.[65] In 1967, the American Council for Nationalities Service and the NCCIJ received substantial grants from the Ford Foundation to train ''priests, teachers, youth leaders, neighborhood level and city-wide ethnic group leaders, in an attempt to build solid bridges on common projects between the Negro community and much of white Cleveland.''[66] (The Ford Foundation also funded a CORE voter registration drive that in effect served as a tool of the Stokes campaign—several observers interpreted the foundation's support for Project Bridge as a way to avoid implications of partisanship.) In discussions with the foundation, the NCCIJ emphasized that the turmoil in Chicago during the previous summer demonstrated the need to work closely with the Catholic Church, since for many citizens, "the local church is their strongest link to the past." The NCCIJ quoted a Cleveland welfare expert to the effect that "The city has become three clenched fists ready to strike at Negroes—one Hungarian, one Polish, one Italian."[67]

Researchers from local colleges and five nuns with doctorates in the social sciences staffed the project. Programs included a pulpit exchange in which priests from the inner-city spoke in white ethnic areas, the formation of a cluster of community organizations, a series of radio programs bringing together Catholics and African-Americans, education programs in the Catholic schools, and the merging of three parochial grade schools into one school which could then serve all residents of "this diverse community."[68]

Successes were limited. Pulpit exchanges worked reasonably well, but few callers participated in the radio program and organizers gave only minimal supervision to teachers involved in school programs. From the beginning, staffers, including the nuns, encountered suspicion that they represented outsiders—even communists—eager to destroy the neighborhood. "This sense of communist infiltration into the civil rights movement," commented Sister Loretta Ann Madden, "is so deeply ingrained into the nationality groups that it overshadows all other possible misreadings within the workings of the minority groups." Community organizers faced the same hurdles. "The Buckeye area," complained Sister Roberta Steinbacher, "was just impenetrable." Overall, residents evinced a "tremendous resistance to becoming involved with Negroes on community problems."[69]

Assumptions underlying the program came under renewed scrutiny. A few of the staffers became less certain that racism in white ethnic parishes was the primary problem. One nun acknowledged awkwardness in discussions between "blue-collar ethnics" and social scientists, while another commented that fear of crime—"the people are very disturbed because they are not able to hold church services in the evening"—dominated local discussions.[70]

The release of the Kerner Report stressing the need to eradicate white racism and the developing separatist ideology within the African-American community also pushed program organizers away from "bridging." In 1968, the Ford Foundation agreed to support a "Summer in Suburbia" program in which seventy-three nuns visited over 30,000 homes asking open-ended questions about racial issues. The nuns deemed half the conversations "positive" and a slightly smaller proportion "negative"; a significant percentage of residents were clearly home but failed to open the door.[71] At the same time, a group of twenty-six Black Catholics formed a Black Task Force within the diocese, and some Project Bridge staffers gave the group enthusiastic support. Attempts at interracial dialogue dwindled. "[I]n the first six months of the second year . . . white staff worked in white suburban communities and black staff in black neighborhoods." Three of the supervising nuns eventually resigned from the project because "they could no longer work within Bridge's institutional framework based on integration."[72]

Staffers' energy quickly moved away from civil rights and toward reform of the institutional Church. At the onset of the project, the five nun-administrators stopped wearing habits in order to diminish the distance between religious and laypeople. " 'Bridging' " one nun concluded, "could not be limited." In January 1969 one staffer warned that only "confrontation" and not "gentle assimilation" could produce changes in the Church structure. The next month, two staffers participated in a protest mass at the city's cathedral. Even as policemen called by nervous chancery officials stormed up the aisles to arrest rebel clergy, one priest read a statement castigating Catholic leadership for supporting the war in Vietnam, along with "imposing [and] also . . . perpetuating white racism." He added that "we are continually preoccupied with its [the church's] failures to implement the instructions of the Second Vatican Council concerning the restructuring of the church at each of its levels." When arrested, the priests sang "We Shall Overcome." The Project Bridge board endorsed the "idealism" of the gesture and staffers proclaimed that an end to "collaboration with the church" meant that the project could finally be "free." By the end of the year, one nun working with the project had left religious life and begun a campaign to discourage Catholics from contributing money to their parishes.[73]

IV

The initial decision of Project Bridge organizers to focus on "white ethnic" communities suggests how the "white ethnic revival"—within both the Church and American society—became the logical corollary to Black Power. A cadre of scholars and activists produced a series of publications documenting the persistence of ethnicity in American life and emphasizing the distinctive quality of the Catholic ethnic experience. The tone was aggrieved. Community organizer Barbara Mikulski, later a Maryland senator, attacked "phony white liberals, pseudo black militants and patronizing bureaucrats" who opposed federal funding for parochial schools and tax reductions for the working class. "These social classes which seem so committed to expiating guilt for injustices done to the blacks," Father Andrew Greeley argued, "were quite unconcerned about injustices and exploitations worked upon white ethnics and their ancestors."[74]

A cluster of organizations—often assisted by foundation grants—emerged to examine white working-class communities. Polish priests in Detroit started a "Black-Polish Conference." Father Greeley's volume on white ethnics became the first publication in a series on ethnicity sponsored by the American Jewish Committee, and he later became a director of the Center for the Study of American Pluralism. Father Geno Baroni left his ministry in Washington's African-American ghetto to supervise a National Center for Urban Ethnic Affairs.

Baroni explained that the Black Power movement had caused him to reflect upon his own Italian roots. "Blacks," he noted, "[had] become increasingly resentful of whites who tried to work with them, and rightly so. I believe the black community should do its own thing."[75] Baroni addressed a general session of the national bishops' conference in 1969. White ethnics, Baroni warned, are "a major source of our vocations and the backbone and support of our Church. . . . We have to go beyond the coalition of the 1960's that we had for the Civil Rights movement." The bishops agreed. One year later, they announced that "if there is to be a resolution of the racial crisis which currently grips our society, a critical role will be played by the white ethnic working-class communities."[76]

Proponents of the revival highlighted the obvious. Identifiable communities of Euro-American Catholics did exist in all of the northern cities, and effective public (and religious) policy needed to take these communities into account. Nevertheless, the assumption of many scholars and activists that ethnic identity would endure with relatively little change—Michael Novak entitled his study *The Rise of the Unmeltable Ethnics*—proved mistaken. The functional predictors—fluency in a foreign language, intermarriage, residential segregation—pointed toward assimilation, if perhaps not as rapidly as anticipated immediately after the Second World War. Even cultural attributes such as taste in food, clothing style, and career choice became less distinctive. By the end of the decade, scholars used the phrase "symbolic ethnicity" to describe a process of literal choice, not a birthright.[77]

The separation of the academic phenomenon known as the "white ethnic revival" from the issues most pertinent to residents of heavily Catholic communities also became evident. On the one hand, academics and activists celebrated a variety of ethnic traditions. While receptive to these developments, residents were more likely to be appalled by the demands made by Black Power advocates than to replicate those demands within their own communities. One correspondent complained to New York's Cardinal Cooke about nuns supporting community control of predominantly African-American schools. "We do not demand our children be taught only by teachers of Irish extraction. We do not demand our children be taught Irish culture, Irish history."[78]

The saga of Chicago priest Francis X. Lawlor illuminates these matters. The son of a New York policeman, Lawlor joined the Augustinian order and spent twenty-two years as a science teacher and disciplinarian at St. Rita's High School in Chicago, as well as engaging in occasional solo campaigns against "smutty" books and immodest dress.[79] By Lawlor's own account, the 1966 marches by Martin Luther King convinced him that areas on the Southwest Side of the city needed protection. Within three years, Lawlor's efforts had yielded 186 block clubs with 10,000 nominal members and two separate offices—all dedicated to "holding the line" at Ashland Avenue for an indeterminate amount of time. "St. Leo's, St. Sabina's, St. Kilians were changing," he later explained, "Sacred Heart had already gone Negro and I came to feel that everybody panics." He added in a 1968 letter to Cardinal Cody that "Unless we hold the line at Ashland Av., it would only be a matter of time till every Catholic parish and school, built at great sacrifice by the people on the Southwest Side of Chicago, would be empty. Non-Catholic whites would suffer the same fate."[80] In a jab at liberal pretensions, he insisted that his program was ecumenical—linking local Protestants and Catholics in a way that "fits in with the Cardinal's diocesan program."[81]

Liberal Catholics viewed Lawlor's program as anathema. One priest took to calling him "Father George Wallace" while other Catholics debated him on local TV and condemned his tactics. Lawlor's activities also failed to please either Cardinal Cody or Lawlor's Augustinian superiors. In the spring of 1968, Cody and the Augustinian order commanded Lawlor to leave the Chicago area. To Cody's disgust, Lawlor returned less than a month later to continue his organization of block clubs, even though he was forced to live in a private home and work without parish affiliation. At every opportunity, Lawlor reiterated his opposition to racism. "I love the Negro as much as Cardinal Cody or anyone," he claimed, "They're all children of God." Instead, he characterized his opponents as racist, since they ignored "time-honored morally legitimate and healthy cultural, social and economic standards of the community."[82] No one else, Lawlor argued in a front-page interview in the *Chicago Daily News,* wanted to help the "forgotten people who are caught between the ghettos and the suburbs."[83]

Through fliers distributed outside of churches, endless rounds of block club speeches, private masses, and neighborhood rallies, Lawlor articulated the fears and frustrations of his constituency. He incessantly invoked what he perceived as the parallel example of the civil rights movement. "Our people are fed up," he told one crowd, "and we will demonstrate our opposition as the late Martin Luther King, Jesse Jackson, Al Raby, and the various black groups have been doing. Our right to survive in our own homes is as important as the black's right to live in their communities."[84]

The ethos was blue-collar and patriotic. Rallies were held in local parks or automobile showrooms, and Lawlor wore a hard hat in a number of advertisements. His newsletter logo contained four interlocking circles representing church, family, community, and unity. Block clubs strongly endorsed Flag Day, and ads in the club newsletter welcomed home individual soldiers from Vietnam. Before one rally of 2,500 people, Lawlor again emphasized that his campaign was not "anti-Negro" but merely a recognition that "All of your work as soldiers, sailors and marines to defend your country becomes meaningless if you cannot fight for your own back yard."[85]

Lawlor's claims were far less "ethnic" than "white." Some of Lawlor's rallies occurred simultaneously with ethnic celebrations, and certainly many of his supporters were second- and third-generation immigrants, but Lawlor framed his complaints in racial idioms. He entitled articles "The White Minority," and stressed the "disintegration of white communities."[86] Distinctions between various "white" ethnicities mattered far less than common grievances. At one Gage Park rally, Lawlor asked all present to raise their hands and signify their country of origin. Lawlor then triumphantly concluded that since all the countries mentioned were European, "our culture is a shared culture. We identify with each other."[87]

Far more than appeals to ethnic solidarity, two elements of Lawlor's analysis gave it persuasive power. First, he correctly identified an enormous problem: white flight and subsequent neighborhood resegregation. When Lawlor announced that "There has been no integration here in Chicago," he voiced a self-evident but rarely addressed truth. In addition, while public (and archdiocesan) officials negotiated with Black Power advocates and sponsored racial education programs, Lawlor zeroed in on the paralyzing fear caused by racial transition. He lamented that "There are no programs sponsored in our behalf by the government or the churches or the civil rights groups. We are the forgotten men and women whose plight is ignored and whose destruction is apparently of no concern to those in positions of power.'' A group of parishioners from Little Flower Parish echoed these concerns. ''Middle-class white people living west of Ashland Avenue are a forgotten minority,'' they claimed. ''We have stood placidly aside while our neighbors from other parishes (St. Sabina's, St. Kilian's, and St. Margaret's) have been forced to move.''[88]

Lawlor was also a priest. Through his very presence, he reminded supporters of the initial foundation of many neighborhood communities, and served as a rallying point for those appalled by changes within the postconciliar church. Supporters conspicuously held crosses at rallies, passed out decals reading ''Pray for Father Lawlor'' and maintained that unlike clergy abandoning the priesthood for secular pursuits, ''Father Lawlor will continue to be a priest forever because he was ordained to do so. In the eyes of God, he is always a priest.''[89]

Of course Father Lawlor's very defiance of authority, despite his adherence to the traditional Catholic forms, was itself only possible in the late 1960s. Conservative Catholics furious over the breakdown of church discipline in other matters now harbored a priest as rebellious—just on different issues—as the Berrigan brothers. In a hint of how quickly even conservatives turned to the language of ''rights,'' Lawlor immediately appealed to the newly formed Association of Chicago Priests when Cody tossed him out of the archdiocese. ''If this type of treatment can be handed to me, it can happen to any and every priest,'' Lawlor warned. ''This is the moment for truth or continued chaos and scandal.'' The association, torn between disgust with Lawlor's views and unease about Cody's disregard for procedure, agreed to consider the matter.[90]

Letters to archdiocesan officials discussing Father Lawlor reveal an extraordinary sense of betrayal. One Lawlor defender complained to Cardinal Cody about ''young priests who can't get by a Sunday sermon without injecting the brotherhood theme or using some 'way out' opening whenever they address the congregation.''[91] Another woman explained that she was second-generation Irish, the mother of seven children, and the wife of a local milkman. She added that she had never used birth control pills or allowed ''hanky panky.'' ''Is it too much to ask,'' she wrote, ''that the next 40 years (God willing) we try and enjoy them in the same neighborhood, with the same wonderful people we have known?'' It was time, she concluded, to start ''bucking all the powers that be, including the Church.'' A longtime South Side resident detailed her painful departures from five parishes because of racial transition, ''beginning with Holy Angels and ending with St. Philip Neri.'' She added that ''I have concluded from your and the Church's very apparent disinterest in us that the Irish hierarchy are continuing their obvious distain [sic] for us, the middle-class backbone of the Catholic Church in the United States.'' Another woman explained that her husband had worked two jobs to send the family's children to Catholic schools during twenty fondly remembered years in St. Sabina's parish. ''We were among the last to leave, but today you would call me a racist because I cannot and will not live in an integrated neighborhood. My knowledge of integration is not from newspapers, or T.V. I lived it. . . . We have only one true friend in the Church, Father Francis X. Lawlor.''[92]

These disputes also provided a glimpse of the shifting political terrain. Launched by his campaign to ''hold the line'' at Ashland Avenue, Father Lawlor fashioned a career in Chicago politics, crushing Democratic machine candidates in a series of campaigns that culminated in his election as alderman for the city's 15th ward.

Following the advice of political seer Kevin Phillips, Republican politicians at the national level also made a conscious play for Catholic votes.[93] Catholics overwhelmingly supported Hubert Humphrey in 1968, but Republican analysts continued to press what they perceived as an advantage on social issues—particularly busing, crime, and aid to parochial schools. In a memo to aides, President Nixon urged that the ''first priority'' be given to ''the blue-collar issues and the Catholic issues where McGovern simply cannot be appealing to them.'' In the spring of 1972, following a glowing introduction from Philadelphia's Cardinal Krol, Nixon endorsed federal aid for parochial schools in a well-publicized address before several thousand Catholic schoolteachers. (Republicans were also delighted by the refusal of Democratic liberals to allow prominent Catholics such as Mayor Daley and AFL-CIO chief George Meany to participate in the 1972 convention.) In the fall, for the first time, a majority of the nation's Catholics voted for the Republican presidential candidate.[94]

V

The tangled connections between the parochial schools and racial issues serve as a particularly revealing road map to end-of-the-decade Catholicism. Following the rejection of the common public school by the American bishops in the late nineteenth century, successive generations of parents, nuns, and priests dedicated themselves to constructing a Catholic educational alternative. Since World War II, the pace of the building programs had increased, as parents moving away from older neighborhoods clamored for Catholic education. Even as the harried pastor began construction of the elementary school—often prior to the church building itself—diocesan officials would stop by the rectory and inform him of the

Richard Nixon, Henry Kissinger, and Philadelphia's John Cardinal Krol
in 1972. AP/Wide World.

levy for the local Catholic high school. In 1958, Catholics spent $175 million on school construction costs alone. (One Brooklyn fundraising drive that began in that year, for five new diocesan high schools, exceeded its $24 million goal by $14 million.) While never enrolling more than 60 percent of all Catholic children, the over 14,000 schools extant in 1964 did enroll 5,662,328 students, 12 percent of the national total. In cities such as Philadelphia, Chicago, and Cleveland, from 20 to 40 percent of the student population attended Catholic schools.[95]

By the early 1960s, however, Catholic school administrators faced daunting challenges. In part, the crisis was financial. Increased public school expenditures, with subsequent jumps in property taxes, made the task of securing donations for parochial schools more difficult. (Conflicts over proposed tax hikes often divided school boards into Catholic and non-Catholic factions.)[96] In addition, the same parents demanding Catholic education also insisted upon facilities roughly comparable to more generously funded public schools, sending parochial building and maintenance costs on an upward spiral.

Crucial to keeping tuition charges at a minimum—and most Catholic schools continued to request only nominal fees—was a supply of women religious willing to work for token wages. As the number of entrants into convents began a long decline, pastors balanced the limited available funds with the need for salaried lay teachers. Added to these financial concerns was a falling birth rate and a logistical mismatch between school buildings located within the city and an increasingly suburban Catholic population. Total enrollment in parochial high schools and grade schools plummeted from over 5.5 million in 1965 to 2.5 million in 1990.[97]

But balance sheets told only part of the story. By the mid-1960s, many Catholic liberals viewed the schools as embarrassing anachronisms. A few analysts compared African-Americans and Catholics. "The walls of segregation which have imprisoned the Negro and the Catholic," argued one author, "are crumbling in the seismic shifts of the day. The Negro leadership is courageously and rightly helping to tear them down. The timid Catholic effort to shore them up seems wrong and worthy of failure."[98] Women religious available for teaching positions viewed such work more skeptically in the wake of the Second Vatican Council. One outspoken nun refused to become a "money-saving device for a middle-class society with middle-class values," while a colleague rejected work with "comfortable Catholics." A few nuns left Catholic schools to accept positions with inner-city public schools.[99]

Mary Perkins Ryan fired the most important salvo in her 1964 manifesto, *Are Parochial Schools the Answer?* In a time of ecumenicism, she asked, how could the Church devote such enormous resources to constructing a "mediocre Catholic milieu"? This "socioreligious segregation and the idea that such segregation is a desirable thing," she concluded, only harmed Catholic children while removing Catholic parents from the larger school polity. Liberals also seized upon the results of two major studies of Catholic schools in the mid-1960s, arguing that the data demonstrated the inefficacy of schools in promoting racial tolerance and commitment to the Church.[100]

Activists voiced similar concerns. One New Jersey priest publicly lambasted his bishop for organizing a $16.5 million dollar fund drive for "our white middle-class children while Negro and Puerto Rican children in the same diocese will continue to receive inadequate schooling, live in inadequate housing, be deprived of opportunities for betterment and bitten by rats." Father Groppi in Milwaukee went so far as to attack traditional Catholic schools as merely a "gimmick to bring people into the Church." The schools, he argued, must be "social-action oriented from the first grade up."[101] One speaker at a National Catholic Social Action Conference meeting urged the Church to simply sell all Catholic school buildings and then hand the money over to the poor. Commented NCCIJ executive secretary Mathew Ahmann, "The Church school serves the middle-class community. The sisters are tied to slavery in service of the middle class rather than being free to serve the slums of the inner city."[102]

Questions concerning the necessity of Catholic schools merged with debates on school integration. If by the mid-1960s Catholic schools enrolled from 20 to 40 percent of the *total* student population within the northern cities, they enrolled a significantly higher percentage of the *white* population. In Philadelphia, almost 60 percent of the city's white high school students were in Catholic schools. In Pittsburgh, 3.6 percent of the Catholic high school students were African-American, compared to 49.8 percent for the city's public schools. Statistics for Pittsburgh elementary schools were equally telling—2.9 percent of St. Lawrence's students were African-American, while four blocks away African-Americans made up 72.6 percent of the student body at Fort Pitt public school. Corpus Christi grade school was 74 percent white, while at neighboring Lemington (a public school) whites made up only 4.3 percent of the student population.[103] Not only did the schools promote a religious siege-mentality, then, their very existence thwarted efforts to integrate urban public schools. "Are Catholic schools," one columnist asked, "aggravating the problem of racial imbalance in American cities, or at least blocking possible solutions?"[104]

Attacks came from Catholic and non-Catholic sources. In Pittsburgh, the CIC lambasted the parochial school system for its failure to integrate at the same rate as the public schools—one 1967 statement mourned the construction in "a heavily Catholic area" of a new high school which would threaten the tenuous integration of a local public high school. Carl Francis, a Philadelphia Urban League representative, informed a group of local priests that Catholic schools are "producing many racists and are giving us some of our greatest problems." (Francis added that "When we go into neighborhoods that are predominantly Catholic, we run into a great deal of racial hostility.")[105] The city's school-board chairman complained that "the public school system would be much happier if the percentage of non-whites in parochial schools was nearer to that in public schools."[106]

Records of the Chicago archdiocesan school board during the 1960s reveal an agonizing minuet. Should facilities at heavily African-American schools be improved despite the inevitable segregation that would result? How could the board persuade African-American parents uneasy about sending their children on long bus rides to participate in busing programs? Would a racial quota system aimed at limiting resegregation hold up in court? Could integration occur in white suburban areas where "It is already impossible to accommodate all students who wish to attend Catholic high schools"?[107]

Forced-busing programs amplified these fears. Given the Supreme Court's rejection of metropolitan busing programs, public school busing efforts generally centered around white and African-American areas within city limits. As the administrators of the largest private school system in each city, Catholic leaders were inexorably drawn into the issue. When the Chicago schools announced plans in 1968 to bus public schoolchildren, the board and Cardinal Cody offered their support through a Catholic (albeit voluntary) busing plan. Cody also requested priests to read a letter supporting the public school plan in all parishes.[108] After prodding by liberal priests, Detroit's Cardinal Dearden ordered schools to scrutinize each new application for enrollment. "Even in the case of application to the school from people in the parish," he warned, "careful examination should be made of some parents' sudden decision to enroll their child in a Catholic school."[109]

The issue predictably widened the split between the two Catholic cultures. In Boston a handful of priests openly defied the archbishop's request and agreed to allow all students into their parish schools—one Dorchester priest helped African-American parents lead joint, biracial protests against the public school busing program. Small groups of Chicago Catholics protested in front of Cardinal Cody's home; one leader of a local taxpayer group urged Catholics to quit contributing to both local parishes and the ongoing school fund drive. (The school fund drive, ironically, was already under fire from Catholic liberals who questioned any financial support for Catholic schools.)[110] When one pastor admonished his parishioners during mass for protesting outside of a public grade school, a handful simply walked out of the church. "I know Vatican II said the Catholic Church is to determine what theology to teach our children," wrote one woman, "but it has not as yet said [that] they must be bused to learn that theology!"[111]

Racial incidents heightened anxieties. In Philadelphia, African-American students at a number of the city's predominantly white Catholic high schools protested what they viewed as unjustified expulsions and cultural insensitivity. The students—in negotiations mediated by NAACP advisors and an African-American Catholic attorney—demanded Black Student Unions, holidays in honor of Martin Luther King, Jr., and Malcolm X, and more focus on African-American history. To the disgust of liberal clergy and laypeople, the principal of Roman Catholic High School reacted to these incidents by instituting what he termed "martial law" and warning of mass expulsions. Following a tense confrontation between several hundred of the school's African-American and white students outside the school building, another teacher blamed "outside political groups" for provoking trouble, asserting that "about 90 percent [of the African-American students] are being controlled by about 10 percent of the black students."[112]

Correspondingly, some Catholic high schools with predominantly white student bodies now existed in largely African-American neighborhoods. As the last redoubts of an older Catholic world, the schools occasionally became battle-grounds for white and African-American students. As one Catholic sociologist complained, "There are parochial schools with at least 95 percent of the enrollment non-Negro existing in the same neighborhood where the public school enrollment is 90 percent or more Negro, a fact of isolation that leads to misconception, hostility, and occasional violence."[113]

In Cleveland, conflicts between students at Cathedral Latin (1,162 male students, 97 percent white) and public school John Hay across the street (1,494 students, 95 percent African-American) resulted in the cessation of sporting events between the two schools in 1968. Chicago's predominantly white Mount Carmel High School was located only a few blocks from heavily African-American Hyde Park High School. On the day following Martin Luther King's death, police intercepted one group of students from Hyde Park marching toward Mount Carmel. Administrators at Mount Carmel responded by dismissing all students early. As buses began to lumber away from the curb, groups of Blackstone Rangers—the city's most feared African-American street gang—rocked the vehicles. Eventually, the leader of the Rangers orchestrated a series of deafening chants—"Blackstone!"—as each bus moved down 64th Street. Groups of African-American gang members also occasionally startled members of the Mount Carmel football team by yelling gang slogans over the fence during practice.[114]

Critics repeatedly voiced a cluster of concerns—that Catholic schools prevented public school integration, served a middle-class constituency, kept nuns trapped in educational servitude, promoted religious separatism, and doomed parishes to a perpetual sea of red ink—at parish council meetings, educators' conferences, and school board discussions. Monsignor John Donahue, S.J., the director of the U.S. Catholic Conference's department of education, demonstrated the ideological distance separating liberals from those advocating a continuation of the parochial building boom in a 1968 article timed to coincide with the annual conference of the National Catholic Educational Association. Monsignor Donahue conceded that Catholic schools had been appropriate for their time. He also argued, however, that Vatican II and the urban crisis necessitated fresh approaches to educational issues. In particular, since Catholic "numbers are greatest in the very areas of the country . . . where the racial crisis is most urgent," administrators should reallocate Catholic resources toward the ghetto poor. All educational monies should be channeled first to the urban ghetto, second toward religious education outside of the schools, and only in the last resort to the maintenance of existing elementary and high schools.[115]

VI

Defenders of Catholic schools responded passionately. Cleveland bishop Clarence Elwell denounced the withdrawal of nuns from the classroom simply because a few critics said the nuns were "wasting their time when they could be out engaging in apostolic activity, teaching liturgy to all and doing social work." All systems of education, Elwell added, are "based on an ideology."[116]

The counterattack was particularly vehement on the issue of integration. Catholic schools clearly pulled substantial numbers of white students out of the public school system. Given that white families tended to abandon a neighborhood when the number of minority students in the public schools increased dramatically, however, the schools also enabled white Catholic families (at least in the short run) to remain in the city longer than their non-Catholic counterparts. The criticism made by one sociologist—that " 'Catholic' realtors have pushed sales to white Catholics in certain neighborhoods using the Church authorities' ability to maintain segregated schools as a selling point"—could also be read as an acknowledgment of the role played by Catholic schools in anchoring a portion of the white population.[117] As a number of studies suggested, Catholic schools fostered neighborhood integration even as they limited potential integration within the public schools.[118]

Catholic spokespersons focused on these points. Even if a strong parochial system did inhibit integrated education, its "very existence," according to a New York archdiocesan official, "has helped to make or to keep many neighborhoods integrated." Pittsburgh bishop John B. McDowell defended the construction of a high school in a heavily Catholic neighborhood in similar fashion. "It was our conviction that a strong Catholic school program had a holding power on a part, at least, of the White population. . . . [this program] came from the people and . . . expressed their desire to stay in Pittsburgh." Detroit's Cardinal Dearden conceded that "Where a fair racial mixture has been achieved, the effect has been due, in large part, to the Catholic schools." Indeed, since Catholic schools were able to maintain white student populations for a much longer period of time, the Catholic schools were often more integrated than already resegregated public schools.[119]

Supporters of the schools also pointed to the extraordinary success of Catholic education in inner-city areas. The Catholic school continued to be the most visible manifestation of the Catholic presence in the inner city, and a symbol of excellence to African-American parents disturbed by declining educational and disciplinary standards in the public schools. Enrollment of African-Americans in parochial schools sharply increased in the 1960s and 1970s, even as total Catholic enrollment declined.[120]

To be sure, predominantly African-American schools did not escape questions raised by competing definitions of Catholic identity. The pioneers of Catholic work in African-American areas during the 1930s and 1940s had clearly viewed the school as one means to a particular end: bringing the gift of the Catholic faith to a largely non-Catholic population. As one Albany nun phrased it in her history of the diocese, "The next attempt to teach the truths of the Faith to members of the Negro race was by means of a Catholic school."[121] The new sensitivity to both local cultures and other religions, as well as the perception among liberals that conciliar teachings emphasized social justice over evangelization, weakened support for conversion programs. Requirements that non-Catholic parents attend instruction classes or weekly mass were quietly dropped, along with once ubiquitous convert classes. Members of the Chicago archdiocesan school board protested that requiring parents to attend instructional classes in the faith "invaded an individual's freedom of conscience," while in Cleveland priests worried about news media reaction to any compulsory program. Nuns cautioned against "theological imperialism." In retrospect, one priest commented, "It seemed to be understood that neither black students in the parochial schools nor their parents were likely to become Catholics."[122]

Nevertheless, African-Americans repeatedly demonstrated their affection for parochial schools. A standard tableau of the late 1960s was the initial decision of archdiocesan administrators to close inner-city schools awash in a sea of red ink, followed by heroic efforts of parents to save the school and tense negotiations between administrators and local African-American Catholics. African-American parents at Newark's St. Charles Borromeo raised a remarkable $115,000 to keep their school open. In Detroit, African-American Catholics coordinated a series of demonstrations and sit-ins to protest the replacement of inner-city schools with after-school religious education programs."The schools are the only things the blacks have in the Church," complained Joseph Dulin, the African-American principal of St. Martin de Porres High School. Parishioners of St. Agnes parish added in a letter to Cardinal Dearden that "We love our schools and also need them desperately because of the inadaquate [sic] and poor quality Public School system in our Black community."[123] Rochester bishop Fulton J. Sheen unilaterally decided to give St. Bridget's parish (including the rectory, church, school, and parish land) to the federal Housing and Urban Development agency in 1967 as a gesture of concern for the inner city. (Sheen carried the repentance imagery to the extent of announcing the decision on Ash Wednesday.) Inner-city residents and priests vigorously protested the decision. "The church school [is] the most important thing in the neighborhood," complained pastor Francis Vogt, "There is enough empty property around without taking down the church and the school." Ultimately, Sheen rescinded the gift.[124]

A series of studies explained this support. African-American graduates of Catholic schools were more likely—regardless of the family's financial background—to attend and graduate from college and score higher on standardized achievement tests. Researchers attributed this success to a variety of factors: the emphasis within Catholic schools on academic subjects for all students (in part, because the schools could not afford expensive shop facilities), the ability to enforce disciplinary measures, a minimal administrative bureaucracy, and small student populations.[125]

Researchers also emphasized the importance of the larger Catholic community, even in schools where many of the students were not Catholic. Successful Catholic schools served both to create and strengthen social bonds between parents, parishioners, students, and teachers. At school fundraisers, liturgies, and other functions, educational and religious communities merged. These powerful connections—quite different from communities formed simply through neighborhood boundaries (public schools) or voluntary choice (elite private schools)—helped create a "social integration" which embedded students in a larger, supportive community.[126]

The best Catholic schools, in other words, provided some of the social structure necessary for the success of disadvantaged students. The paradigmatic example was Chicago's Holy Angels elementary school. By the early 1970s, the school enrolled a remarkable 1,282 students in its fortress-like seventy-five-year-old building a few blocks from the city's largest and most notorious public housing project, the Robert Taylor Homes. (Ironically, the parish school had been notably unwilling to accept African-American children in the early 1940s). Pastor George Clements and administrators stressed African-American pride, Catholic doctrine, weekly mass attendance (even for non-Catholics) and sustained parental involvement in the operation of the school through fund drives and frequent meetings between parents and teachers. Despite constant financial worries, the success of the school's pupils drew national attention.[127]

Even as the schools themselves closed, the experience at Holy Angels and educational research seemed to confirm the value of traditional Catholic educational policies. Operating the nation's largest private school system had always been extraordinarily demanding. Until the 1960s, however, Catholics generally accepted—indeed demanded—schools as one piece of a larger set of social structures. Working in concert with the family, Catholic societies, and the parish, Catholic schools were perceived as playing a crucial role in the formation of Catholic adults.

This set of assumptions supporting not only the schools but distinctly Catholic organizations became embattled during the 1960s. Just as liberals had joined with other like-minded individuals in the battle against segregation, so too, according to this view, should Catholics abandon parochialism. What value, liberals asked, came from the separation of Catholics from American society? "[T]here is no reason," argued one theologian, "why the Church should maintain and, what is worse, continue to build institutions of health, education, and welfare."[128] Other Catholics added that the value of groups such as the Catholic Interracial Council was only marginal, given that "Catholics, instead of belonging to their own ghetto

Father George Clements with students from Holy Angels elementary school. (1971). AP/Wide World.

society, [should] join with others of good will in a common organization."[129]

A group of miniature autobiographies by Catholic intellectuals published in 1965 revealed a remarkable distaste for the parish, the Catholic school, the Catholic college, and various Catholic societies. "The parishes that I know in New York, New Haven, and Providence," commented one academic, "are utterly deficient." A bitter appraisal of St. John's University in New York emphasized the community's "Catholic ethnocentrism, with its related defensiveness, puritanism, fundamentalist interpretation of religious matters, and political reaction."[130]

Significantly, one of the most discussed innovations of the era in liberal Catholic circles was the "floating" or "underground" parish. According to participants, these small, individualistic groups produced a "truly meaningful kind of Christian community" outside the boundaries of institutional structures. (One group specifically advocated a "disengagement from existing institutions.") The beauty of the underground church, one enthusiast concluded, was its devotion to "liberative personal and communal love-relationships." Again, these developments were routinely placed in the context of contemporary race relations. A New Jersey priest formed an experimental community after deciding that "just like the Negro situation—we just can't wait forever while people make minor changes in the law." In Philadelphia, the hesitant archdiocesan response to racial incidents became the genesis of one of the largest nontraditional groups.[131]

In effect, Catholic liberals—like American liberals generally—became enamored of a rights-based language that emphasized the relationship of the individual to society (or the Church) instead of the ways in which societal structures shaped individual development. At its best, sensitivity to individual rights shaped the response of Catholic activists to pressing issues—marching through heavily Catholic neighborhoods, for example, in support of the opportunity to purchase a home in any area of the city. Unfortunately, few issues possessed the moral clarity of the causes initially championed by civil rights activists. The dilemmas presented by metropolitan segregation and African-American poverty, for example, defied legal solutions. Even the passage of open housing laws, while important, failed to eliminate either African-American ghettos or white ethnic fears.[132]

When extended to ecclesiastical matters, the language of "rights" also stressed individual concerns. Perhaps nothing so characterized religious thought of the period as the belief that the institution—be it the hierarchical Church, an unresponsive parish, or a mediocre school—denied individual fulfillment. In this view, the endless fundraisers for Catholic schools, the bingo games to renovate the parish church, and the acclaim for pastors with "good business sense" masked a hollow spirituality which relegated religion to the margins of contemporary life. Father Richard McBrien matter-of-factly posed the question in a 1969 volume entitled *Do We Need the Church?* McBrien's answer—a tentative "yes" which included an attack on the Church's connections to the "residential community"—hinted at the chasm separating liberals from traditional Catholic notions.[133]

Only these intersections—between liberals uneasy about standard parochial institutions and practices, and traditionalists shocked to see those same institutions and practices disappear—explain the disarray evident in American Catholicism by the end of the 1960s. To a new generation of Catholic activists, the Church of their youth seemed spiritually bankrupt. One Brooklyn priest meditated on the meaning of his vocation after fruitlessly trying to calm white Catholic parishioners following a racial disturbance.

> And it dawned on me as it never had before how little Jesus Christ had to do with a person's being Catholic. Here were many Church-going adults and youths, crying out against the Negroes opposite them, challenging the authority of the police and wondering why a priest would be concerned. The Gospel of Jesus meant nothing here, the Church of Jesus had no place here. Perhaps more startling even [was] this: I was a priest, and the priest had no place here—with the layman and his life and concerns in the street.[134]

From a more traditional perspective, the prosaic task of maintaining the communal structures that had once sustained Catholic spirituality now seemed strangely unfashionable. Even the commitment of priests and nuns to the religious life seemed uncertain, as waves of younger, often more activist, clergy and religious decided to pursue other endeavors. (In 1970, over four thousand nuns left their religious orders; by this point, 4.6 percent of priests aged 29–34 were resigning *annually*.)[135] An Omaha Catholic noted that the archdiocesan priest most closely connected to civil rights activities had left the priesthood for marriage "by a J.P." From now on, the disgruntled communicant warned, "we will refuse to attend a church, or a Mass, or to support a church which is presided over by a civil rights activist priest."[136] One woman informed interviewer Robert

Coles of her misgivings about the younger clergy. "They're talking as if we did something wrong for being white. . . . Priests never used to talk about the Negro when I was a child. Now they talk to my kids about them all the time. I thought the Church is supposed to stand for religion and eternal things."

Internecine battles over these issues caused one commentator to fear for the health of the institution itself. One would hope, he concluded, that Catholics will maintain "what has seemed worth preserving for almost 2,000 years."[137]

CONCLUSION

Ambition was never wanting. In a North Philadelphia neighborhood just beginning in the late nineteenth century to distinguish itself from rural Pennsylvania, a group of Jesuits began the task of building a parish church in conjunction with the foundations for a Jesuit high school and college. The church, Gesu, was modeled on its Roman namesake, with towers soaring to twelve stories, 8,000-pound church bells and ten-foot-thick stone walls. The absence of interior pillars ensured a breathtaking expanse within the church building, culminating in a seventy-two-foot-high main altar. By 1895, the stone cross perched atop the church roof could be seen "from distant parts of the city."[1]

Within a generation, parishioners built a five-story rectory for the fifty-odd Jesuit priests, brothers, and scholastics in residence, a convent for the Sisters of the Immaculate Heart of Mary, and two separate parochial grade-school buildings. St. Joseph's college moved to a more spacious campus on the edge of the city in 1927, but St. Joseph's Prep remained in buildings next to the Gesu. With several thousand members living within the compact area from 15th Street to 20th Street on the west, Columbia to Fairmount on the north, the parish became one of the city's most prominent. Local politicians and business people proudly claimed membership, occupying the most elegant homes around the church in what had become a predominantly working-class, Irish-American neighborhood. Glossy booklets were produced in honor of various parish societies and in commemoration of annual parish festivals. Upon the dedication of the new parochial school building in 1917, 1,200 men from the parish Holy Name Society, followed by members of the local Ancient Order of Hibernians chapter, marched around the neighborhood in a solemn procession. Huge crowds requiring seven separate daily services squeezed into the church for special novenas. A touch of elegance was evident each year at the 5:00 a.m. Christmas mass, when white-gloved ushers escorted parishioners to their pews as a full choir filled the church with Latin hymns.[2]

And yet even those halcyon days were marked with tension. From at least the turn of the century, small pockets of African-Americans lived in and around the neighborhood, provoking some uneasiness concerning the future of the parish. Between 1907 and 1910, a Jesuit missionary recently returned from Jamaica, Father A. J. Emerick, organized a "national" parish, Our Lady of the Blessed Sacrament, for African-American Catholics in North Philadelphia. The response from his fellow religious at Gesu was guarded. As Emerick put it, some parishioners and priests feared that "too close proximity of a larger number of the colored people might injure the Gesu Church." Eventually, he established the church on Broad Street, just beyond the Gesu parish limits, since it "was thought that if I was stationed at the Gesu, my presence would draw too many colored people around there."[3]

Twenty years later the problems presented by a racially changing neighborhood had grown more pressing. From the end of the First World War until the 1940s, African-Americans poured into once predominantly Jewish areas to the north and west of the parish. One Philadelphia housing official pointedly observed as early as 1931 that those families "whose children have benefitted from public school education" (i.e. not families with children attending parochial schools) tended to first leave changing neighborhoods.[4] By this time leaders of white Protestant congregations had already begun planning the closing of church facilities.[5]

The contrast with Catholic populations and parishes was noteworthy. Irish, German, and Polish families, and this was particularly true in the 47th ward in which Gesu was located, tended to have high rates of home-ownership and low rates of foreclosure.[6] One resident maintained that realtors consciously began their blockbusting efforts away from the Gesu parish buildings because "If they had started right in front of the parish they [the neighbors] probably would not have succumbed."[7]

Even so, the future of the parish neighborhood was clearly uncertain. Twenty-one residents of Gratz Street contributed to a typical Gesu collection in 1933, but only one resident remained on the rolls seven years later, a signal of how neighborhood change threatened parish life.[8] Between 1931 and 1937, the assessed value of property in the neighborhood plunged 42 percent, almost twice the average rate of decline in depression-era Philadelphia.[9]

In 1936, Father James Maguire, a Jesuit living in the Gesu rectory, resolved to stop the area's seemingly inexorable "decline." A native of County Mayo, Ireland, Maguire had become a member of Gesu soon after his arrival in Philadelphia as a young man. After entering the priesthood he, too, served as a Jamaican missionary, among other endeavors, before returning to his home parish in 1935.[10]

A year after his return, Maguire spearheaded the formation of the Gesu Parish Neighborhood Improvement Association, whose very name suggested the Catholic propensity for equating parish and neighborhood. The organization's expressed purpose was to "keep and bring into the parish respectable home-owners and tenants and to prevent the further influx of undesireables into the neighborhood."[11] Strenuous efforts were necessary, since "We can truly say that more good for souls, for the community and for the Church has been accomplished within the buildings erected on the site bounded by Stiles and Thompson Streets, 17th and 18th Streets, than in any other set of buildings of like size in our city."[12]

The next several years saw a flurry of activity. Gesu parish still counted four thousand members, but African-Americans now predominated on some of the area's side streets, and members of the GPNIA perceived time as running short. Parishioners were urged to fix up their apartments and homes. Volunteers on each block within the parish reported any vacancies to the GPNIA, which kept lists of "desireable applicants for various homes in this vicinity." Father Maguire and GPNIA members badgered city officials to place traffic lights at dangerous intersections and stop business enterprises which "destroy[ed] the peace of the surrounding NEIGHBORHOOD."[13] GPNIA members founded the 18th and Thompson Federal Credit Union in 1939, making home loans to over 318 parishioners in two years. Most important, GPNIA members declared an all-out assault upon "unscrupulous speculators who seek to hasten the decline of the parish and gain financial advantage from this depreciation." All of these tasks were deemed necessary in order to restore the parish "to its former enviable standing."[14]

Maguire became notorious for personally berating real estate agents interested in profiting from racial turnover and landlords willing to divide apartments into multiple units. Referring to his work in Jamaica, he emphasized his lack of racial prejudice. But he also stressed his belief that "our colored brethren are happier and more content in their own little neighborhoods," and of course the organization's frequent invocation of the term "undesireables" was shorthand for the attempt to stop African-Americans, of whatever class or religion, from entering the neighborhood. The various declarations of the GPNIA—such as "WE SHALL STAND OUR OWN GROUND"—also needed little translation, as did a campaign to persuade parishioners to sign restrictive covenants barring the sale of their homes to African-Americans. Maguire and allies within the GPNIA even organized a fund drive in order to purchase an abandoned property near the church, adding an attractive lot for the Gesu parochial school while simultaneously fending off a purchase attempt by an African-American group.[15]

These increasingly feverish efforts sparked public controversy. African-American Catholics asking to become members of Gesu parish, or send their children to its parochial school, continued to be directed toward Our Lady of the Blessed Sacrament.[16] Reporters for the local African-American newspaper confronted Maguire's Jesuit superior, Father Thomas Love, in the spring of 1941. "We have nothing against Negroes," argued Love, "but we cannot stand by and see property which represents the life-savings of some of our oldest parishioners become almost valueless or go to ruin. I recognize that there is a conflict of rights here, but my interest is primarily in the question of preserving the unity of my parish."[17]

II

The terms "unity" and "rights" are suggestive. As Father Love knew even then, formal Catholic thinking on racial issues was moving toward an endorsement of interracialism, a position which would make the theological "right" of Catholic property owners to discourage African-Americans from destroying parish "unity" less clear. That Father Maguire could organize the GPNIA even as another Jesuit, John LaFarge, conferred with Pius XI on the necessity of interracialism, suggests the distance between working as an editor for a Jesuit magazine in Manhattan and parish ministry in North Philadelphia, but the gap between theory and practice was closing. LaFarge, himself, was aware of the Philadelphia situation. He knew most of the priests working in the city's African-American parishes, as well as parishioners and fellow Jesuits at Gesu, and he also maintained contact with a prominent African-American attorney living two blocks from the Gesu church. The "only course," LaFarge noted in a private memo, "is for Father Love and Father Maguire to pull out of the Gesu Improvement Association and turn their attention towards integrating Negroes into the parish by zealous missionary and educational effort."[18]

LaFarge added that there was "Every reason to believe that the Gesu Action is most distasteful to Cardinal Dougherty." He was correct. While hardly a crusader for African-American civil rights, Dougherty nonetheless viewed these matters from a different vantage point. During the 1920s and early 1930s, Dougherty had worked to ensure that Irish pastors welcomed Italian children into parochial schools and parishes, and he expected African-Americans to receive the same courtesies. As early as 1926, one Philadelphia priest working in an African-American parish had warned that while "separate national churches have the aspects of favors and privileges" for European immigrants, African-American Catholics interpreted separate churches as evidence that they were not "wanted in the ordinary Catholic churches, schools, colleges, etc." Dougherty also knew that a handful of Philadelphia parishes had established successful convert programs. One parish, St. Ignatius, registered 862 African-American baptisms in 1939, another such parish, St. Elizabeth's, shared a border with Gesu.[19]

African-American Catholics involved in these conversion programs wondered why clerics at Gesu parish were so unwelcoming. One anguished African-American Catholic eager to see the "work of the Church among the Negro here" succeed, described the situation as "increasingly painful."[20] Even conservative bishops such as New York auxiliary McIntyre feared that news from Philadelphia might create obstacles for convert programs in Harlem.[21]

Dougherty's acceptance of an invitation to speak before a group of African-American Elks indicates his own desire to make a public statement. To waves of applause, Dougherty declared that "Any Catholic who despises anyone for any reason, particularly on such baseless foundation as race or color, is not a loyal Catholic."[22] A local pastor excitedly predicted that Dougherty's remarks would aid "those of us within the Catholic [Church] who have for years been working to INTEGRATE the Catholic Negro."[23]

When Father Love of Gesu parish asked Dougherty to support the effort to keep a lot near the parish church out of African-American hands, then, Dougherty responded with caution. He first noted his admiration for the missionary work of St. Peter Claver and Philadelphia's Mother Katherine Drexel. He then observed that "Just at present a special campaign is being made by our priests and many of our zealous Catholics to convert the Negroes of Philadelphia; if they thought that I am doing them an injustice in any way, and particularly by an endeavor to prevent them to acquire property, I fear that the campaign would be, if not frustrated, at least somewhat impeded."[24]

Calls for change could also be heard from within Gesu. One member of the parish was Anna McGarry, a remarkable second-generation Irishwoman who since the early 1930s had worked to create a genuinely interracial parish community. When blockbusting real estate agents moved up and down her predominantly Irish block, McGarry refused to sell and welcomed her new African-American neighbors. At the time, McGarry recalled, "I believed that my husband had given his life to protect democracy [in World War I] but here in my own neighborhood I saw actions that nullified his sacrifice."[25]

McGarry and a friend welcomed members of the Federated Colored Catholics at a Gesu parish communion breakfast, provoking raised eyebrows among other parishioners. She also invited John LaFarge to give a lecture on Catholics and racial prejudice. When African-Americans joined her and LaFarge for a dinner at her house, neighbors commented that "she'll even bring niggers into her own house for a priest."[26] The fledgling Philadelphia Catholic Interracial Council, organized in part by McGarry, joined with representatives of the NAACP, the National Negro Congress, and local Jewish groups in public criticism of the Gesu clergy. Ironically, reports of the issue in the African-American *Philadelphia Tribune* appeared just above McGarry's own column on Catholic interracialism.[27]

Perhaps most important, neighborhood turmoil occurred in a particular political context. The issue became public in part because opponents of segregation knew that American opposition to the Axis powers guaranteed an audience for their claims. When African-Americans termed the actions of the Gesu priests "Hitler Fascism in Philadelphia" they struck a resonant chord. Members of the GPNIA might still sing "God Bless America" (along with "When Irish Eyes are Smiling") at the conclusion of meetings, but the language of American nationalism, obtained with such effort by their immigrant ancestors, had moved beyond their control.[28]

The combination of these forces—Jesuit intellectuals rethinking racial issues, African-American Catholics and a Cardinal worried about derailing local evangelical efforts, the work of activists like Anna McGarry within the parish, and the ideological window opened by the war—made the reconstruction of boundaries within Gesu inevitable. It did not make it easy. Direct approaches to Cardinal Dougherty were brushed off. "No appointment was granted with the Cardinal; we sent special messages and everything," observed one African-American leader. "Finally, we received word that the Cardinal had left for an extended trip." As late as 1946, Jesuits still considered "public opinion" when pondering whether to sell a neighborhood property owned by the parish to an African-American buyer.[29] African-American parishioners were hardly welcomed with open arms; one African-American recalled Gesu churchgoers moving out of the pew when he sat down near them. When a friend asked him if the people at Gesu "accepted" him, he responded "God accepts me."[30]

Still, reports on the GPNIA in the Gesu parish bulletin, and perhaps

Anna McGarry in 1948. Photo located at Urban Archives, Temple University.

the organization itself, ceased in 1942, and Father Maguire became viewed by his colleagues as an embarrassment.[31] By 1944 the new pastor of Gesu parish was also a member of a national Jesuit Committee on Interracial Justice chaired by LaFarge. An anonymous correspondent accused the new pastor of "coddling the colored people and trying to force them down the throats of us Irish Catholics." The correspondent added that "The men wanted to march on the Cardinal about LaFarge and expose him for what he is[.] If he loves the colored people so much why does he not go down to the south." The Gesu school was opened to African-Americans in 1945. Some priests encouraged African-American families to ignore parishioner hostility, and Anna McGarry worked to ensure African-American integration into parish life. "Irish Night" continued, but with growing numbers of participants tracing their origins to the Caribbean or the rural South.[32]

By 1950, only 6 percent of the residents of Ward 47 lived on all-white blocks.[33] Over the course of the next decade, the parish slowly became involved in neighborhood life, instead of solely ministering to the last white holdouts. The Jesuit official making the annual inspection of the parish in 1954 belatedly recognized that "the parish is definitely assuming the character of a mission. This calls for serious study of missionary activity among the colored." Educated by their African-American parishioners, priests became more involved in the housing, welfare, and educational issues so important to area residents. One priest walked up and down the blocks of the parish each day, meeting families, inquiring about problems, using these contacts to launch new programs for neighborhood youth.[34]

The Jesuits also resolved to keep St. Joseph's Prep in North Philadelphia despite the fact that the vast majority of potential Catholic high school students now lived some distance from the neighborhood. In 1951, one realtor bluntly advised the priests that they should move the high school as quickly as possible since "the neighborhood will never return to its former status as a good middle-class section." Instead, alumni and teachers recruited pupils from Northeast and South Philadelphia, as well as providing scholarships for local, African-American students. After much of the school burned to the ground in 1965, an immediate rebuilding campaign produced new structures on the same site.[35]

In the context of the civil rights movement and the Second Vatican Council, neighborhood ministries assumed greater responsibilities. Nuns and parishioners founded a day nursery for the children of working mothers. Resources were channeled toward the parish grade school. Parish-sponsored community centers offered local residents legal and economic

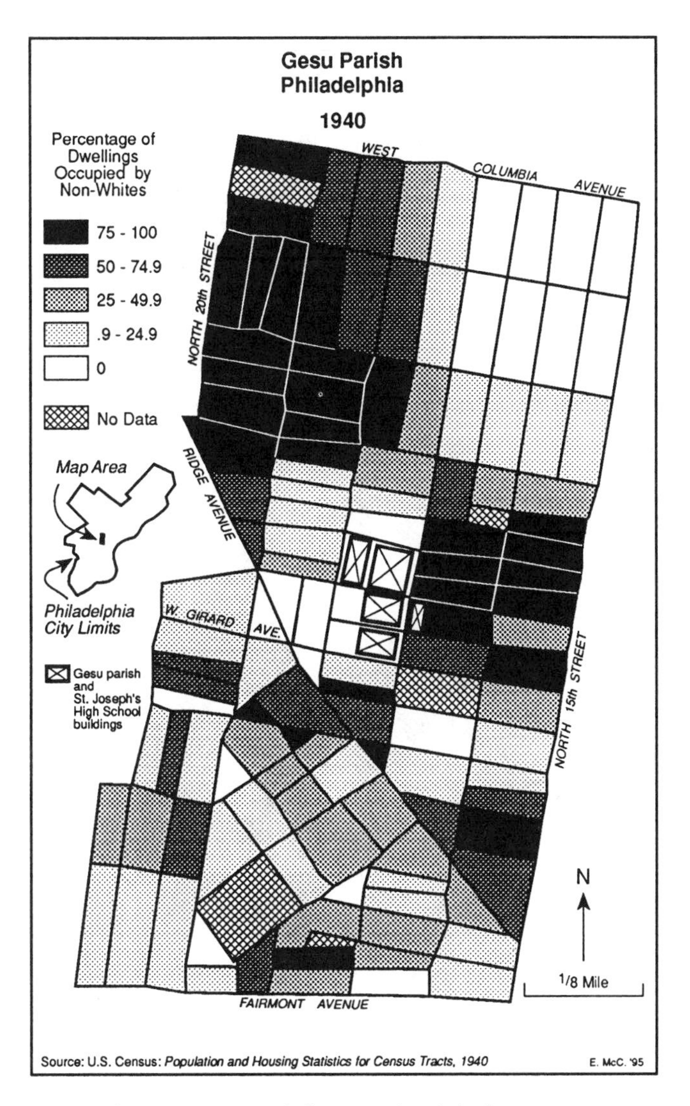

Source: U.S. Census: *Population and Housing Statistics for Census Tracts, 1940*

advice, and priests represented the area in civic improvement organizations. A parish group called "Our Neighbors' Association" worked to find employment for area residents, and Gesu Catholic Youth programs became among the most successful in the city.[36]

The neighborhood deteriorated. One pastor noticed as early as 1963 that parish leaders, primarily middle-class African-Americans, frequently left the area for safer surroundings. The area of North Philadelphia in which the parish was located lost over 55 percent of its population between 1950 and 1980. By that point, two out of every five families lived below the poverty line. Unemployment hovered at around 20 percent.[37]

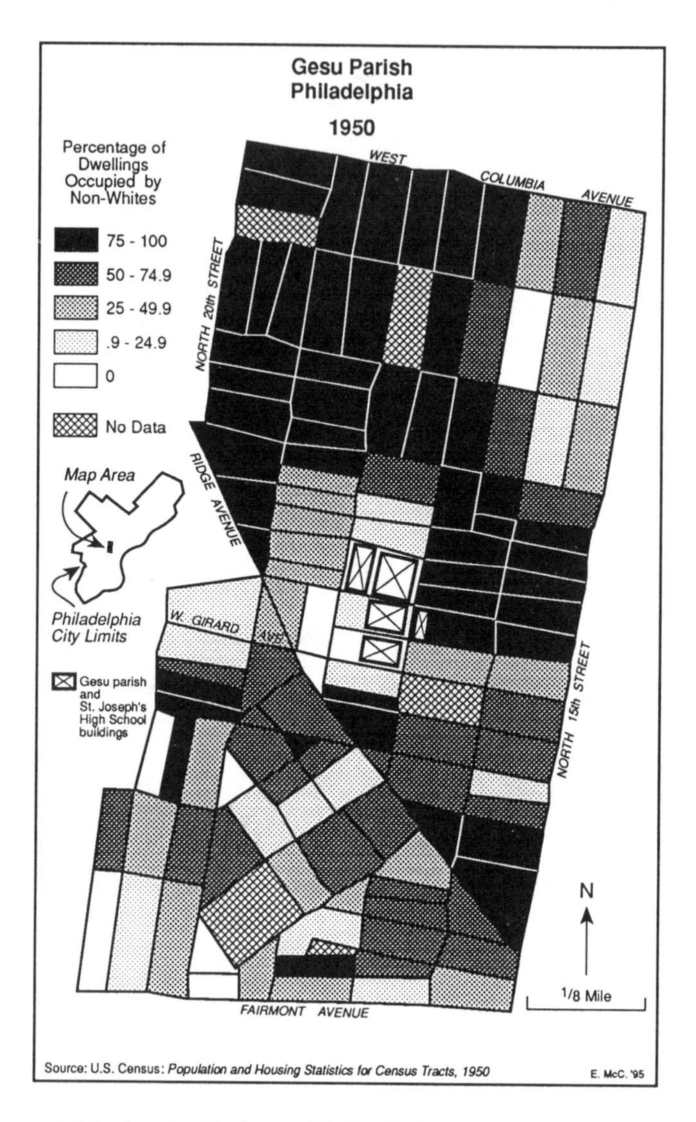

In a neighborhood with few stable institutions, the Gesu school, with a history of sustaining high academic standards, played an important role. Ninety-five percent of the school's 400 students in 1991 were African-American, mostly poor and living within two miles of the Gesu church.[38] When these students walk to school, they pass homes and a church building constructed by Catholic immigrants whose dreams now seem irrevoca-

(*facing page*) A 1976 procession in front of the Gesu Church, Philadelphia. Photograph by Robert S. Halvey, reprinted with permission.

bly distant. Only when the students gather within the still imposing church, dressed in uniforms similar to those worn by generations of Gesu schoolchildren before them, do threads of continuity between that era and our own become imaginable.

III

The power of the Gesu drama lies in its familiarity. Hundreds of similar productions were staged in the streets and parishes of the urban North during the twentieth century, productions that illuminate the shifting web of connections between religion, race, and community. The language with which Catholics spoke of "races" within the church, the persistent vision of the world as a series of geographical and racial enclaves, and the scale of the enormous churches, schools, convents, and rectories testified to a Catholic sensibility at odds with the population movements of twentieth-century America.

This vision proved impossible to sustain. By the 1940s, two "Catholic" traditions offered competing views of the Church's role. Catholics such as Father Maguire pictured tightly knit, homogeneous parishes and schools, even as more cosmopolitan members of Maguire's own parish, notably Anna McGarry, worked to eradicate the bigotry implicit in such narrow definitions of community.

The moment of triumph for Anna McGarry's vision was brief. In retrospect, the Second Vatican Council and the 1960s were the opening acts in a remarkably turbulent, unfinished drama. A generation after the opening of the council, unprecedented disagreement with Church teaching in all areas connected to gender and sexuality, a steep drop in vocations to the religious life, and marked declines in mass attendance suggested an institution less capable of making claims on Catholic lives. Opinion surveys also discovered a startling, widespread ignorance concerning fundamental matters of Catholic doctrine. Sociologists continued to marvel at the ability of Catholic parishes to stabilize urban neighborhoods, but the unity of purpose necessary to sustain such parishes seemed less certain.[39]

The agonizing disputes over school busing that wracked Boston reflected this new situation. As in the housing disputes described throughout this volume, groups of white Catholics proclaimed their opposition to liberal notions of racial progress, particularly notions that failed to extend beyond city lines. That many of the public officials advocating busing programs were themselves Catholic, including federal judge Arthur Garrity and Senator Edward Kennedy, only heightened the tension. Brutal attacks on buses carrying frightened African-American schoolchildren recalled the rocks lobbed at Martin Luther King by residents of Chicago's Southwest Side.

And yet the similarities obscured important differences. The moral imperative supporting busing programs was never as obvious, even to African-Americans, as the right to purchase a home or rent an apartment. And Chicago's Catholic Church faced racial tension at a moment of unusual strength, with parochial organizations at the height of their influence, and many clergy, nuns, and parishioners energized by Vatican II. Broad programs of racial education were inaugurated, noble promises were made. By the mid-1970s, even in heavily Catholic Boston, Church influence had narrowed. Catholic leaders worked as best they could to lessen violence in white neighborhoods, but those responsible for the violence seemed increasingly distant from an institution once notable for its hold on working-class males.[40]

Even the territorial structure of the parish system became increasingly fragile, as more Catholics became accustomed to "shopping" for a parish whose liturgy and programs matched their own inclinations. Here the diversity in liturgical practice permitted by the Second Vatican Council—along with the widening divide between liberal and conservative Catholics—logically took the form of individual parishes appealing to different segments of a broader Catholic market. A de facto congregationalism became dimly evident.[41]

The new era in American Catholic history eventually forced bishops and parishioners to reckon with the scores of almost empty churches, convents, rectories, and auditoriums dotting all of the northern cities—a somber legacy of the extended encounter between Euro-American Catholics and African-Americans. Beset by high maintenance costs on aging physical plants, these parishes are often valiantly supported by small groups of elderly white holdouts and African-American newcomers. Complicating this picture is a pressing clergy shortage and a steadily growing suburban Catholic population, which from an episcopal perspective makes the relatively low priest-to-parishioner ratios in the central cities an unaffordable luxury. In the 1990s, for the first time in American Catholic history, bishops closed scores of once vibrant parishes in the northern cities, leaving grafitti-scarred buildings topped by crosses as poignant relics of another era. In Philadelphia, in 1993, Gesu was one of fourteen parishes to shut its doors, although the Jesuits promised to continue supporting the parish school and to use the church as a place of worship for students at Gesu and St. Joseph's Prep.[42]

IV

Other markers were more hopeful. Ten African-American bishops worked in dioceses across the country by the mid-1980s, providing a heretofore absent voice in Church leadership circles.[43] John Paul II's special meeting with African-American Catholics during his 1987 visit to the United States also suggested a growing recognition of African-American Catholics as a distinct people. The number of African-American Catholics steadily expanded. By 1990, 9.2 percent of the nation's African-Americans identified themselves as Catholic, making Catholicism the nation's second largest African-American religious community.[44]

As the African-American migration to the northern cities slowed to a trickle after 1970, Catholic attention shifted to immigrants from Asia, the Caribbean, and Latin America.[45] A significant number of these new Americans were Catholic, leading to a novel patchwork of Vietnamese, Latino, and Haitian parishes in each northern city. Even as they revitalized moribund parishes, these immigrants reminded now affluent Euro-American Catholics of their own history, and provided an unusual opportunity for contact across lines of class and ethnicity.

In all of the northern cities, underpaid teachers, heroic parents, overworked administrators, and diocesan officials continued to make extraordinary sacrifices—only grudgingly acknowledged by public officials and scholars—in order to maintain Catholic schools as signals of concern for the nation's cities. Genuinely integrated parish communities throughout the North belied the old canard that 11:00 Sunday morning was the most segregated hour of the week. Each city also contained a handful of vibrant African-American Catholic parishes, often located in deeply impoverished areas.[46]

The focus shifted from "integration" to "community." While careful to stress that Catholic parishes must welcome all newcomers, activists emphasized the importance of local groups as a means of creating a more equitable society. The record of Catholic-sponsored, Alinsky-style community organizations has been uneven, but Alinsky's ideas have maintained their popularity in Catholic circles because of the congruence between the Catholic concern for neighborhood and Alinsky's vision of small-scale, nonpartisan community organizations. Through the annual Campaign for Human Development fund drive and various Catholic Charities projects, the Church became the nation's largest sponsor of community organization efforts. In San Antonio, Los Angeles, Brooklyn, and elsewhere, Catholic parish organizations continue to anchor community groups dedicated to improving neighborhood living conditions.[47]

A renewed interest in Catholic social teaching complemented this work, especially as the economic and cultural individualism of the postwar era came under closer scrutiny. Both socialist and some free-market societies, in the Catholic view, mistakenly assumed that individuals emerged onto the societal stage as blank slates, able to make rational decisions based upon their evaluation of objective criteria. Neither system placed any special value upon private associations or volunteer organizations. By contrast, Catholic social thought stressed the importance of "mediating institutions" as incubators of civic and moral virtue. The 1989 revolutions in Eastern Europe—as well as concurrent events in Brazil, El Salvador, and the Phillipines—were particularly encouraging on these matters, especially as it became evident that Catholic organizations had provided much of the cultural breathing space necessary for civil society.[48] The term "social capital" (and the realization that religious organizations were its most prominent source) became familiar in a variety of intellectual settings. Robert Bellah's influential work, in particular, drew upon Catholic social teaching in its recommendations for American society.[49]

What also became more clear in retrospect is that Catholics mourning the collapse of traditional urban parish communities defended a vision with its own moral content. That Gesu parish in the 1940s proved unable to welcome African-American Catholics is both a tragedy and one chapter in a complicated, ongoing story. An alternative reading of the parish history might stress the ways in which parochial structures, at their best, embedded both Euro- and African-Americans in worlds larger than themselves, connecting moral tradition to personal virtue.[50] Indeed, some of the strongest African-American and Latino parishes of the post–Vatican II era have forms—especially the merging of school, parish, and cultural communities—familiar to American Catholic historians.

These developments might best be termed a rediscovery. Contemporary Catholic intellectuals sound uncannily like immigrant pastors asking parents to donate hard-earned wages to Catholic churches and schools or like the first Euro-American Catholic workers in the African-American community. Today's community organizer, like an early twentieth-century proponent of Catholic schools, views "community" as a set of practices distinct from either economic self-interest or bureaucratic obligation. Both share a belief in the value of institutions as sources of the good society. Both understand their work not merely as socially useful but as part of a gospel obligation.

Clearly, a literal replication of older Catholic traditions is undesirable. The most discouraging lesson gained from the extended Catholic grappling with racial issues in the twentieth century is that parish communities frequently proved incapable of manifesting concern for those outside their boundaries. Discrimination against even African-American Catholics painfully displayed the frailty of the line between community and fortress. To borrow from an older rhetoric, acts of contrition are necessary.

Measured against this often disheartening history is the need for ways to strengthen family, faith, and community. For all their faults, the Catholic parish worlds described throughout this study performed precisely this task. Dusty piles of parish histories now gathered in diocesan archives illuminate this point. Here lie row upon row of photographs displaying proud clerics standing in front of just-completed buildings, worshippers streaming into Sunday mass, nuns assisting children with schoolwork, and parish athletic teams clutching recently acquired trophies.

Viewed from this angle, Catholics sustained faith while structuring a genuine community life. And in turn, this weaving of geography, education, and church produced a more resilient social fabric. For today's Catholics, comparable achievements in a more inclusive context would seem a worthy goal.

ACKNOWLEDGMENTS

I am indebted to many individuals and institutions. The Pew Program in Religion and American History awarded me a faculty research grant, as did Harvard University's William F. Milton Fund. A postdoctoral fellowship from the Lilly Program in Arts and the Humanities at Valparaiso University introduced me to an unusually thoughtful group of scholars, and now, friends. Stanford University and the Mellon Foundation provided support in the project's earliest stages. An earlier version of chapters 7 and 8 was published in *Religion and American Culture: A Journal of Interpretation* 4 (Summer 1994), 221–54, © by The Center for the Study of Religion and American Culture.

Archivists and librarians across the country also provided vital assistance. Special thanks go to the staff at Manuscripts and Archives, University of Notre Dame, Georgetown University Special Collections, Joseph Coen at the Archives of the Diocese of Brooklyn, Roman Godzak at the Archives of the Archdiocese of Detroit, Archie Motley at the Chicago Historical Society, Christine Krosel at the Archives of the Diocese of Cleveland, Ron Patkus at the Archives of the Archdiocese of Boston, Phil Runkel and Charles Elston at Marquette University Special Collections, Nancy Sandleback at the Archdiocese of Chicago Archives and Record Center, Shawn Weldon at the Philadelphia Archdiocesan Historical Research Center, and Anthony Zito and John Shepherd at Catholic University. The dedication of Father Peter Hogan, S.S.J., to the wonderful collections on Catholicism and race relations assembled by the Josephite Fathers in Baltimore is matched only by the welcome he gives novice researchers.

Father Steven Avella, Dominic Capeci, Father Donald Crosby, S.J., Philip Gleason, Father Paul Jervis, Ira Katznelson, Gretchen Knapp, J. Anthony Lukas, Sister Mary Nona McGreal, O.P., Father Philip Murnion, Father Thomas O'Gorman, Dominic Pacyga, Ellen Skerrett, David Southern, Brother Thomas W. Spalding, Tom Sugrue, Leslie Woodcock Tentler, and Father Anthony J. Vader shared material from their own research. Hope Brophy worked above and beyond the call of duty to make my visits to Detroit profitable; Father George Bur, S.J., and Father Thomas Clifford, S.J., did the same in Philadelphia.

Many volunteers—including Susan Aylward, Sheila McGreevy-Barry and Joseph Barry, Todd Graff, Doug Griffiths, Joe and Mary Keating, the Koreman family, Rebecca Lowen, Jim and Diane Luke, Charlie Manak, Tom McCaffery, Helen McGreevy, Kevin McGreevy, Mary McGreevy and Ernest Grumbles, Jack and Carol McManus, Molly Ochsner, Peggy and Greg Pehl, Mike Rataczak, Brian and Vera Roney, William and Mary Tobin, and Roger Wu—extended food, shelter, and good cheer during my travels. Tim Milford served as a valued research assistant. Eliza McClennen constructed the maps; Monroe Horn helped assemble the data. Bill Tobin provided friendship and thoughtful readings of the manuscript in equal measure. Meeting, and marrying, Jean McManus made the final steps of the journey the most rewarding.

My debt to the students, parents, faculty, and staff of Hales Franciscan High School in Chicago is humbling. Audience members and colleagues at Stanford University, Valparaiso University, the University of Notre Dame, Yale University, St. Louis University, Harvard University, College of the Holy Cross, the University of Chicago, and various professional conferences graciously responded to portions of this project. The patient criticism of George Fredrickson and Mark Schwehn was especially appreciated. David Kennedy's careful evaluation of my work and support throughout graduate school allowed me to stay the course.

My parents, Patrick and Kathleen McGreevy, have been my greatest resource and blessing.

NOTES

Sources and organizations frequently cited in the notes are identified by the following abbreviations.

AAB	Archives of the Archdiocese of Boston.
AAC	Archives of the Archdiocese of Chicago.
AAD	Archives of the Archdiocese of Detroit.
AAM	Archives of the Archdiocese of Milwaukee.
ACHR	Archdiocesan Commission on Human Relations Papers, Archives of the Archdiocese of Detroit.
ACHREC	Archdiocesan Commission on Human Relations and Ecumenicism, Archives of the Archdiocese of Chicago.
ACLU	American Civil Liberties Union Papers, University of Chicago Special Collections.
ACUA	Archives of the Catholic University of America.
ADB	Archives of the Diocese of Brooklyn.
ADC	Archives of the Diocese of Cleveland.
AMC	Anna McGarry Papers, Marquette University Special Collections.
AUS	Association of Urban Sisters Papers, Archives of the Archdiocese of Boston.
BCIM	Bureau of Catholic Indian Missions Collection, Marquette University Special Collections.
CHS	Chicago Historical Society.
CIC	Catholic Interracial Council Papers, Chicago Historical Society.
EB	Eugene Bleidorn Papers, State Historical Society of Wisconsin at Milwaukee.
EGAN	Msgr. John J. Egan Papers, Manuscripts and Archives, University of Notre Dame.
ESF	Emil Schwarzhaupt Foundation Papers, University of Chicago Special Collections.
HPD	H. Paul Douglass Collection on Microfiche, Andover-Newton Library, Harvard University.
JAL	J. Anthony Lukas Papers, State Historical Society of Wisconsin at Madison.
JG	James Groppi Papers, State Historical Society of Wisconsin at Milwaukee.

JLF	John LaFarge Papers, Georgetown University Special Collections.
MAUND	Manuscripts and Archives, University of Notre Dame.
MUSC	Marquette University Special Collections.
NAACP	National Association for the Advancement of Colored People Papers, Library of Congress.
NCCC	National Conference of Catholic Charities Papers, Catholic University of America Archives.
NCCIJ	National Catholic Council for Interracial Justice Papers, Marquette University Special Collections.
NCWC	National Catholic Welfare Conference Papers, Archives of the Catholic University of America.
NFCCS	National Federation of Catholic College Students Papers, Manuscripts and Archives, University of Notre Dame.
NYCIC	New York Catholic Interracial Council Papers, Catholic University of America Archives.
PAHRC	Philadelphia Archdiocesan Historical Research Collection.
RMP	Robert Moses Papers, New York Public Library.
SA	Saul Alinsky Papers, University of Illinois at Chicago Special Collections.
SAD	Social Action Department, National Catholic Welfare Conference Papers, Archives of the Catholic University of America.

Introduction

1. Chicago Commission on Race Relations, *The Negro in Chicago: A Study of Race Relations and a Race Riot* (Chicago: University of Chicago Press, 1922), 656, 663–64, 666.

2. Ibid., 40, 115; James R. Grossman, *Land of Hope: Chicago, Black Southerners, and the Great Migration* (Chicago: University of Chicago Press, 1989), 118.

3. *The Negro in Chicago,* 16, 19, 155.

4. Ibid., 64–67. My account is pieced together from the slightly varying descriptions in "Mob Besieges Church," *Chicago Tribune,* September 21, 1920, p.1/2; "Clash Zone Quiet; 600 Police on Duty," *Chicago Daily News,* September 21, 1920, p.1/3; "Priest Calms Angry Crowd; Saves 3," *Chicago Defender,* September 25, 1920, p.1.

5. William A. Osborne, *The Segregated Covenant: Race Relations and American Catholics* (New York: Herder and Herder, 1967) is the most relevant monograph. Especially thoughtful on early twentieth-century Protestantism is Ralph E. Luker, *The Social Gospel in Black and White: American Racial Reform, 1885–1912* (Chapel Hill: University of North Carolina Press, 1991). For the later period, see James F. Findlay, Jr., *Church People in the Struggle: The National Council of Churches and the Black Freedom Movement, 1950–1970* (New York: Oxford University Press, 1993).

6. Exceptions to this statement include J. Anthony Lukas, *Common Ground: A Turbulent Decade in the Lives of Three American Families* (New York: Alfred A. Knopf, 1985) and Jonathan Reider, *Canarsie: The Jews and Italians of Brooklyn Against Liberalism* (Cambridge: Harvard University Press, 1985).

7. The work of Barbara Fields is crucial. See especially her "Ideology and Race in American History," in *Region, Race, and Reconstruction: Essays in Honor of C. Vann Woodward,* J. Morgan Kousser and James M. McPherson, eds. (New York: Oxford University Press, 1982), 143–77 and Fields, "Slavery, Race and Ideology in the United States of America," *New Left Review,* 181 (May/June 1990), 95–118. For an effort to emphasize the importance of "status" or "race" without viewing it as transhistorical, see David R. Roediger, *The Wages of Whiteness: Race and the Making of the American Working Class* (London: Verso, 1991), 3–17; George M. Fredrickson, *The Arrogance of Race: Historical Perspectives on Slavery, Racism, and Social Inequality* (Middletown: Wesleyan University Press, 1988), 154–61 and Fredrickson, "Reflections on the Comparative History and Sociology of Racism" (unpublished paper in author's possession).

8. For example, James R. Barrett, "Americanization from the Bottom Up: Immigration and the Remaking of the Working Class in the United States, 1880–1930," *Journal of American History,* 97 (December 1992), 1001–2; Roediger, *The Wages of Whiteness,* esp. 43–92, 133–63.

9. Herbert Hill makes the most powerful case for immigrant racism as a constitutive element of the American labor movement, especially in his "Race, Ethnicity and Organized Labor," *New Politics* 1 (Winter 1987), 31–82. Even Hill, however, when looking for illustrations of violent twentieth-century racial conflicts between working-class "whites" and African-Americans, repeatedly refers to disputes over housing, not labor issues. One of Hill's key examples, assaults on African-American workers marching against the Allen-Bradley Company in Milwaukee, indirectly makes this point. The march was led by Father James Groppi, who engendered such bitter hostility precisely because of his position as both a Catholic priest and a leader in the local open housing movement. See Hill, pp. 72–75.

10. See Douglas S. Massey and Nancy A. Denton, *American Apartheid: Segregation and the Making of the Underclass* (Cambridge: Harvard University Press, 1993), 40–42, 48–49; Arnold R. Hirsch, "With or Without Jim Crow: Black Residential Segregation in the United States," in *Urban Policy in Twentieth Century America,* Arnold R. Hirsch and Raymond Mohl, eds. (New Brunswick: Rutgers University Press, 1993), 68–69.

11. Ira Katznelson makes this observation in *City Trenches: Urban Politics and the Patterning of Class in the United States* (New York: Pantheon, 1981), 16.

12. David Roediger states that the nineteenth-century Catholic Church did nothing more than "reflect the racial attitudes of its members." *Wages of Whiteness,* 140. On the absence of Catholicism from standard twentieth-century historical accounts, see Leslie Woodcock Tentler, "On the Margins: The State of American Catholic History," *American Quarterly,* 45 (March 1993), 104–27.

13. Robert Wuthnow, "Understanding Religion and Politics," *Daedalus,* 120 (Summer 1991), 1–20.

14. Since the U.S. Census Bureau refuses to ask questions about religious affiliation, historical estimates of the American Catholic population are notoriously vague. One sample survey conducted by the Census in 1957 demonstrated that 47 percent of the population in the northeastern states, and 37.8 percent of the population in all urban areas with greater than 250,000 residents, identified themselves as Catholic. Summarized in Milton M. Gordon, *Assimilation in American Life: The Role of Race, Religion, and National Origins* (New York: Oxford University

Press, 1964), 195–96. Two national studies that chart religious affiliation by county, using pastoral membership estimates, are *Churches and Church Membership in the United States* (New York: National Council of Churches, 1956) and Douglas W. Johnson, Paul R. Picard, Bernard Quinn, *Churches and Church Membership in the United States: An Enumeration by Region, State, and County* (Washington: Glenmary Research Center, 1974). These numbers were possibly low. Not all Catholics regularly attending church registered with a parish. Also, since the parish system was territorial and national (and not congregational) pastors had an incentive to underestimate parish populations as a means of avoiding larger diocesan taxes and fund drive assessments. Parish censuses conducted in the 1930s invariably discovered many "unregistered" Catholics. On parish censuses, see James Peterson, "Knowing Our Own," *American Ecclesiastical Review,* 88 (1933), 291–301; Thomas F. Coakley, "Revelations of a Parish Census," *American Ecclesiastical Review,* 90 (May 1934), 529–39.

15. Philip Gleason, "Forum," *Religion and American Culture,* 2 (Winter 1992), 13.

Chapter 1

1. Cyprian Davis, *The History of Black Catholics in the United States* (New York: Crossroad, 1990), 216; Cyprian Davis, "The Holy See and American Black Catholics: A Forgotten Chapter in the History of the American Church," *U.S. Catholic Historian,* 7 (1988), 179; John Tracy Ellis, *The Life of James Cardinal Gibbons: Archbishop of Baltimore, 1834–1921* (Milwaukee: Bruce, 1952), 402–4.

2. Steven J. Ochs, *Desegregating the Altar: The Josephites and the Struggle for Black Priests, 1871–1960* (Baton Rouge: Louisiana State Press, 1990), 141.

3. John T. Gillard, *The Catholic Church and the American Negro* (Baltimore: St. Joseph's Society Press, 1929), 45, 47; Ochs, *Desegregating the Altar,* 58, 141–43, On Drexel, Katherine Burton, *Golden Door: The Life of Katharine Drexel* (New York: P. J. Kenedy and Sons, 1957), and Sister Consuela Marie Duffy, S.B.S., *Katherine Drexel: A Biography* (Philadelphia: Peter Reilly, 1965).

4. Gillard, *The Catholic Church and the American Negro,* 47–54. Parish totals from *The Official Catholic Directory, 1920* (New York: P. J. Kenedy & Sons, 1920), 43–52, 249–51, 465–66, 575.

5. Marilyn Nickels, *Black Catholic Protest and the Federated Colored Catholics, 1917–1933* (New York: Garland, 1988), 32

6. Ochs, *Desegregating the Altar,* 221–29.

7. Ibid., 237–38, 243.

8. *The Papal Encyclicals: 1903–1939,* Claudia Carlen, I.H.M., ed. (Raleigh: The Pierian Press, 1990), 287; Carlo Falconi, *The Popes of the Twentieth Century* (Boston: Little,Brown, 1967), 145–47; Ochs, *Desegregating the Altar,* 283–84.

9. Fr. John LaFarge to Fr. John Gillard, July 1, 1930, file 55, box 55, JLF.

10. Ochs, *Desegregating the Altar,* 283–88. The apostolic delegate, Archbishop Pietro Fumasoni-Biondi, repeatedly emphasized his interest in African-American Catholics. See the 1931 account of his appearance in Milwaukee in *Silver Anniversary, St. Benedict the Moor Mission* (Milwaukee: 1934), 31.

11. Charles Shanabruch, "The Catholic Church's Role in the Americanization of Chicago's Immigrants, 1833–1928" (University of Chicago, Ph.D. diss., 1975), 564; Gary Wray McDonough, *Black and Catholic in Savannah, Georgia*

(Knoxville: University of Tennessee Press, 1993), 219–20. Minutes of the Annual Meetings of the Bishops of the United States, September 22, 1920, ACUA; Ochs, *Desegregating the Altar,* 258–61, 287; Davis, *The History of Black Catholics in the United States,* 230–31.

12. Robert H. Lord, John E. Sexton, and Edward T. Harrington, *History of the Archdiocese of Boston in the Various Stages of Its Development, 1604–1944* (New York: Sheed and Ward, 1943), 725; John Daniels, *America Via the Neighborhood* (New York: Harper & Brothers, 1920), 242–46.

13. Stephen Joseph Shaw, "Chicago's Germans and Italians, 1903–1939: The Catholic Parish as a Way-Station of Ethnicity and Americanization" (University of Chicago, Ph.D. diss., 1981), 214–15.

14. Andrew Greeley discusses Bridgeport churches in *Neighborhood* (New York: the Seabury Press, 1977), 27–30.

15. Thomas Jablonsky, *Pride in the Jungle: Community and Everyday Life in Back of the Yards Chicago* (Baltimore: Johns Hopkins University Press, 1993), 103–4.

16. Agnes Meyer, "Orderly Revolution," *Washington Post,* June 4, 1945, p.9; Saul Alinsky, *Reveille For Radicals* (Chicago: University of Chicago Press, 1945), 122. Catholic percentage in Edward Kantowicz, "Church and Neighborhood," *Ethnicity,* 7 (1980), 349–66; also see Bernard I. Kahn, "The Catholic Church and Its Relationship to the Back of the Yards Neighborhood Council: A Study in Community Behavior" (University of Chicago, M.A. thesis, 1949), 13.

17. James Sanders, *The Education of an Urban Minority: Catholics in Chicago, 1833–1965* (New York: Oxford University Press, 1977), 48; Robert Slayton, *Back of the Yards: The Making of a Local Democracy* (Chicago: University of Chicago Press, 1986), 50.

18. Joseph E. Ciesluk, *National Parishes in the United States* (Washington, D.C.: Catholic University of America Press, 1944), 12–23; Rev. Anthony Bernard Mickells, *The Constitutive Elements of Parishes* (Washington, D.C.: Catholic University of America Press, 1950), 8–9; *Webster's Third New International Dictionary: Unabridged* (Springfield: G. & C. Merriam, 1966), 1642, notes that the definition can be either "the residents of such area or the members of one church."

19. Ciesluk, *National Parishes,* 23.

20. Leslie Woodcock Tentler, *Seasons of Grace: A History of the Catholic Archdiocese of Detroit* (Detroit: Wayne State University Press, 1990), 3; Richard Linkh, *American Catholicism and European Immigrants, 1900-1924* (New York: Center for Migration Studies, 1975), 109; Mary Cantwell, *American Girl: Scenes From a Small-Town Childhood* (New York: Random House, 1992), 90.

21. For example, Lizabeth Cohen, *Making a New Deal: Industrial Workers in Chicago, 1919–1939* (New York: Cambridge University Press, 1990), 88–93.

22. Richard S. Sorrell, "The Survivance of French Canadians in New England (1865–1930): history, geography and demography as destiny," *Ethnic and Racial Studies,* 4 (January 1981), 89–105; Elliot Robert Barkon, "French Canadians," in *The Harvard Encyclopedia of American Ethnic Groups,* Stephan Thernstrom, ed. (Cambridge: Harvard University Press, 1980), 395–99.

23. Rev. John P. Gallagher, *A Century of History: The Diocese of Scranton: 1868–1968* (Scranton: The Diocese of Scranton, 1968), 401–2.

24. Tentler, *Seasons of Grace,* 308; Joseph John Parot, *Polish Catholics in*

Chicago, 1850–1920: A Religious History (Dekalb: Northern Illinois University Press, 1981), 231.

25. Parot, *Polish Catholics in Chicago, 187–88,* 207–14.

26. Sanders, *Education of an Urban Minority,* 70; Rev. John V. Tolino, "The Future of the Italian-American Problem," *American Ecclesiastical Review,* 101 (September 1939), 221–32; idem, "Solving the Italian Problem," *American Ecclesiastical Review,* 99 (September 1938), 246–56; J. Zorili, "Some More Light on the Italian Problem," *American Ecclesiastical Review,* 79 (1928), 256–68.

27. Silvano M. Tomasi, *Piety and Power: The Role of the Italian Parishes in the New York Metropolitan Area, 1880–1930* (Staten Island: Center for Migration Studies, 1975), 45.

28. Philip Klein, *A Social Study of Pittsburgh: Community Problems and Social Services of Allegheny County* (New York: Columbia University Press, 1938), 251; James Groppi, unfinished autobiography, folder 4, box 14, JG; Tomasi, *Piety and Power,* 76.

29. John O'Grady, *Catholic Charities in the United States: History and Problems* (Washington, D.C.: National Conference of Catholic Charities, 1930), 277; John C. Murphy, *An Analysis of the Attitudes of American Catholics Toward the Immigrant and the Negro, 1825–1925* (Washington D.C.: Catholic University of America Press, 1940), 92–93.

30. Edward R. Kantowicz, *Corporation Sole: Cardinal Mundelein and Chicago Catholicism* (Notre Dame: University of Notre Dame Press, 1983), 82; Klein, *A Social Study of Pittsburgh,* 719–20.

31. Theodore Abel, *Protestant Home Missions to Catholic Immigrants* (New York: Institute of Social and Religious Research, 1933), 24, 41; also see Graham Taylor, *Pioneering on Social Frontiers* (New York: Arno, 1976), 196.

32. Marian and Charles Leber, "1948–1949 Survey of the Newark Ironbound Section," 10–11, 13–14, #2395, HPD; Brand Blanshard, *The Church and the Polish Immigrant* (1920), copy in Widener Library, Harvard University, 31–32. On Detroit, see Gerhard Lenski, *The Religious Factor: A Sociological Study of Religion's Impact on Politics, Economy and Family Life* (Garden City: Doubleday, 1961), 33–36.

33. W. Lloyd Warner and Leo Srole, *The Social Systems of American Ethnic Groups* (New Haven: Yale University Press, 1945), 173–76.

34. Eileen McMahon, "What Parish Are You From? A Study of the Chicago Irish Parish Community and Race Relations, 1916–1970" (Loyola University, Ph.D. diss., 1989), 92; Gary Ross Mormino, *Immigrants on the Hill: Italian-Americans in St. Louis, 1882–1982* (Urbana: University of Illinois Press, 1986), 164.

35. "Painting of Church is a Last Tribute to Msgr. Garstka," *Buffalo Evening News,* August 20, 1941, p.8.

36. Ambrose Kennedy, *Quebec to New England: The Life of Monsignor Charles Dauray* (Boston: Bruce Humphries, 1948).

37. Report of Nicholas Von Hoffman, November 17, 1956, folder 6, box 34, ESF.

38. Sanford Horwitt, *Let Them Call Me Rebel: Saul Alinsky—His Life and Legacy* (New York: Alfred A. Knopf, 1989), 61; Edward Kantowicz, "Church and Neighborhood," 349–66.

39. Suggestive on Catholics and American sports is Frank Deford, "Religion in Sport," *Sports Illustrated,* 44 (April 19, 1976), 96–97.

40. Alfred Kazin, *A Walker in the City* (New York: Harcourt, Brace and Jovanovich, 1951), 167–68; John R. Commons, *Races and Immigrants in America* (New York: Macmillan, 1920), 217.

41. *The Official Catholic Directory* (New York: P. J. Kenedy and Son, 1940), 138–69, 265–79.

42. Fr. Walter, O.S.B., "Census-Taking and Its By-Products," *Ecclesiastical Review,* 71 (1924), 457; Rev. William Francis Fitzgerald, *The Parish Census and the Liber Status Animarum* (Washington, D.C.: Catholic University of America Press, 1954), 84–90; Joseph H. McMahon, "The Organization of a City Parish," *Ecclesiastical Review,* 63 (1920), esp. 359–60.

43. For example, *St. Luke Parish Golden Jubilee, 1917–1967* (Gary: n.p., 1967), 15; Robert J. Mastrion, "History of Epiphany Roman Catholic Church" (St. Francis College, B.A. thesis, 1961), 28–29.

44. Ernest W. Burgess, "Residential Segregation in American Cities," *The Annals of the American Academy of Political and Social Science,* 140 (November 1928), 112. Robert Park, "The Urban Community as a Spacial Pattern and Moral Order," in *The Urban Community,* Ernest Burgess, ed. (Chicago: University of Chicago Press, 1926), 18. On immigrants and the African-American ghetto, Richard L. Morrill and O. Fred Donaldson, "Geographical Perspectives on the History of Black America," *Economic Geography,* 48 (January 1972) 15; Dennis Clark, *The Ghetto Game: Racial Conflicts in the City* (New York: Sheed and Ward, 1962), 107–34; Harold M. Rose, "Social Processes in the City: Race and Urban Residential Choice," Commission on College Geography Resource Paper #6 (Washington, D.C.: Association of American Geographers, 1969), 16.

45. Slayton, *Back of the Yards,* 31; Olivier Zunz, *The Changing Face of Inequality: Urbanization, Industrial Development, and Immigrants in Detroit* (Chicago: University of Chicago Press, 1982), 152–61. On Newburyport, Stephan Thernstrom, *Poverty and Progress: Social Mobility in a Nineteenth Century City* (Cambridge: Harvard University Press, 1964), 201. On Italian-Americans, see Phyllis H. Williams, *South Italian Folkways in Europe and America* (New York: Russell & Russell, 1969), 45–46, and Walter Firey, *Land Use in Central Boston* (Cambridge: Harvard University Press, 1947), 214.

46. Otis Dudley Duncan and Stanley Lieberson, "Ethnic Segregation and Assimilation," *American Journal of Sociology,* 64 (January 1959), 371.

47. Joseph B. Schuyler, S.J., *Northern Parish: A Sociological and Pastoral Study* (Chicago: Loyola University Press, 1960), 109.

48. Burgess, "Residential Segregation," 112. Also see Louis Wirth, "The Ghetto," in *Louis Wirth: On Cities and Social Life,* Albert J. Reiss, Jr., ed. (Chicago: University of Chicago Press, 1964), 95; Oscar Handlin, *The Newcomers: Negroes and Puerto Ricans in a Changing Metropolis* (Cambridge: Harvard University Press, [1st ed. 1959] 1965), 34–35, 80; Thomas Kessner, *The Golden Door: Italian and Jewish Immigrant Mobility in New York City, 1880–1915* (New York: Oxford University Press, 1977), 151–60; Yona Ginsberg, *Jews in a Changing Neighborhood: The Study of Mattapan* (New York: The Free Press, 1975), 51.

49. "Msgr. John W. Keough Dies; Total Abstinence Leader," undated clipping, file 1961, newsclippings, box 48, CIC; Fr. Walter, O.S.B., "Census-Taking and Its By-Products," 455; Brother Gerald J. Schnepp, *Leakage from a Catholic Parish* (Washington, D.C.: Catholic University of America Press), 314.

50. "Territorial or National Parish?" *Ecclesiastical Review*, 61 (1919), 97–98; "National Pastors and Assistance at Marriage," *Ecclesiastical Review*, 80 (1929), 92.

51. H. Paul Douglass, "The 'South End,' Newark, New Jersey and Its Churches," 5–6 (1945), #2393, HPD.

52. Samuel C. Kincheloe, "The Behavior Sequence of a Dying Church," in *The Church in the City: Samuel C. Kincheloe and the Sociology of the City Church*, Yoshio Fukuyama, ed. (Chicago: Exploration Press, 1989), 51–100; Ross W. Sanderson, *The Strategy of City Church Planning* (New York: Institute of Social and Religious Research, 1932), 12–13. H. Paul Douglass explicitly contrasts Protestants and Catholics in *The Saint Louis Church Survey* (New York: George H. Doran, 1924), 64–77, and *The Springfield Church Survey* (New York: George H. Doran, 1926), 44. Also see H. Paul Douglass and Edmund deS. Brunner, *The Protestant Church as a Social Institution* (New York: Harper and Brothers, 1935), 70–73.

53. For "splendidly organized" see John J. Gilkerson, Jr., *Middle-Class Providence, 1820–1940* (Princeton: Princeton University Press, 1986), 280–81. For "fragmentation of their membership" see Samuel C. Kincheloe, *The American City and Its Church* (New York: Friendship Press, 1938), 95.

54. H. Paul Douglass, "The Churches of Greater Detroit," (1943) #2270, HPD.

55. James Weldon Johnson, *Negro Americans, What Now?* (New York: The Viking Press, 1938), 25.

56. On Orthodox synagogues, David A. Wallace, "Residential Concentration of Negroes in Chicago" (Harvard Ph.D. diss., 1953), 135. H. Paul Douglass briefly notes similarities between Protestant churches and synagogues in "Churches of Buffalo and Vicinity," (1947), 81, #2431, HPD.

57. Gerald Gamm, "Neighborhood Roots: Exodus and Stability in Boston, 1870–1990" (Harvard University, Ph.D. diss., 1994). Andrew M. Greeley, *The Catholic Myth: The Behaviors and Beliefs of American Catholics* (New York: Scribners, 1990), 156. Also see Rev. Robert G. Howe, "Baltimore Parish Study" (September 1967), file 3, box 5, John Cardinal Dearden Papers, MAUND; Lenski, 194–97; Mathew Edel, Elliot D. Sclar, and Daniel Luria, *Shaky Palaces: Homeownership and Social Mobility in Boston's Suburbanization* (New York: Columbia University Press, 1984), 134.

58. Robert A. Woods and Albert J. Kennedy, *The Zone of Emergence: Observations of the Lower, Middle and Upper Working Class Communities of Boston, 1905–1914* (Cambridge: M.I.T. Press, 1962), 131–34.

59. Paul Wrobel, *Our Way: Family, Parish, and Neighborhood in a Polish-American Community* (Notre Dame: University of Notre Dame Press, 1979), 39; Sister M. Martina Abbot, S.C., *A City Parish Grows and Changes* (Washington, D.C.: Catholic University of America Press, 1953), 77.

60. Kathleen Neils Conzen makes the same point slightly differently by pointing out that residential concentrations of any one national group were initially neces-

sary to create a sense of group identity, but that the community institutions themselves then helped shape group and neighborhood identities. See Conzen, "Immigrant Neighborhoods and Ethnic Identity," *Journal of American History*, 66 (December 1979), 603–15. In my view, Thomas Philpot's otherwise excellent study of neighborhood change in Chicago, by focusing solely on ethnicity, slights the even more important religious connections between residents of Chicago's "white" neighborhoods. It was these connections, and their physical and cultural enactment in both buildings and rituals, that sustained neighborhood identity for many area residents. See Thomas Lee Philpot, *The Slum and the Ghetto: Neighborhood Deterioration and Middle-Class Reform, Chicago, 1880–1930* (New York: Oxford University Press, 1978), 131–35.

61. Roger S. Ahlbrandt, Jr., *Neighborhoods, People and Community* (New York: Plenum, 1984), 26, 63–65, 187; Edward O. Laumann, *Bonds of Pluralism: The Form and Substance of Urban Social Networks* (New York: John Wiley & Sons, 1973), 106, 121, 202.

62. For examples of newspaper advertisements, see *The New World*, April 18, 1958, 27; John Lukacs, *Philadelphia: Patricians and Philistines, 1900–1950* (New York: Farrar, Straus and Giroux, 1981), 332; McMahon, "What Parish Are You From?" 92, 96, 130; Eugene Edward Obidinsky, *Ethnic to Status Group: A Study of Polish Americans in Buffalo* (New York: Arno, 1980), I-8.

63. Msgr. Edward Hawks, *History of the Parish of St. Joan of Arc, Harrowgate, Philadelphia* (Philadelphia: Peter Reilly, 1937), 102; John E. Walsh to Rev. Hugh Lamb, May 23, 1930, Dougherty correspondence, PAHRC; Colleen McDannell, *The Christian Home in Victorian America, 1840–1900* (Bloomington: Indiana University Press, 1986), 58.

64. H. Paul Douglass, "Churches of Buffalo and Vicinity," 88–89, 96.

65. Claris Edwin Silcox and Galen M. Fisher, *Catholics, Jews, and Protestants: A Study of Relationships in the United States and Canada* (New York: Harper and Brothers, 1934), 69–70.

66. Caroline Golab, *Immigrant Destinations* (Philadelphia: Temple University Press, 1977), 154–55.

67. John Deedy, *American Catholicism: And Now Where?* (New York: Plenum, 1987), 152–53; Paul Wilkes, "As the Blacks Move In, the Ethnics Move Out," *New York Times Magazine*, January 24, 1971, 9+; Kenneth Wilson Underwood, *Protestant and Catholic: Religious and Social Interaction in an Industrial Community* (Boston: Beacon Press, 1957), 189; John Gregory Dunne, *Harp* (New York: Simon and Schuster, 1989), 47; Patricia Hampl, "Parish Streets," (St. Joseph, Minn.: College of St. Benedict, 1990), 1; McMahon, "What Parish Are You From?" 130–31; Taylor, *Pioneering on Social Frontiers*, 195. On naming as signifying community, see Alasdair MacIntyre, *Whose Justice? Which Rationality?* (Notre Dame: University of Notre Dame Press, 1988), 378–79.

68. Mary C. Waters, *Ethnic Options: Choosing Identities in America* (Berkeley: University of California Press, 1990), 99.

69. One 1951 study found that 45 out of the 50 Catholic families in one midwestern parish had had their homes blessed. See Rev. Lucan R. Freppert, O.F.M., "A Study of Religious Rituals in Families of an Urban Parish" (Catholic University of America, M.A. thesis, 1951), 35; Margaret Byington, *Homestead: The Households of a Mill Town* (New York: Charities Publication, 1910), 152; on

Catholic homes, McDannell, *The Victorian Home,* 66–70. The practice of venerating a "traveling" statue of Mary occurred in both Polish and Irish parishes. For example, McMahon, "What Parish Are You From?" 137–40, and Wrobel, *Our Way,* 104–5; Taylor, *Pioneering on Social Frontiers,* 196.

70. Also see R. Laurence Moore, "Religion, Secularization, and the Shaping of the Culture Industry in Antebellum America," *American Quarterly,* 41 (June 1989), 235.

71. Stephen L. Cabral, *Tradition and Transformation: Portuguese Feasting in New Bedford* (New York: AMS Press, 1989), 41–44; McMahon, "What Parish Are You From?" 81.

72. Sister Mary Joan, O.P., and Sister Mary Nona, O.P., *Guiding Growth in Christian Living: A Curriculum for the Elementary School,* volume 1: *Primary Grades* (Washington, D.C.: Catholic University of America Press, 1957), 273–74; Robert Orsi, "The Center Out There, In Here, and Everywhere Else: The Nature of Pilgrimage to the Shrine of St. Jude, 1929–1965," *Journal of Social History,* 25 (Winter 1991), 216; Ralph L. Woods and Henry L. Woods, *Pilgrim Places in North America: A Guide to Catholic Shrines* (New York: Longmans, Green, 1939).

73. Rev. Charles Bruehl, D.D., "Pastoral Work in the Large City," *Homiletic and Pastoral Review,* 22 (1922), 604; "A Code of Domestic and Neighborhood Ethics," *Ecclesiastical Review,* 77 (1927), 521–22.

74. Quoted in R.P. Chenu, O.P, "Catholic Action and the Mystical Body," *Restoring All Things: A Guide to Catholic Action,* John Fitzsimons and Paul McGuire, eds. (New York: Sheed & Ward, 1938), 4. Also see passages from Aquinas cited in Michael Novak, *The Catholic Ethic and the Spirit of Capitalism* (New York: The Free Press, 1993), 301–2.

75. "Catholic Social Teaching Through the Regular Curriculum," *Institute of Social Order Bulletin,* 1 (September 1943), 6.

76. Chenu, "Catholic Action," 8–9; Bruehl, "Pastoral Work," 601. The work of David Hollenbach, S.J., is illuminating on these matters. See his "A Communitarian Reconstruction of Human Rights: Contributions from Catholic Tradition," in *Catholicism and Liberalism: Contributions to American Public Philosophy,* R. Bruce Douglass and David Hollenbach, eds. (Cambridge: Cambridge University Press, 1994), esp. 127–28. Also see David Hollenbach, S.J., *Claims in Conflict: Retrieving and Renewing the Catholic Human Rights Tradition* (New York: Paulist Press, 1979), 41–106.

77. Woods and Kennedy, *The Zone of Emergence,* 213; Bruehl, "Pastoral Work," 361.

78. Karl Rahner, S.J., "The Theology of the Parish," in *The Parish: From Theology to Practice,* Hugo Rahner, S.J., ed., Robert Kress, trans. (Westminster: Newman Press, 1958), 25–32.

79. I am indebted to Andrew M. Greeley's exploration of some of the implications of David Tracy's work in Greeley's *The Catholic Myth,* 34–64. Also see Michael J. Himes and Kenneth R. Himes, O.F.M., *Fullness of Faith: The Public Significance of Theology* (New York: Paulist Press, 1993), 82–83; David Tracy, *The Analogical Imagination: Christian Theology and the Culture of Pluralism* (New York: Crossroad, 1981), esp. 405–44.

80. Allen Luke, "The Secular Word: Catholic Reconstructions of Dick and

Jane,'' in *The Politics of the Textbook,* Michael W. Apple and Linda K. Christian-Smith, eds. (New York: Routledge, 1991), 176.

81. *Silver Jubilee of Our Lady of the Angels Church, Cleveland Ohio* (Cleveland, n.p. 1947), 30.

82. Cited in Gamm, ''Neighborhood Roots,'' 71.

83. McMahon, ''What Parish Are You From?'' 38, 40–41, 81, 84, 178–84, 294–95.

84. Joseph Chalaskinski, ''Polish and Parochial School Among Polish Immigrants,'' [1935?] folder 2, box 33, Community Areas Project Collection, cited in Cohen, *Making a New Deal,* 87; William I. Thomas and Florian Znaniecki, *The Polish Peasant in Europe and America* (Boston: Richard G. Badger, 1920), 5: 41.

85. John Joseph Bukowczwk, ''Steeples and Smokestacks: Class, Religion and Ideology in the Polish Immigrant Settlements in Greenpoint and Williamsburg, Brooklyn, 1880–1929'' (Harvard University, Ph.D. diss., 1980), 216–21; Sister M. Martina Abbot, S.C., *A City Parish Grows and Changes* (Washington, D.C.: Catholic University of America Press, 1953), 24.

86. Thomas and Znaniecki, *The Polish Peasant in Europe and America,* 5: 48–52, 63–64.

87. Polly Kline, ''Red Hook Procession to Re-enact Aged Rite,'' *New York Daily News,* March 26, 1964, p. B7.

88. Robert Anthony Orsi, *The Madonna of 115th Street: Faith and Community in Italian Harlem, 1880–1950* (New Haven: Yale University Press, 1985).

89. Nicholas John Russo, ''Three Generations of Italians in New York City: Their Religious Acculturation,'' in *The Italian Experience in the United States,* Silvano Tomasi and Madeline H. Engel, eds. (Staten Island: Center for Migration Studies, 1970), 195–209; Humbert S. Nelli, *Italians in Chicago, 1880–1930: A Study in Ethnic Mobility* (New York: Oxford University Press, 1970), 199; Stephen Joseph Shaw, ''Chicago's Germans and Italians, 1903–1939,'' 242; Orsi, *The Madonna of 115th St.,* 182–84, 220.

Chapter 2

1. John T. Gillard, S.S.J., *Colored Catholics in the United States* (Baltimore: The Josephite Press, 1941), 163–65, 167; Gilbert Osofsky, *Harlem: The Making of a Ghetto: Negro New York, 1890–1930* (New York: Harper and Row, 1966), 134.

2. John T. Gillard, S.S.J., *The Catholic Church and the American Negro* (Baltimore: St. Joseph's Society Press, 1929), 105.

3. See Kenneth A. Scherzer, *The Unbounded Community: Neighborhood Life and Social Structure in New York City, 1830–1875* (Durham: Duke University Press, 1992), 50–52; Alan Burstein, ''Immigrants and Residential Mobility: The Irish and Germans in Philadelphia, 1850–1880,'' in *Philadelphia: Work, Space, Family, and Group Experience in the Nineteenth Century,* Theodore Hershberg, ed. (New York: Oxford University Press), 1981, 199–200; Stephanie W. Greenberg, ''Industrial Location and Ethnic Residential Patterns in an Industrializing City: Philadelphia, 1880,'' in *Philadelphia,* Hershberg, ed., 204–29.

4. Data from *Fourteenth Census of the United States*, vol. 2, *Population* (Washington, D.C.: Government Printing Office, 1922).

5. Mathew Frye Jacobson, *Special Sorrows: The Diasporic Imagination of Irish, Polish, and Jewish Immigrants in the United States* (Cambridge: Harvard University Press, 1995), 185.

6. See Edward McSweeney to W.E.B. DuBois, May 10, 1922, frame 1199, reel 10, W.E.B. DuBois Papers, microfilm copy.

7. Leslie Woodcock Tentler, *Seasons of Grace: A History of the Catholic Archdiocese of Detroit* (Detroit: Wayne State University Press, 1900) 496; John Clayton, ''Catholics of Many Nations Hold Meetings,'' *Chicago Tribune*, June 21, 1926, p.3.

8. James F. Connelly, *The History of the Archdiocese of Philadelphia* (Philadelphia: Archdiocese of Philadelphia, 1976), 386–89; *Statuta Provincialia et Diocesana* (Philadelphia, 1934), 41. On Irish and Italian tension in Philadelphia, see Richard A. Vanbero, ''Philadelphia's South Italians and the Irish Church: A History of Cultural Conflict,'' in *The Religious Experience of Italian-Americans*, Silvano Tomasi, ed. (Staten Island: The Italian American Historical Association, 1975), 50–51.

9. Dominic Pacyga, *Polish Immigrants and Industrial Chicago Workers on the South Side,1880–1922* (Columbus: Ohio State University Press, 1991), 212–27; Dominic Pacyga, ''Race, Ethnicity, and Acculturation: Chicago's Poles and Irish and the 1919 Race Riot,'' paper in author's possession.

10. Elizabeth B. Sweeney to Cooper, March 6, 1930, correspondence and notes, Fr. John Cooper Papers (unprocessed), ACUA.

11. Rev. Maurice S. Sheehy, *National Attitudes in Children* (Washington, D.C.: Catholic Association for International Peace, 1932), 4, 7, 8; ''Widening Group Friendship,'' *Commonweal*, 16 (August 10, 1932), 359.

12. Louis Tordella to Brooklyn Catholic Action Council (n.d.), carton 1, box 1, NYCIC.

13. ''Constitution of the Federated Colored Catholics of the United States,'' *St. Elizabeth's Chronicle*, 2 (November 1929), 27; Marilyn Nickels, *Black Catholic Protest and the Federated Colored Catholics, 1917–1933* (New York: Garland, 1988), 47, 83.

14. David James, interview with author, January 11, 1991.

15. Oliver G. Waters to Bishop Joseph Schrembs, March 31, 1922, Our Lady of Blessed Sacrament File, ADC.

16. John Wooten to Archbishop William O'Connell, August 3, 1920, St. Richard's parish file sp-129, AAB; John Daniels, *In Freedom's Birthplace* (New York: Arno [1914] 1969), 229–30, cited in William C. Leonard, ''The History of Black Catholics in Boston, 1790–1840'' (Northeastern University, M.A. thesis, 1993), 16.

17. Gertrude Walsh, ''The Negro Apostolate in Brooklyn, 1920–1950'' (Saint John's University, M.A. thesis, 1951), 5; Sr. M. Adeline Lemanska, H.F.N., ''The Catholic Negroes in the Diocese of Brooklyn'' (Villanova University, M.A. thesis, 1951), 34; Oliver G. Waters to Bishop Joseph Schrembs, March 31, 1922, Our Lady of Blessed Sacrament File, ADC.

18. Fr. E. R. Dyer to Fr. Stephen, December 10, 1921, folder 7, box 6, series

5/2, Bureau of Catholic Indian Missions Collection; Francis B. Cassily, "Why Colored Churches?" *The Chronicle,* 3 (1930), 39–40.

19. Joe Chatham to John LaFarge, n.d., carton 1 box 2, NYCIC; Anselm M. Townsend, O.P. to John LaFarge, February 19, 1936, folder 6, box 52, JLF.

20. Francis J. Gilligan, "The Color Line Considered Morally," *Ecclesiastical Review,* 81 (1929), 487.

21. Rev. Francis J. Gilligan, S.T.L., *The Morality of the Color Line: An Examination of the Right and the Wrong of the Discriminations Against the Negro in the United States* (Washington, D.C.: Catholic University Press, 1928), 97, 100–104, 113–14, 134–37, 20. In 1946, John LaFarge urged that Gilligan's work not be reprinted because of the "dynamic and explosive problems" which might ensue. LaFarge to Fr. Raymond McGowan, March 20, 1946, Social Action Division file, box 39, NCWC.

22. Useful on the term "ethnic" is Werner Sollors, *Beyond Ethnicity: Consent and Descent in American Culture* (New York: Oxford University Press, 1986), esp. 20–26. On the scientific literature, Carl N. Degler, *In Search of Human Nature: The Decline and Revival of Darwinism in American Social Thought* (New York: Oxford University Press, 1991), 48–55.

23. "National Pastors and Assistance at Marriage," *Ecclesiastical Review,* 80 (1929), 89–90.

24. Mundelein in foreword to "Report of the American Board of Catholic Missions, From July 1, 1928 to July 1, 1929," BCIM; Henry William Francis, "The Dying National Parish and Compulsory Membership Registration," *Ecclesiastical Review,* 95 (1936), 380.

25. David R. Roediger, *Towards the Abolition of Whiteness: Essays on Race, Politics, and Working Class History* (London: Verso, 1994), 181–98; also David R. Roediger, *Wages of Whiteness: Race and the Making of the American Working Class* (London: Verso, 1991), 133–63, and Robert Orsi, "The Religious Boundaries of an In-between People: Street *Feste* and the Problem of the Dark-Skinned Other in Italian Harlem, 1920–1990," *American Quarterly,* 44 (September 1992), 313–48.

26. Minutes of the Northeastern Clergy Conference for the Negro Welfare, March 20, 1934, folder 12, box 52, JLF.

27. Anselm M. Townsend, O.P., to John LaFarge, February 19, 1936, folder 6, box 2, series II, JLF

28. Gustave B. Aldrich, "Another View of Separate Churches," *The Chronicle,* 3 (1930), 56–57; address by Fr. William J. Walsh at Catechetical Conference, October 29, 1936, folder 15, box 2, series II, JLF.

29. Joseph Schrembs to Most Rev. P. Fumasoni-Biondi, February 25, 1927, Our Lady of the Blessed Sacrament file, ADC; James W. Sanders, *The Education of an Urban Minority: Catholics in Chicago, 1833–1965* (New York: Oxford University Press, 1977), 209.

30. W.E.B. DuBois, "The Catholic Church and Negroes," *The Crisis,* 30 (July 1925), 121.

31. *Diamond Jubilee: All Saints Roman Catholic Church* (Brooklyn, 1942), 10; Stephen J. Shaw, "An Oak Among Churches: St. Boniface Parish, Chicago, 1864–1990," in *American Congregations,* vol. 1, *Portraits of Twelve Religious*

Communities, James P. Wind and James W. Lewis, eds. (Chicago: University of Chicago Press, 1994), 354.

32. C. J. Nuesse, "Empirical Problems for Social Research in the Parish," in C. J. Nuesse and Thomas J. Harte, eds. *The Sociology of the Parish: An Introductory Symposium* (Milwaukee: Bruce, 1951), 223. John O'Rourke, "A Study of the Adjustment of a City Parish to Changing Social Conditions in the Surrounding Community" (Fordham, M.S.W. thesis, 1941), 70; Shaw, "An Oak Among Churches," 367–70.

33. James T. Farrell, *Studs Lonigan: A Trilogy Comprising Young Lonigan, The Young Manhood of Studs Lonigan, and Judgment Day* (Urbana: University of Illinois Press, 1993), 402, 375.

34. Ibid., 412, 434, 560.

35. Ibid., 402, 408. Also insightful on Farrell is Carla Cappetti, *Writing Chicago: Modernism, Ethnography, and the Novel* (New York: Columbia University Press, 1993), esp. 134–36.

36. "Church of the Nativity" (1938), Nativity of Our Blessed Lord File, ADB.

37. Joseph J. McCarthy, "History of Black Catholic Education in Chicago, 1871–1971" (Loyola University, Ph.D. diss., 1973), 82–3; *A History of the Parishes of the Archdiocese of Chicago,* Msgr. Harry Koenig, ed. (Chicago: Archdiocese of Chicago, 1980), vol. 1, 72–75.

38. Sanders, *Education of an Urban Minority,* 212; Morgan S. Odell, "United Religious Survey of the Greater Grand Crossing Area," Dec. 1929–Jan. 1930, folder 1, box 134, Ernest Burgess Papers, cited in Thomas Philpot, *The Slum and the Ghetto: Neighborhood Deterioration and Middle-Class Reform, Chicago, 1880–1930* (New York: Oxford University Press, 1978), 199.

39. *Second Annual Report of the Detroit Housing Commission* (Detroit, 1935), 33–34.

40. Fr. Francis Beecherini to Rt. Rev. Msgr. John Doyle, November 6, 1929, San Francesco Parish file 27.4, AAD; Tentler, *Seasons of Grace,* 503–12.

41. *Santa Maria Parish Silver Jubilee, 1919–1944,* Santa Maria parish file 1.8, AAD; Ciarocchi to Mooney, April 20, 1943, Santa Maria parish file 1.5, AAD.

42. William Foote Whyte, "Race Conflicts in the North End of Boston," *New England Quarterly,* 12 (December 1939), 623–42.

43. Farrell, *Studs Lonigan,* 618. I became aware of this passage through Robert Butler, "Farrell's Ethnic Neighborhood and Wright's Urban Ghetto: Two Visions of Chicago's South Side," *Melus,* 18 (Spring 1993), 163; History of West Englewood, December 1925, in Vivian Palmer Papers, CHS.

44. John LaFarge, *The Manner is Ordinary* (New York: Harcourt, Brace, 1954), 1–105; Edgar Alexander, "Church and Society in Germany: Social and Political Movements and Ideas in German And Austrian Catholicism, 1789–1950," Toni Stolper trans., in *Church and Society: Catholic Social and Political Thought and Movements, 1789–1950,* Joseph N. Moody, ed. (New York: Arts, 1953), 487–490.

45. LaFarge, *The Manner is Ordinary,* 157, 168–69.

46. Nickels, 224; John LaFarge, "The Immediate Negro Problem,"*America,* 26 (December 24, 1921), 222.

47. Gilbert J. Garraghan, *The Jesuits of the Middle United States* (New York: America Press, 1938), 3: 564; William Markoe, "An Interracial Role," box 4,

series 103, Fr. William Markoe Papers, Marquette University Special Collections; also see Nickels, *Black Catholic Protest,* 136–75.

48. Copy of vow in file 1917–1946, box 3, series 103, Markoe Papers.

49. William Markoe, "A Solution to the New Race Problem," *America,* 24 (November 27, 1920), 125–26; Nickels, *Black Catholic Protest,* 156; to Claude Barnett, [1931,] Reel 5, Series J, Claude Barnett Papers, microfilm copy.

50. William Markoe, "Negro Higher Education," *America,* 26 (April 1, 1922), 558–60; "The Negro and Catholicism," *America,* 30 (February 23, 1924), 449–50; "The Importance of Negro Leadership," *America,* 29 (October 13, 1923) 605–6.

51. Nickels, *Black Catholic Protest,* 133, cites Thomas W. Turner, "Does the Roman Catholic Church Need a National Federation of Colored Communicants?," address delivered at the ninth annual convention of the F.C.C., August, 27, 1933, Turner Papers, Howard University.

52. Philip Gleason, *Keeping the Faith: American Catholicism Past and Present* (Notre Dame: University of Notre Dame Press, 1980), 136–51; William Halsey, *The Survival of American Innocence: Catholicism in an Era of Disillusionment, 1920–1940* (Notre Dame: University of Notre Dame Press, 1987), 148. On the larger context of the neo-Thomistic revival, see Halsey, 138–68.

53. A. H. Clemens, "The Catholic Sociologist Faces a New Social Order," *American Catholic Sociological Review,* 4 (October 1943), 159.

54. Cited in David O'Brien, *American Catholics and Social Reform: The New Deal Years* (New York: Oxford University Press, 1968), 67.

55. On the 1919 episcopal statement, Joseph M. McShane, *"Sufficiently Radical": Catholicism, Progressivism, and the Bishop's Program of 1919* (Washington, D.C.: Catholic University of America Press, 1986), 169.

56. On Catholics and early Roosevelt initiatives, George Q. Flynn, *American Catholics and the Roosevelt Presidency, 1932–1936* (Lexington: University of Kentucky Press, 1968), 78–102; Philip Gleason, *Keeping the Faith,* 141; O'Brien, *American Catholics and Social Reform,* 17–18; on Catholic Action, *Transforming Parish Ministry,* Jay P. Dolan, R. Scott Appleby, Patricia Byrne, and Debra Campbell, eds. (New York: Crossroad, 1989), 222–51.

57. Msgr. Fulton Sheen, *The Mystical Body of Christ* (New York: Sheed and Ward, 1935), 359. For an overview, see F. X. Lawlor, "Mystical Body of Christ," *New Catholic Encyclopedia,* vol. 10 (New York: McGraw Hill, 1967), 166–171.

58. *Restoring All Things: A Guide to Catholic Action,* John Fitzsimons and Paul McGuire, eds. (New York: Sheed & Ward, 1938), 8.

59. Sheen, *Mystical Body,* 368.

60. Stritch quoted in Joseph P. Chinnici, O.F.M., *Living Stones: The History and Structure of Catholic Spiritual Life in the United States* (New York: Macmillan, 1989), 155.

61. For example, Klemens von Klemperer, *Ignaz Seipel: Christian Statesman in a Time of Crisis* (Princeton: Princeton University Press, 1972), 54–65; Joseph N. Moody, "Anti-Semitism," *Conference Bulletin of the Archdiocese of New York,* 14 (1936), 87.

62. Carlton J. H. Hayes, *Essays on Nationalism* (New York: Macmillan, 1926), 237; idem, "Obligation to America: Part II," *Commonweal,* 1 (January 7, 1925),

227. Also see idem, *Patriotism, Nationalism, and the Brotherhood of Man* (Washington, D.C.: Catholic Association for International Peace, 1937), 19–23.

63. Rev. Bernard J. Sheil, D.D., "The Negro—Our Dispossessed," *Interracial Review*, 15 (October 1942), 151; Betty McCormack to NFCCS Office, April 4, 1944, file "Interracial Justice Commission," box 7, National Federation of Catholic College Students Papers, MAUND.

64. Harry Sylvester, *Moon Gaffney* (New York: Henry Holt, 1947), 96–97.

65. John LaFarge, "The Cardinal Gibbons Institute," *Commonweal*, 15 (February 17, 1932), 433–35; LaFarge, "What Do the Colored Catholics Want?" *America*, 43 (September 20, 1930), 565–67.

66. Linna Bresette, "Report of Social Action Activities with Colored Catholics," November 20, 1943, "Negro in Industry" file, box 39, NCWC; William M. Markoe, "Our Jim Crow Federation," *The Chronicle*, 3 (July 1930), 149; idem, "Better Race Relations," *The Chronicle*, 5 (October 1932), 196–97.

67. Nickels, *Black Catholic Protest*, 107; Hazel McDaniel Teabeau, "Federated Colored Catholics Make History in New York City Convention," *The Chronicle*, 5 (October 1932), 199.

68. W.E.B. DuBois, "The Negro and the Catholic Church," *The Crisis*, 40 (March 1933), 68–69. Ironically, LaFarge and Markoe viewed themselves as progressives on the question of lay participation since they did urge laypeople to become involved in social questions. The difference was that they viewed "solid doctrinal instruction"—by priests—as a necessary first step. See John LaFarge, "How Catholic Action Triumphs through Conquest," *America*, 61 (February 4, 1939), 412–13.

69. See Rev. John Burke to Bishop John McNicholas, January 19, 1933, file "Social-Action—Race Relations, 1929–1947," box 89, NCWC.

70. Teabeau, "Federated Colored Catholics," 200.

71. George W. B. Conrad to Claude Barnett, January 13, 1933, reel 5, series J, Claude Barnett Papers, microfilm copy; Nickels, *Black Catholic Protest*, 162.

72. Nickels, *Black Catholic Protest*, 98; John LaFarge, "The Crux of the Mission Problem," *America*, 36 (November 6, 1926), 80–81.

73. George K. Hunton, *All of Which I Saw, Part of Which I Was: The Autobiography of George Hunton as told to Gary MacEoin* (New York: Doubleday, 1967), 58; John LaFarge, *Interracial Justice: A Study of the Catholic Doctrine of Race Relations* (New York: America Press, 1937), 172–74.

74. See O'Brien, *American Catholics and Social Reform*, for an overview of Catholic activism during the 1930s.

75. The most comprehensive study of the New York Catholic Interracial Council is Martin Zielinski, " 'Doing the Truth': The Catholic Interracial Council of New York, 1945–1965" (Catholic University of America, Ph.D. diss., 1989), esp. 1–46; Guichard Parris, interview with Martin Zielinski, January 12, 1988, copy in possession of Martin Zielinski.

76. Ed Marciniak, interview with Steven Avella, December 14, 1983; "Catholic Action for the Negro," *America*, 49 (June 10, 1933), 221; Tracy Mitrano, "The Rise and Fall of Catholic Women's Higher Education in New York State, 1890–1985" (State University of New York, Binghamton, Ph.D. diss., 1988), 143.

77. John LaFarge, "Racial Truth and Racist Error," *Thought,* 14 (1939), 19–35.

78. John Higham, *Send These to Me: Jews and Other Immigrants in Urban America* (New York: Atheneum, 1975), 128.

79. Wallace Stegner, "Who Persecutes Boston?" *The Atlantic Monthly,* 174 (July 1944), 45–52; Hilda Polacheck, "The Ghost of the Convent," January 26, 1939, file "Hilda Polacheck," box A708, Federal Writers Project Papers, Library of Congress; James Weschler, "The Coughlin Terror," *The Nation,* 149 (July 22, 1939), 93; Ronald H. Bayor, *Neighbors in Conflict: The Irish, Germans, Jews, and Italians of New York City, 1929–41* (Baltimore: Johns Hopkins University Press, 1978), esp. 87–108.

80. John Cogley, "Peddling Anti-Semitism," *Work,* 2 (October 1944); Bayor, *Neighbors in Conflict,* 107–8; John F. Stack, Jr., *International Conflict in an American City: Boston's Irish, Italians, and Jews, 1935–1944* (Westport: Greenwood Press, 1979), 92–140.

81. Rt. Rev. John A. Ryan, D.D., *American Democracy vs. Racism, Communism* (New York: Paulist Press, 1939); John LaFarge, *The Race Question and the Negro* (New York: Longmans, Green, 1943), xiv.

82. Ryan to Daniel Lord, December 5, 1938, box 16, Ryan Papers (unprocessed collection), ACUA.

83. "Christian American Jew Baiting," *Christian Social Action,* September 1939, box 22, American Catholic Trade Unionist Papers, Archives of Labor and Urban Affairs, Wayne State University; William Miller, *A Harsh and Dreadful Love: Dorothy Day and the Catholic Worker Movement* (New York: Liveright, 1972), 152.

84. Memo of the Auxiliary Bishop of Chicago, n.d. [1939], *Actes et Documents du Saint Siege Relatifs a la Seconde Guerre Mondiale* (Libreria Editrice Vaticana, 1972), vol. 6, 123–24; Notes from Mr. Hurley of the Secretary of State, March 8, 1940, in ibid. (1965), vol. 1, 381–82.

85. Suggestive on this topic are Elazar Barkan, *The Retreat of Scientific Racism: Changing Concepts of Race in Britain and the United States Between the World Wars* (Cambridge: Cambridge University Press, 1992), 279–346; Degler, *In Search of Human Nature,* 59–211; and George Stocking, *Race, Culture, and Evolution: Essays in the History of Anthropology,* rev. ed. (Chicago: University of Chicago Press, 1982), 270–307.

86. LaFarge to Franz Boas, May 10, 1939, file 3, box 10, JLF; also see Philip Gleason, "Minorities (Almost) All: The Minority Concept in American Social Thought," *American Quarterly,* 43 (September, 1991), 392–424.

87. Anthony Rhodes, *The Vatican in the Age of the Dictators, 1922–1945* (London: Hodder and Stoughton, 1973), 161–210; Donald J. Dietrich, *Catholic Citizens in the Third Reich: Psycho-Social Principles and Moral Reasoning* (New Brunswick: Transaction Books, 1988), 132–79.

88. *The Papal Encyclicals: 1903–1939,* Carlen, ed., 527.

89. John LaFarge, "The Pope Deals With Nazi Persecution," *Catholic Mind,* 35 (May 8, 1937), 211; John LaFarge, "Racism and Social Unity," *Catholic Mind,* 37 (January 8, 1939), 491–492; Jacques Maritain, "The Menace of Racialism," *Interracial Review,* 10 (May 1937), 70; on Maritain and racialism, Bernard

Doering, *Jacques Maritain and the French Catholic Intellectuals* (Notre Dame: University of Notre Dame Press, 1983), 139, 151–53.

90. John LaFarge, "Racism and Social Unity," 494; letter to Clergy Conference of Negro Welfare, December 8, 1938, folder 6, box 52, JLF.

91. "Symposium To be Published on 'Race: Nation: Person:'" *Interracial Review,* 17 (May 1944), 78; Bishop Joseph W. Corrigan, Preface, in *Race: Nation: Person: Social Aspects of the Race Problem,* G. Barry O'Toole, ed. (New York: Barnes and Noble, 1944), vi-vii, 69; *Scientific Aspects of the Race Problem* (Washington, D.C.: Catholic University of America Press, 1941).

92. Johannes Schwarte, *Gustav Gundlach, S.J., (1892–1963)* (Munich: Verlag Ferdinand Schöningh, 1975), 74–77; LaFarge, *The Manner is Ordinary,* 272–274.

93. Jim Castelli, "Unpublished Encyclical Attacked Anti-Semitism," *National Catholic Reporter,* 9 (December 15, 1972), 1/15; Schwarte, *Gustav Gundlach,* 78–80; LaFarge, *The Manner is Ordinary,* 272–274. The refusal of Pius XII to issue the more emphatic statement co-authored by LaFarge is, of course, one small part of a long-running debate over the role of the Church during the Nazi persecution of the Jews. For a recent article on the matter, see Conor Cruise O'Brien, "A Lost Chance to Save the Jews?," *New York Review of Books,* 36 (April 27, 1989), 27–28+.

94. Ruth Fox, "Catholicism and Racism," *Interracial Review,* 17 (February 1944), 25; "The Baroness Jots it Down," *Harlem Friendship House News* (April 1944), 6; *The Papal Encyclicals: 1939–1958,* Claudia Carlen, I.H.M., ed. (Raleigh, North Carolina: Pierian Press, 1990), 24, 40; Francis E. McMahon, *A Catholic Looks at the World* (New York: Vanguard Press, 1945), 278.

95. Elizabeth Ogg, *Longshoremen and Their Homes* (New York: Greenwich House, 1939), 27–28; Caroline Ware, *Greenwich Village, 1920–1930: A Comment on American Civilization in the Post-War Years* (Berkeley: University of California Press, 1994), 135. "Racial Catholics" cited in Bayor, *Neighbors in Conflict,* 108.

96. Gillard, *Colored Catholics in the United States,* 32, 221.

97. Bishop Molloy to LaFarge, August 18, 1942, folder 25, box 17, JLF; Earl Boyea, "The National Catholic Welfare Conference: An Experience in Episcopal Leadership, 1935–1947" (Catholic University, Ph.D. diss., 1987), 342.

98. Compare the letter sent by Chicago bishop Bernard Sheil to the directors of the United Jewish Appeal, December 29, 1939 in *Actes Du Saint Siege Relatifs a la Guerre Mondiale* (Libreria Editrice Vaticana, 1972), vol. 6, 211–12, with Sheil, "The Negro—Our Dispossessed," *Interracial Review,* 15 (October 1942), 151.

Chapter 3

1. Reynolds Farley and Walter R. Allen, *The Color Line and the Quality of Life in America* (New York: Oxford University Press, 1989), 113.

2. John T. Gillard, S.S.J., *Colored Catholics in the United States,* (Baltimore: St. Joseph's Society Press), 31-33; Reverend Joseph F. Eckert, S.V.D., "The S.V.D. Negro Mission Report," *St. Augustine Messenger,* 23 (May 1945), 102–104.

3. Gertrude Guidry and John Sherwood, interview with author, July 29, 1991; Sr. Mary Ita O'Donnell, "Monsignor John Belford (1861–1951): His Social and

Political Criticisms'' (Saint Francis College, B.A. thesis, 1966), 1–30; ''A Manifest Duty,'' *Commonweal,* 39 (December 10, 1943), 196.

4. Msgr. John K. Sharp, *An Old Priest Remembers, 1892–1978* (Hicksville, N.Y.: Exposition Press, 1978), 97; W.E.B. DuBois to Rev. Alcuin Knecht, February 14, 1930, frame 560, reel 33, W.E. B. DuBois Papers, microfilm copy.

5. George K. Hunton, *All of Which I Saw,* 128; ''Hayes Criticism of Negro Views Irks Dr. Belford,'' *Brooklyn Daily Eagle,* October 18, 1929, p.3; ''The NAACP Battle Front,'' *The Crisis,* 36 (December 1929), 409; John F. Moore, *Will America Become Catholic?* (New York: Harper & Row, 1931), 201; Gustave B. Aldrich, ''Another View of Separate Churches,'' *The Chronicle,* 3 (March 1930), 55–56; Harold Connolly, *A Ghetto Grows in Brooklyn* (New York: New York University Press, 1977), 63–64, 77, 139. Also see the correspondence between the NAACP and a representative of the New York archdiocese in St. Mathew's file, box 269, series C, group I, NAACP.

6. *The Church Bulletin* (St. Peter Claver), 49 (April 1940), 4; John F. Moore, *Will America Become Catholic?,* 201. Quinn did openly disagree with Belford's ideas on formal segregation. See ''Would Bar Negroes If Number Increased At His Church,'' *New York Amsterdam News,* October 23, 1929, p.10.

7. Joseph J. McCarthy, ''History of Black Catholic Education in Chicago, 1871–1971'' (Loyola University, Ph.D., diss., 1971), 76; Anna M. McGarry, ''Parish in A Changing Neighborhood,'' *Interracial Review,* 27 (February 1954), 26; Roy Wilkins to Archbishop Spellman, February 8, 1940, ''Catholics-1940–1955'' file, box 167, series A, group II, NAACP; Roy Wilkins with Tom Mathews, *Standing Fast: The Autobiography of Roy Wilkins* (New York: Viking Press, 1982), 81; Willie Agatha Backus, ''A Case Study of Twenty Negro Catholic Families in Saint Charles Borromeo Parish, Harlem, Receiving Aid to Dependent Children because of Absence of the Father from the Home'' (Fordham University, M.S.S. thesis, 1952), 49; Rev. William J. Culleen, ''The Monsignor—Who Could Forget Him?'' *Stray Notes,* 37 (April 1960), 65.

8. For example, Vance Packard, *The Status Seekers* (New York: D. McKay, 1959), 204.

9. St. Clair Drake and Horace Cayton, *Black Metropolis: A Study of Life in a Northern City* (Chicago: University of Chicago Press, 1993), 413–15.

10. Cecily Mara O'Connor, ''A Negro Catholic Parish: A Study of the Parish of St. Mark the Evangelist in Its Religious, Educational and Social Aspects'' (Fordham University, M.S.S. thesis, 1937), 33.

11. E. J. Fitzmaurice, bishop of Wilmington, to Tenelly, November 1, 1939, folder 4, box 13, series 5/2, BCIM.

12. Quoted in Sr. Mary Caroline Hemesath, ''A History of the Whites' Policy Toward the Chicago Negro from 1865 to the Present Time with Particular Reference to Religious Factors'' (Catholic University of America, M.A. thesis, 1941), 96; Rev. Henry Thiefels, C.S.P., ''Report'' [1939], Sacred Heart Parish file, AAD, cited in Tentler, *Seasons of Grace,* 573; Fr. Philip Steffes to Archbishop Kiley, April 11, 1944, St. Benedict the Moor file, AAM.

13. *Information In Regard to the Proposed South Park Gardens Housing Project* (Chicago: Chicago Housing Authority, 1938), 60. The most thorough treatment of work in Holy Angels parish is Fr. Patrick Curran, ''Missionary Apostolate Among the non-Catholic Negroes in Chicago,'' *Christ to the World,* 5 (1960),

295–308. Also see Avella, 283–88; Fr. Joseph Richards, interview with Thomas Joyce, December 10, 1974, AAC; Dempsey J. Travis, *An Autobiography of Black Chicago* (Chicago: Urban Research Institute, 1981), 32.

14. Sr. Mary Clarice Sobczyk, O.S.F., "A Survey of Catholic Education for the Negro of Five Parishes in Chicago" (DePaul University, M.A. thesis, 1954), 23–50; Fr. Joseph Richards, interview with Thomas Joyce, December 10, 1974; Rev. Joseph M. Barry, O.M.I., to Bishop John Noll, May 14, 1949, folder 1, St. Monica's parish file, Archives of the Diocese of Gary; William Hogan, interview with author, January 4, 1991.

15. Fr. Joseph Richards, interview with Thomas Joyce, December 10, 1974; Curran, "Missionary Apostolate," 299–300; Sobczyk, "Survey," 49–50.

16. Drake and Cayton, *Black Metropolis,* 413–15, 685; "Catholics and Color," *Ebony,* 1 (November 1945), 21; Harold Fey, "Catholicism and the Negro," *The Christian Century,* 61 (December 20, 1944), 1477.

17. On conversion statistics, Stephen Ochs, *Desegregating the Altar: The Josephites and the Struggle for Black Priests* (Baton Rouge: Louisiana State University Press, 1990), 461; Robert McClory, "Church Loses Ground," *National Catholic Reporter,* 13 (February 4, 1977), 1. One 1963–64 survey estimated New York's African-American population to be 10.7 percent Catholic, with much higher percentages for migrants from the Caribbean. See Nathan Kantrowitz, *Ethnic and Racial Segregation in the New York Metropolis: Residential Patterns Among White Ethnic Groups, Blacks, and Puerto Ricans* (New York: Praeger, 1973), 28. On converts as a percentage of parish membership, see Rev. Clarence Williams, C.PP.S., "Black Catholicism in Chicago: An Oral History," *Journal of Urban Ministry,* 1 (Winter 1979), 27–28; Frank Fahey, "The Sociological Analysis of a Negro Parish," (University of Notre Dame, Ph.D. dissertation, 1959), 87.

18. "417 are Confirmed at Two Churches," *New York Times,* June 19, 1939, p. 8; Lester Gaither, "A Negro Youth Center: Saint Charles Borromeo Youth Project, Harlem, New York, 1950–1953" (Fordham University, M.S.S. thesis, 1954), 37–41.

19. Dorothy Ann Blatnica, V.S.C., " 'In Those Days': African-American Catholics in Cleveland, 1922–1961" (Case Western Reserve University, Ph.D. diss., 1992), 69–71, 75.

20. Arthur G. Falls, M.D., "How Would Negroes Vote?" *America,* 50 (October 7, 1933), 21; Claude Barnett to Dr. Harry V. Richardson, July 16, 1949, reel 5, series J, Claude Barnett Papers, microfilm copy; Drake and Cayton, *Black Metropolis,* 685; Taylor Branch, *Parting the Waters: America in the King Years 1954–1963* (New York: Simon and Schuster, 1988), 366.

21. "Respect the Church," *The Council Review,* 6 (June 1927), 3–4; Fr. Joseph Richards, interview with Thomas Joyce, December 10, 1974.

22. *Celebrating Our Restoration: St. Peter Claver R.C. Church* (Brooklyn, 1991), copy in ADB. Joseph G. McGroarty, "Negro Sanctity," *Catholic World,* 164 (October 1946), 32–40; Curran, "Missionary Apostolate," 306–7.

23. Ted Le Berthon, "Why Jim Crow Won at Webster College," *Pittsburgh Courier,* February 5, 1944, p.1; "Negro Expounds Catholic Doctrine at Health Camp," *Buffalo Star,* July 26, 1946, p.1. On African-American conversion to Catholicism, see Daniel F. Collins, "Black Conversion to Catholicism: Its Implica-

tions for the Negro Church,'' *Journal for the Scientific Study of Religion,* 10 (1971), 208–18.

24. Dorothy Ann Blatnica, V.S.C., *''At the Altar of Their God'': African-American Catholics in Cleveland, 1922–1961* (New York: Garland, 1995), 26; E. Franklin Frazier, *The Black Bourgeoisie: The Rise of a Negro Middle Class in the United States* (New York: The Free Press, 1957), 79--80, 110–11; Negro Scholarship applications for academic year 1958, file December 16–31, 1957, box 20, CIC.

25. Cornelius J. Ahern to LaFarge, March 25, 1935, folder 6, box 52, JLF.

26. Rev. Capistran J.Haas, O.F.M., *History of the Midwest Clergy Conference on Negro Welfare* (Chicago: Franciscan Press, [c. 1963]), 3–15.

27. ''Report Presented to the Commission for the Catholic Missions Among the Colored People,'' folder 6, box 8, series 5/2, BCIM; McIntyre to LaFarge, October 16, 1946, folder 22, box 17, JLF.

28. Msgr. Daniel Cantwell, interview with Steven Avella, December 14, 1983, AAC; Ed Marciniak, interview with Steven Avella, December 14, 1983, AAC; Stritch to Msgr. Daniel M. Cantwell, October 18, 1946, box 2969, Chancery correspondence, AAC.

29. Thomas J. Harte, C.Ss.R., *Catholic Organizations Promoting Negro-White Race Relations in the United States* (Washington, D.C.: Catholic University of America Press, 1947), 56-57; idem, ''Leadership Techniques of a White Pastor in a Negro Parish'' (Catholic University of America, M.A. thesis, 1944), 21; Margaret Roach, interview with author, December 13, 1990.

30. Ethel Lee to Brother Anselm, December 3, 1937, Dougherty correspondence, PAHRC. Also see Cecily Mara O'Connor, ''A Negro Catholic Parish: A Study of the Parish of St. Mark the Evangelist in Its Religious, Educational, and Social Aspects'' (Fordham University, M.A. thesis, 1937), 61.

31. Richard J. Roche, *Catholic Colleges and the Negro Student* (Washington, D.C.: Catholic University of America Press, 1948), 123; Blatnica, *''At the Altar of Their God,''* 116–20.

32. Atkielski to Fr. Louis Pastorelli, January 12, 1940, St. Benedict the Moor file, AAM; Rev. Vincent A. Dever to Schrembs, August 8, 1942, Our Lady of the Blessed Sacrament file, ADC.

33. *Divini Redemptoris* in Carlen, *The Papal Encyclicals: 1903–1939,* 537.

34. ''On Bringing Modern Society Back to Christ,'' *Acta Romana* of 1938, volume 9, reprinted in *Institute of Social Order Bulletin,* 1 (December 1943), 1.

35. LaFarge, *The Manner is Ordinary,* 246; Anna McGarry to Arthur Fauset, November 2, 1938, CIC correspondence file, box 1, AMC. General discussions of Catholic anti-communism can be found in Robert L. Franck, ''Prelude to Cold War: American Catholics and Communism,'' *Journal of Church and State,* 34 (Winter 1992), 39–56; David O'Brien, *American Catholics and Social Reform: The New Deal Years* (New York; Oxford University Press, 1968), 70–96; Donald F. Crosby, S.J., *God, Church and Flag: Senator Joseph R. McCarthy and the Catholic Church, 1950–1957* (Chapel Hill: University of North Carolina Press, 1978), 3–26.

36. ''Bolshevism and the Negro,'' *The Chronicle,* 3 (May 1930), 109.

37. John T. Gillard, S.S.J., *Christ, Color and Communism* (Baltimore: The Josephite Press, 1937); Norbert Georges, O.P., ''Pitching in Pittsburgh,'' *The*

Torch (January 1943), 3–4; Joseph F. Thorning, S.J., "Communism in the United States," *America,* 65 (September 21, 1935), 559–61.

38. Preuss in *The True Voice,* October 8, 1937, clipping enclosed in "Report on Negro Work and Application for Aid," September 23, 1937, folder 4, box 9, series 5/2, BCIM.

39. Ethel L. Lee to Cardinal Dougherty, January 9, 1938, Dougherty correspondence, PAHRC.

40. Ochs, *Desegregating the Altar,* 396; "Minutes of the Chicago-Missouri Province Meeting on Communism and Atheism," West Baden College, West Baden, Ind., June 22–26, 1935, folder 32, box 4, JLF; John LaFarge, "Catholics and the Negro," *Interracial Review,* 10 (September 1939), 139.

41. The most comprehensive source is Elizabeth Louise Sharum, "A Strange Fire Burning: A History of the Friendship House Movement" (Texas Tech University, Ph.D. diss., 1977); also see Catherine de Hueck, *Friendship House* (New York: Sheed and Ward, 1947), 8, 111.

42. Allan A. Archibald, "Many are Called . . . but," *Harlem Friendship House News,* 3 (October 1943), 1/4; David James, interview with author, January 11, 1991; "Friendship House Clarifies," *Harlem Friendship House News,* 2 (March 1943), 2; de Hueck, *Friendship House,* 66.

43. Leo V. Miller to LaFarge, August 28, 1946, folder 2, box 29, JLF. John Lewis Gaddis concludes that Catholics provided the "most vocal center of skepticism concerning Soviet ideological intentions" during the war. See Gaddis, *The United States and the Origins of the Cold War, 1941–1947* (New York: Columbia University Press, 1972), 52–54.

44. John F. Cronin, S.S., *Catholic Social Principles: The Social Teachings of the Church Applied to American Economic Life* (Milwaukee: Bruce, 1950), 82; idem, "The Problem of American Communism in 1945," iii, file 26, box 1, Fr. John F. Cronin Papers, MAUND; Garry Wills, *Bare Ruined Choirs: Doubt, Prophecy, and Radical Religion* (Garden City: Doubleday, 1972), 234; Joshua B. Freeman and Steve Rosswurm, "The Education of an Anti-Communist: Father John F. Cronin and the Baltimore Labor Movement," *Labor History,* 33 (Spring 1992), 217–47; Mary C. Golden to John LaFarge, March 14, 1948, "1946–1948" file, box 1, AMC.

45. John LaFarge to Mrs. D.P. McDonald, July 30, 1943, file 9, box 29, JLF; Mrs. D.P. McDonald to LaFarge, July 19, 1943, file 9, box 29, JLF.

46. Gunnar Myrdal, *An American Dilemma: The Negro Problem and Modern Democracy* (New York: Harper & Brothers, 1944), xlvii; John LaFarge, *The Race Question and the Negro: A Study of the Catholic Doctrine on Interracial Justice* (New York: Longmans, 1943), 290.

47. Myrdal, *An American Dilemma,* xlviii; Walter A. Jackson, *Gunnar Myrdal and America's Conscience: Social Engineering and Racial Liberalism, 1938–1987* (Chapel Hill: University of North Carolina Press, 1990), 135–85.

48. R. M. MacIver, "Group Images and Group Realities" and "Summation," in R. M. MacIver, ed., *Group Relations and Group Antagonisms: A Series of Addresses and Discussions* (New York: Harper & Brothers, 1944), 7, 216; Robin M. Williams, Jr., *The Reduction of Intergroup Tensions: A Survey of the Research on Problems of Ethnic, Racial, and Religious Group Relations* (New York: Social Science Research Council, 1947), 7. Much of this paragraph is indebted to Philip

Gleason, "Americans All: World War II and the Shaping of American Identity," *Review of Politics,* 43 (October 1981), 483–518, and John Higham, *Send These to Me: Immigrants in Urban America* (Baltimore: Johns Hopkins University Press, 1984), 198–232.

49. George L.A. Reilly, *Sacred Heart in Bloomfield: A Diamond Jubilee History of the Parish, 1878–1953* (Bloomfield, N.J.: Sacred Heart Church, 1953), 169–71; "The Cross and the Flag," from St. Mary of the Annunciation parish, copy in St. John's Seminary Library, Boston, Mass. Seventeen hundred and fifty-nine men of Brooklyn's Our Lady of Perpetual Help served in the conflict. Twenty had died by 1943. See John F. Byrne, C.Ss.R., *A Golden Harvest: Church of Our Lady of Perpetual Help* (Brooklyn, 1943), 69–70, 133–42.

50. Francis J. Spellman, "Bigotry Is Un-American," *Catholic Digest,* 9 (December 1944), 1–6; "Archbishop Spellman Urges Every True American to Fight Bigotry," *Interracial Review,* 18 (November 1945), 158–59. On the Lincoln Memorial ceremony, J. M. Hayes to Msgr. Carroll, February 10, 1943, "Social Action—Race Relations, 1929–1947" file, box 89, NCWC.

51. Harte, *Catholic Organizations Promoting Negro-White Relations in the United States,* 22, 131; "A Manifest Duty," *Commonweal,* 39 (December 10, 1943), 196; Margaret Mead, "How Religion Has Fared in the Melting Pot," in *Religion and Our Racial Tensions,* Willard L. Sperry, ed. (Cambridge: Harvard University Press, 1945), 61–62.

52. J. L. O'Sullivan, Foreword, in *A Catholic Church in America: Or One Priest to Another* (Milwaukee: St. Benedict's Press, 1942), 4; Guichard Parris, "The Negro Worker in the Postwar World," *Stray Notes,* 7 (June 1945), n.p.; Albert Sidney Foley, "The Catholic Church and the Washington Negro" (University of North Carolina, Ph.D. diss., 1950), 144, copy in ACUA.

53. Marion Bruce to Executive Secretary, October 19, 1945, Social Actions—Race Relations file, box 89, NCWC. Ms. Bruce reacted to the use of Ryan's speech in a 1945 address by Archbishop Rummel. On the Howard speech, Mr. Jacob Billikaph to Ryan, March 4, 1943, Ryan correspondence (unprocessed) ACUA; Rt. Rev. John Augustine Ryan, "The Place of the Negro in American Society," March 2, 1943, copy in "Social Action—Race Relations, 1929–1947" file, box 89, NCWC.

54. John LaFarge, S.J., *The Race Question and the Negro* (New York: Longmans, Green, 1943), 274.

55. W.L. Clayton, "In The Course of Events," *The Chronicle,* March 10, 1945; "New Parish for Colored Catholics," *Boston Post,* May, 1945—clippings in Josephite Archives, Baltimore, Maryland.

56. Cassius J. Foster to Edward Marciniak, January 16, 1945, folder 1, box 1, CIC.

57. From 1942 course taught by LaFarge entitled "Racism in America," file 3, box 10, JLF. Charles L. Rawlings, "Detroit Interracial Council," *Interracial Review,* 19 (November 1946), 166–67; Archibald F. Glover, "The Negro Looks at St. John's," *Interracial Review,* 11 (June 1938), 90–92; "The Brooklyn Catholic Interracial Council," *Interracial Review,* 18 (July 1945), 102–3.

58. Mr. Charles Houston to Mr. Roy Wilkins, March 25, 1938, #0501, reel 4, series B, part 11, NAACP, microfilm copy; George Hunton to Charles Houston, April 20, 1938, #0505, reel 4, series B, part 11, NAACP, microfilm copy.

59. Rev. Frederick McTernan to Rev. Raymond McGowan, "Negro Seminar" file, box 39, SAD. Vatican officials also requested periodic updates on racial issues from staffers at the National Catholic Welfare Conference. See, for example, A. G. Cicognani to Msgr. Howard Carroll, January 22, 1948, "Social Action—Civil Rights, 1938–1949" file, box 85, NCWC. Haas, *History of Midwest Clergy Conference on Negro Welfare,* 18–19; *Minutes of the Annual Meetings of the Bishops of the United States, 1936–1946,* November 15, 1946; Rev. Paul R. E. Francis, S.T.D., "Outflanking Prejudice," *Interracial Review,* 29 (February 1956), 23.

60. NCWC news release, November 26, 1945, "Negro Seminar" file, box 39, SAD.

61. "Archbishop Rummel's Address at Xavier," *Colored Harvest* (April-May 1945), 12.

62. On "racial clannishness," see Daniel M. Cantwell, "Race Relations—As Seen By A Catholic," *American Catholic Sociological Review,* 7 (December 1946), 244.

63. For a variety of perspectives on these matters, see Donald J. Kemper, "Catholic Integration in St. Louis, 1935–1947," *Missouri Historical Review,* 73 (October 1978), 1–22; George H. Dunne, S.J., *King's Pawn: The Memoirs of George H. Dunne,* (Chicago: Loyola University Press, 1990), 80–81; Jeffrey Smith, *From Corps to Core: The Life of John P. Markoe, Soldier, Priest, and Pioneer Activist* (Florissant, Mo.: St. Stanislaus Historical Museum, 1977), 75–76; William Markoe, "Reflections of My Experience in the Field of Race Relations," box 4, series 103, Fr. William Markoe Papers, Marquette University Special Collections. A summary of events in St. Louis and their implications for the Jesuits is contained in Peter McDonough, *Men Astutely Trained: A History of the Jesuits in the American Century* (New York: The Free Press, 1992), 181–97; William Barnaby Faherty, S.J., "Breaking the Color Barrier," *Universitas,* 13 (Autumn 1987), 18–21.

64. George H. Dunne, "The Sin of Segregation," *Commonweal,* 42 (September 21, 1945), 542–45; idem, "Racial Segregation Violates Justice," *America,* 74 (October 20, 1945), 64–65; also see John E. Coogan, S.J., "Christian Untouchables?" *Review for Religious,* 5 (1946), 107–13. On the discovery of a "new" sin, Thomas H. Clancy, "Feeling Bad About Feeling Good," *Studies in the Spirituality of the Jesuits,* 11:1 (January 1979), 19; Dunne, *King's Pawn,* 130–31.

65. Foley, "The Catholic Church and the Washington Negro," 221; Harte, *Catholic Organizations Promoting Negro-White Relations in the United States,* 22; Detroit Catholic Women's Interracial Council Newsletter, (November 1945) in "Social Action—Race Relations, 1929–1947" file, box 89, NCWC.

66. Charles S. Johnson, "Review of the Month," *A Monthly Survey of Events and Trends in Race Relations,* 1 (August 1943), 1; Michael J. Ready to LaFarge, December 11, 1941, folder 24, box 51, JLF.

67. Frank Fahey, "The Sociological Analysis of a Negro Catholic Parish" (University of Notre Dame, Ph.D. diss., 1959), 40–59. Unpublished autobiography of Fr. Vincent Thilman, C.S.C, Thilman file, Archives of the Congregation of the Holy Cross, Indiana Province, Notre Dame, Indiana.

68. "White Kids Rebuff Cleric, Elect Negro Boy as Mayor," *Chicago Defender* (March 13, 1943), and *Defender* clippings in file 9, box 28, Msgr. Daniel

Cantwell Papers, CHS; Peter D'Agostino, "Missionaries in Babylon: The Adaptation of Italian Priests to Chicago's Church, 1870–1940" (University of Chicago, Ph.D. diss., 1993), 394.

69. Niles Carpenter and Daniel Katz, "The Cultural Adjustment of the Polish Group in the City of Buffalo: An Experiment in the Technique of Social Investigation," *Social Forces,* 6 (September 1927), 81.

70. William L. Evans, *Race Fear and Housing in a Typical American Community* (New York: National Urban League, 1946), 20–22.

71. "Buffalo Residents Protest New Site of Housing Project," *Buffalo Evening News,* February 20, 1942, p.25.

72. "Mead, Butler Asked to Help Stop Plans for Negro Housing," *Buffalo Evening News,* August 14, 1941, p.23; Evans, *Race Fear,* 22.

73. Dominic Capeci, *Race Relations in Wartime Detroit: The Sojourner Truth Housing Controversy of 1942* (Philadelphia: Temple University Press, 1984), 89–92, 94–101; Allen Clive, *State of War: Michigan in World War II* (Ann Arbor: University of Michigan Press, 1979), 133, 144–49.

74. Office of War Information: Special Report on Negro Housing, Bureau of Intelligence, March 5, 1942, box 9, Rensis Lickert Papers, Bentley Library, University of Michigan. The Catholic councilman who pledged to work for white occupancy of the homes was a former professional baseball player. The African-American minister leading the protest promised to send him "back to the minors where he belongs." See John Wood, "I Cover the Town," *Michigan Chronicle* (February 14, 1952), 4, notes on article supplied by Dominic Capeci.

75. *Golden Jubilee volume of St. Louis the King Parish* (Detroit, 1973), St. Louis the King parish file, AAD; "Canonical Visitation of 1940," St. Louis the King parish file, AAD; Interviews with Fr. Djuik and Miss Clara Swieczkowska in "Survey of Racial and Religious Conflicts in Detroit" (1943), Civil Rights Congress of Michigan Collection, box 71, Archives of Labor and Urban Affairs, Wayne State University.

76. Interview with Jack Raskin, [1942] Sojourner Truth File, Detroit File 44–49, Federal Bureau of Investigation.

77. Djiuk in Charles S. Johnson and Associates, *To Stem This Tide,* 53.

78. Swieczkowska, "Survey of Racial and Religious Conflicts in Detroit."

79. LaFarge to John Ryan, November 16, 1943, folder 23, box 51, JLF.

80. Swieczkowska, "Survey of Racial and Religious Conflicts in Detroit," (1943); LaFarge to John N. May, August 4, 1943, folder 23, box 51, JLF.

81. Interview with Louis Martin, [1942] Sojourner Truth File, Detroit File 44–49, Federal Bureau of Investigation.

82. Mooney to LaFarge, folder 21, box 51, June 15, 1943, JLF; *Our Bishops Speak: National Pastoral Letters and Annual Statements of the Hierarchy, 1919–1951* (Milwaukee: Bruce, 1952), 119–20; CTF to Roy Wilkins, February 7, 1942, file "Housing General—Jan-Feb 1942," box 233, series A, Group II, NAACP; Leo J. Trese, "Saint Philip, Pray for Us," *Commonweal,* 41 (January 2, 1945), 404–5.

83. Mooney to LaFarge, folder 21, box 51, June 15, 1943, JLF; LaFarge to John N. May, August 4, 1943, folder 23, box 51, JLF.

84. "Protest Letters Call Housing Fight Poor Democracy," *Buffalo Evening News,* August 20, 1941, p.8.

85. "Catholic Churches Amalgamated!" *Buffalo Star,* July 19, 1946, p.11; "Catholics Set the Pace," *Buffalo Star,* July 19, 1946, p.11.

86. A. J. Smitherman in "Catholic-Negro Relations: An Editorial Roundtable," *Opportunity,* 25 (Summer 1947), 139.

87. Roi Ottley, *'New World A-Coming': Inside Black America* (Boston: Houghton Mifflin, 1943), 266; Robert Bradley, Jr., letter to editor, *Michigan Chronicle* (February 14, 1942), 14, from notes given to author by Dominic Capeci; Eleanor Roosevelt to Msgr. Robert F. Keeganin, January 21, 1942, file 4.1, box 3, Cardinal Mooney Papers, AAD; Capeci, *Race Relations in Wartime Detroit,* 134.

88. "The UnGodly," *Michigan Chronicle* (February 14, 1942), from notes given to author by Dominic Capeci.

89. Marie Oresti, interview with author, September 19, 1990; Marie Conti, "Racism and Religion," *The Catholic Worker,* 9 (March 1942), 3; Bette Smith Jenkins, "The Racial Policies of the Detroit Housing Commission and Their Administration" (Wayne State University, M.A. thesis, 1950), 95; Capeci, *Race Relations in Wartime Detroit,* 134; Coogan to Fr. Donovan, May 31, 1944, folder 23, box 51, JLF.

90. "Vatican Fights Racism," *Interracial Review,* 11 (July 1938), 100. Marie Conti, "Racism and Religion," 3. A discussion of this issue in a general sense is found in Peter Hebblethwaite, "The Popes and Politics: Shifting Patterns in 'Catholic Social Doctrine,' " *Daedalus,* 111 (Winter 1982), 85–98.

91. Rev. Francis J. Gilligan, *The Morality of the Color Line: An Examination of the Rights and the Wrongs of the Discrimination Against the Negro in the United States* (Washington, D.C.: Catholic University of America Press, 1928), 134–37.

92. See Lizabeth Cohen, *Making a New Deal: Industrial Workers in Chicago, 1919–1939* (New York: Cambridge University Press, 1990).

93. On Catholic chaplains, Donald F. Crosby, S.J., *Battlefield Chaplains: Catholic Priests in World War II* (Lawrence: University Press of Kansas, 1994).

94. William L. Evans, *Race Fear and Housing in a Typical American Community* (New York: National Urban League, 1946), 22.

Chapter 4

1. Will Herberg, *Protestant, Catholic, Jew: An Essay in American Religious Sociology* (Garden City: Anchor Books, 1960 [rev. ed.]), 153–54; Mary Gordon, "Getting Here from There: A Writer's Reflection on a Religious Past," in *Spiritual Quests: The Art and Craft of Religious Writing,* William Zinsser, ed. (Boston: Houghton Mifflin, 1988), 34; Andrew Greeley, *The American Catholic: A Social Portrait* (New York: Basic Books, 1977), 214–15.

2. John Thomas, S.J., *The American Catholic Family* (Englewood Cliffs: Prentice-Hall, 1956), 8, cited in Peter McDonough, *Men Astutely Trained: A History of the Jesuits in the American Century* (New York: The Free Press, 1992), 435.

3. Liston Pope, "Religion and the Class Structure," *The Annals of the American Academy of Political and Social Science,* 256 (March 1948), 86–90; Hadley Cantril, "Educational and Economic Composition of Religious Groups: An Analysis of Poll Data," *American Journal of Sociology,* 48 (1942–43), 574–79.

4. Brother Gerald J. Schnepp, *Leakage From a Catholic Parish* (Washington, D.C.: Catholic University of America Press, 1942), 169, 284, 309; Stephan Thernstrom, *The Other Bostonians: Poverty and Progress in The American Metropolis, 1880–1970* (Cambridge: Harvard University Press, 1973), 138, 153; Patrick Blessing, "The Irish," in *The Harvard Encyclopedia of American Ethnic Groups*, Stephen Thernstrom, ed. (Cambridge: Harvard University Press, 1980), 539–40.

5. William P. Leahy, S.J., *Adapting to America: Catholics, Jesuits, and Higher Education in the Twentieth Century* (Washington, D.C.: Georgetown University Press, 1991), 126; Ronald W. Schatz, "Connecticut's Working Class in the 1950s: A Catholic Perspective," *Labor History*, 25 (1984), 90; Greeley, *The American Catholic*, 43–47, 53–67.

6. Herberg, *Protestant, Catholic, Jew*, 157–58. The classic statement on the triple melting pot is Ruby Jo Kennedy, "Single or Triple Melting Pot? Intermarriage Trends in New Haven, 1870–1950," *American Journal of Sociology*, 58 (July 1952), 56–59. For a summary of debate sparked by the article, see Michael Hout and Joshua R. Goldstein, "Demographic and Subjective Aspects of Ethnic Identity," *American Sociological Review*, 59 (1994), 70–72.

7. Thaddeus C. Radzialowski notes how one order of nuns abandoned Polish in the schools after receiving letters from former students serving in the military during World War II. To the nuns' dismay, their former students could not advance to higher positions, or even go to confession, because of poor English skills. Radzialowksi, "Reflections on the History of the Felicians in America," *Polish American Studies*, 32 (Spring 1975), 26–27. On problems with the catechism and Polish language issues more generally, see Sr. Marie Ellen Kuznicki, "An Ethnic School in American Education: A Study of the Origins, Development, and Merits of the Educational System of the Felician Sisters in the Polish-American Catholic Schools of Western New York" (Kansas State University, Ph.D. diss., 1972), 167.

8. Agnes E. Meyer, "Orderly Revolution," *Washington Post*, June 5, 1945, p. 9.

9. Rev. Thomas J. Harte, C.Ss.R., "Racial and National Parishes in the United States," in *The Sociology of the Parish*, C. J. Nuesse and Rev. Thomas J. Harte, C.Ss.R., eds. (Milwaukee: Bruce, 1950), 164; W. Lloyd Warner and Leo Srole, *The Social System of American Ethnic Groups* (New Haven: Yale University Press, 1945), 294–95.

10. "Self-Segregation and Negroes," *Interracial Review*, 32 (June 1959), 110.

11. Stephen Joseph Shaw, "Chicago's Germans and Italians, 1903–1939: The Catholic Parish as a Way-Station of Ethnicity and Americanization" (University of Chicago, Ph.D. diss., 1981), 257.

12. *Spiritual Care of Puerto Rican Migrants: Report on the First Conference, Held in San Juan, Puerto Rico, April 11th to 16th, 1955*, William Ferree, Ivan Illich, Joseph P. Fitzpatrick, eds. (Cuernavaca, Mexico: Centro Intercultural de Documentation, [1955] 1970), 7, 11, 41–45; Jaime R. Vidal, "The Rejection of the Ethnic Parish Model," in *Puerto Rican and Cuban Catholics in the U.S., 1900–1965*, Jay P. Dolan and Jaime R. Vidal, eds. (Notre Dame: University of Notre Dame Press, 1994), 70–87; David A. Badillo, "The Catholic Church and the Making of Mexican-American Parish Communities in the Midwest," in *Mexican-Americans and the Catholic Church, 1900–1965*, Jay P. Dolan and Gil-

berto Hinojosa, eds. (Notre Dame: University of Notre Dame Press, 1994), 254–57, 264–65; Allan Figueroa Deck, "The Crisis of Hispanic Ministry: Multiculturalism as an Ideology," *America*, 163 (July 14–21, 1990), 34–35.

13. "Los Caballeros de San Juan," [c.1957], no author, file 151, SA.

14. R. M. MacIver, *The More Perfect Union* (New York: Macmillan, 1948), 12.

15. John T. McGreevy, "Catholicism in the American Intellectual Imagination, 1935–1955," unpublished essay. On the popularity of the Pope with American troops and journalists during the war, see William L. Vance, *America's Rome*, vol. 2, *Catholic and Contemporary Rome* (New Haven: Yale University Press, 1989), 52.

16. On development within the city limits after the war, Jon C. Teaford, *The Rough Road to Renaissance: Urban Revitalization in America, 1940–1985* (Baltimore: Johns Hopkins University Press, 1990), 127–28; on the Catholic propensity to remain in neighborhoods, see Rev. Robert G. Howes, "Baltimore Parish Study" (September 1967), file 3, box 5, John Cardinal Dearden Papers, MAUND; Andrew Greeley, *The Catholic Myth: The Behavior and Beliefs of American Catholics* (New York: Collier Books, 1990), 156–157; Gerhard Lenski, *The Religious Factor* 194–97. On the Catholic birthrate, James D. Cowhig and Leo F. Schnore, "Religious Affiliation and Church Attendance in Metropolitan Centers," *American Catholic Sociological Review*, 23 (Summer 1962), 113–27; Ronald Freedman, Pascal K. Whelpton, John W. Smit, "Socio-Economic Factors in Religious Differentials in Fertility," *American Sociological Review*, 26 (August 1961), 608–14; Charles Westoff, "The Blending of Catholic Reproductive Behavior," in *The Religious Dimension: New Directions in Quantitative Research*, Robert Wuthnow, ed. (New York: Academic Press, 1979), 236–39; on Catholics and suburbs, Paul Brindel, "A Pox on Suburbia," *The Voice of St. Jude*, 22 (April 1957), 7; for the nine million figure, Kenneth T. Jackson, *Crabgrass Frontier: The Suburbanization of the United States* (New York: Oxford University Press, 1985), 238.

17. *Welcome to Boston* (Boston, 1962), 13; Herberg, *Protestant, Catholic, Jew*, 50; Andrew Greeley, *The Church and the Suburbs* (New York: Paulist Press 1963), 31, 58.

18. Herbert Gans, *The Levittowners: Ways of Life and Politics in a New Suburban Community* (New York: Columbia University Press, 1982), 72.

19. See Robert Orsi, "The Center Out There, In Here, and Everywhere Else: the Nature of Pilgrimage to the Shrine of St. Jude," *Journal of Social History*, 25 (Winter 1991), 213–32. On the block rosary, "Block Rosary," *Sign*, 31 (November 1951), 29–30; "Block Rosary in Youngstown," *Ave Maria*, 69 (February 19, 1949), 229; Maria McSherry, "Why Don't You Start the Block Rosary?" *Ave Maria*, 72 (August 19, 1950), 248.

20. Rev. John O'Grady, "The Changing Parish" (Washington, D.C.: National Conference of Catholic Charities, 1955, n.p.). On the general decrease in residential mobility after the war, see Ronald Tobey, Charles Wetherell and Jay Brigham, "Moving Out and Settling In: Residential Mobility, Home Owning, and the Public Enframing of Citizenship, 1921–1950," *American Historical Review*, 95 (1990), 1395–1422.

21. Timothy Richard Allan, "Roman Catholicism and Inner Urban Demographic Change in Buffalo, New York, 1960–1980: A Quantitative and Interpretive Overview" (SUNY Buffalo, Ph.D. diss., 1992), 68–70, 78; Joseph P. Barnett, *The Sesquicentennial History of Saint Denis Parish, 1825–1975* (Havertown, Pa., 1975), 120, 137, 163; Joseph Casino, "From Sanctuary to Involvement: A History of the Catholic Parish in the Northeast," in *American Catholic Parish: A History From 1850 to the Present,* vol. 1, Jay P. Dolan, ed. (New York: Paulist Press, 1987), 78, 99.

22. James Downs, "Neighborhood Decay: Its Causes and Cures, Part II," *America,* 88 (February 7, 1953), 507.

23. Allan, "Demographic Change in Buffalo," 354.

24. James Sheehan, interview with author, September 9, 1990; Samuel Cardinal Stritch, "Neighborhood Conservation," *Catholic Charities Review,* 36 (November 26, 1952), 34.

25. On Shriver, Nicholas Lemann, *The Promised Land: The Great Black Migration and How It Changed America* (New York: Alfred A. Knopf, 1991), 146; John McDermott and Dennis Clark, "Helping the Panic Neighborhood," *Interracial Review,* 28 (August 1955), 135; Hope Brophy, interview with author, September 19, 1990.

26. "Preliminary Reports and Resolution," file August 26–31, 1958, box 24, CIC; CIC Board Minutes, August 12, 1958, file August 1–25, 1958, box 24, CIC; Zielinksi, 139–182; Notes on NCCIJ Planning Conference, December 30, 1957, folder 7, box 29, Msgr. Daniel Cantwell Papers, CHS.

27. For example, Gerald Kelly, S.J., "Some Moral Aspects of the Negro Question," *Institute of Social Order Bulletin* 3 (December 1945), 9–11. For the argument that integration was a moral obligation, Fr. John Markoe, S.J., "Justice for the Negro," *Institute of Social Order Bulletin,* 4 (June 1946), 13–14.

28. Rev. Yves M.-J. Congar, O.P., *The Catholic Church and the Race Question* (Paris: UNESCO, 1953), 40. John Courtney Murray, "Memorandum" [1945], file 585, box 8, John Courtney Murray Papers, Georgetown University Special Collections; Murray to Mr. Theodore M. Lewis, April 12, 1956, file 139, box 2, Murray Papers.

29. For example, "Unchurched Negroes," *Interracial Review,* 21 (May 1948), 69; "Africa and the United States," *Interracial Review,* 24 (April 1951), 51; Ochs, *Desegregating the Altar,* 422–23, 433–38; Stephen Ryan, "African Bishop in the Deep South," *Interracial Review,* 23 (May 1950), 69.

30. "Vatican Rejects Segregation Bid," *New York Times,* August 9, 1957, p.1; Owen McGrath O.P., "The Theology of Racial Segregation," *Catholic Mind,* 55 (November-December 1957), 486. On the Church and South Africa, *Catholics in Apartheid Society,* Andrew Prior, ed. (Capetown, South Africa: David Philip, 1982). Talks given at the New York Catholic Interracial Council offices during the 1950s reflect interest in African issues. Titles included "Interracial Effort Here and in Africa," and "News of Racial Prejudice in United States Used as Communist Propaganda in Africa." See Sister M. Frieda Gobel, "The Educational Program of the Catholic Interracial Councils" (DePaul University, M.A. thesis, 1955), 47.

31. Unpublished autobiography of Rev. Vincent Thilman, C.S.C., p. 6, Thil-

man file, Archives of the Congregation of the Holy Cross, Indiana Province, Notre Dame, Indiana. Also see Nancy J. Weiss, *Whitney M. Young, Jr. and the Struggle for Civil Rights* (Princeton: Princeton University Press, 1989), 50–51.

32. "Panic Cuts Property Value—Msgr. Navin," *Cleveland Catholic Universe Bulletin* (July 24, 1953); Robert T. Stock, "Attack on Home of Negro Spotlights Housing Need," *Cleveland Catholic Universe Bulletin* (November 13, 1953); E. E. White to Bishop Hoban, November 9, 1943, Our Lady of the Blessed Sacrament file, ADC.

33. Fr. Hubert Roberge, "Integration—It's Here," *Integrity,* 8 (April 1954), 11–18; Leslie Woodcock Tentler, *Seasons of Grace: A History of the Catholic Archdiocese of Detroit* (Detroit: Wayne State University Press, 1990), 509; Alden V. Brown, *The Grail Movement and American Catholicism, 1940–1975* (Notre Dame: University of Notre Dame Press, 1989), 87; Roberge to Rt. Rev. John Donavan, September 13, 1954, St. Leo Parish file, 28.13, AAD.

34. Walter White to Paul Robeson, file "School Incidents—Gary," box 520, Series A, Group II, NAACP; Also see Dr. Marlon Edman, "An Analysis of Factors Causing the School Strike at Gary, Indiana," *A Monthly Summary of Events and Trends in Race Relations,* 3 (December 1945), 148–49.

35. Bishop John Noll to Rev. Paul Tanner, December 1, 1945, file "Social Action—Race Relations, 1929–1947," box 89, NCWC. Also see James H. Tipton, *Community in Crisis: The Elimination of Segregation From a Public School System* (New York: Teachers College, Columbia University, 1953), esp. 39, 108, and Ronald D. Cohen, *Children of the Mill: Schooling and Society in Gary, Indiana, 1906–1960* (Bloomington: Indiana University Press, 1990), 178–90.

36. Tipton, *Community in Crisis,* 139.

37. Vice-Chancellor [Rev. Bernard Kearns] to Roberge, June 28, 1956, file 13, box 28, St. Leo Parish File, AAD; Tentler, *Seasons of Grace,* 508–9.

38. Anthony J. Vader, "Racial Segregation Within Catholic Institutions in Chicago: A Study in Behavior and Attitudes" (University of Chicago, M.A. thesis, 1962), 62; Stritch to Mattimore, January 8, 1957, file 1, box 4, Cardinal Stritch Personal Papers, AAC; Stritch to Mattimore, January 4, 1957, file 1, box 4, Stritch Personal Papers; Mattimore to Rt. Rev. Msgr. E.M. Burke, December 29, 1956, file 1, box 4, Stritch Personal Papers.

39. Sr. Patricia Margaret Judge, "The Negro in Queen of the Angels Parish, Newark, N.J.: A Study of the Development of the Religious and Social Services Extended to the Negro by a Catholic Parish, 1930–1964" (Fordham University, M.S.W. thesis, 1965), 159–61; "Queen of Angels Fire," *Advocate,* July 18, 1958, p.8; Anthony DePalma, "Catholicism Is Embracing More Blacks," *New York Times,* January 24, 1989, II, p.1; Roy M. Gasnick, O.F.M., "Where There Is Hatred: The New St. Benedict the Moor Apostolate," *Interracial Review,* 33 (April 1960), 100–101; "The Task Ahead," *Interracial Review,* 32 (July 1959), 132.

40. Steven M. Avella, *This Confident Church: Catholic Leadership and Life in Chicago, 1940–1965* (Notre Dame: University of Notre Dame Press, 1992), 274.

41. *Pastoral Letters of the American Hierarchy,* Hugh J. Nolan, ed. (Huntington, Ind.: Our Sunday Visitor, 1971), 506–510. On the pastoral letter, Fr. Martin Zielinski, " 'Doing the Truth': The Catholic Interracial Council of New York,

1945–1965'' (Catholic University of America, Ph.D. diss., 1989), 212–21; Rev. John F. Cronin, S.S., interview with Rev. Thomas Blantz, March 17, 1978, MAUND; John F. Cronin to ''Family,'' November 15, 1958, file ''Race,'' box 10, SAD; John F. Cronin, ''Religion and Race,'' *America*, 150 (June 23–30, 1984), 472.

42. Bishop of Richmond [John J. Russell] to Archbishop O'Boyle, October 25, 1958, file ''Race,'' box 10, SAD.

43. Archbishop O'Boyle also apparently requested three separate orders of contemplative nuns to pray for approval of the statement. John F. Cronin to ''Family,'' November 15, 1958, file ''Race,'' box 10, SAD.

44. John F. Cronin to Most Reverend Thomas K. Gorman, D.D., September 10, 1958, file ''Race,'' box 10, SAD; Cronin to Cardinal Mooney, September 4, 1958, file ''Race,'' box 10, SAD.

45. ''What We Have Said Stands: Segregation is Wrong,'' *Michigan Catholic* (May 10, 1956).

46. Anne Klejment, ''In the Lion's Den: The Social Catholicism of Daniel and Phillip Berrigan, 1955–1965'' (SUNY-Binghamton, Ph.D. diss., 1980), 170; ''Catholic-Negro Relations: An Editorial Roundtable,'' *Opportunity*, 25 (Summer 1947), 146.

47. Dennis Clark diary, February 29, 1956, box 1, Clark Papers, MAUND; Hannah Lees, ''How Philadelphia Stopped a Race Riot,'' *The Reporter*, 12 (June 2, 1955), 26–29; John F. Bauman, *Public Housing, Race, and Renewal: Urban Planning in Philadelphia, 1920–1974* (Philadelphia: Temple University Press, 1987), 128–29, 165.

48. ''A Report of Racial Tensions and Disturbances in Levittown, Pa., July-October 1957,'' file ''1957,'' box 1, AMC; David Popenoe, *The Suburban Environment: Sweden and the United States* (Chicago: University of Chicago Press, 1977), 124; Marvin Bressler, ''The Myers' Case: An Instance of Successful Racial Invasion,'' *Social Problems*, 8 (Summer 1960), 126–42.

49. Memo to Intergroup Agencies and Community Leaders from Commission on Human Relations (Philadelphia), September 27, 1960, file September 21–30, box 38, CIC; ''Incident at 2608 South Street—October 1960,'' Philadelphia Commission on Human Relations, file Nov. 1–10, box 38, CIC; Joseph Clark and Jay Apt, ''Priest Disperses Crowd Protesting Negro Move,'' *Philadelphia Daily News*, September 27, 1960; ''Priest Calms Mob at Negro's House,'' *New York Times*, September 28, 1960, p.13.

50. ''Commission Confers with Cardinal,'' Chicago Commission on Human Relations monthly bulletin, [1946] folder 1946–1947, box 1, CIC.

51. Bob Senser, ''Story of a Colorful Parish,'' 1951 reprint from *Work*, copy in file 1951, box 2, CIC.

52. Frank Dorey, ''The Church and Segregation in Washington D.C. and Chicago, Illinois'' (University of Chicago, Ph.D. diss., 1950), 288–89.

53. Avella, *This Confident Church*, 261–63; *The Ambrosian*, 7 (July 2, 1950), folder 3, box 29, Msgr. Daniel Cantwell Papers, CHS; Samuel Stritch to Casey, [1950] Chancery Correspondence, AAC; Mr. and Mrs. Martin McLaughlin to Stritch, July 6, 1950, Chancery Correspondence, AAC; Lucy Chapelle, to Stritch, July 1, 1950, folder 3, box 29, Msgr. Daniel M. Cantwell Papers, CHS.

54. Avella, *This Confident Church*, 266–67; Casmir Marcevicius, '' 'Prudence'

and Our Negro Catholics,'' *Interracial Review,* 24 (January 1951), 9; Fr. Kenneth Brigham, interview with Steven Avella, March 11, 1989, AAC; Daniel Cantwell, ''Housing: Seminar on Negro Problems in the field of Social Action,'' (August 1949) folder April-December, 1949, box 1, CIC; Untitled CIC report, [n.d.] folder April-December, 1949, box 1, CIC; Memo on ''Interracial Disturbances at 7407–7409 South Parkway and 5643 South Peoria Street,'' folder 5, box 7, ACLU; Daniel M. Cantwell, interview with Steven Avella, September 22, 1987, AAC; on St. Columbanus as a locus for opposition to racial change, see Philip A. Johnson, *Call Me Neighbor, Call Me Friend: The Case History of the Integration of a Neighborhood on Chicago's South Side* (Garden City: Doubleday, 1965), 14, 43.

55. *Visitation Parish: Diamond Jubilee* (Chicago, 1961); Alice Harper Collins, *Vis Revisited* (Oak Park, Ill.: JOAL Press, 1990); Rev. Martin Farrell, interview with Steven Avella, September 30, 1983, AAC; Robert Slayton, *Back of the Yards,* 157.

56. Collins, *Vis Revisited,* 42, 152–53, 231–78; *Visitation Parish: Diamond Jubilee* (Chicago, 1961).

57. Arnold Hirsch, *Making the Second Ghetto: Race and Housing in Chicago, 1940–1960* (New York: Cambridge University Press, 1983), 55–56.

58. Commission on Human Relations, ''Interracial Disturbances at 7407–7409 South Parkway and 5643 South Peoria Street,'' December 10, 1949, folder 5, box 7, ACLU; Eileen M. McMahon,''What Parish Are You From? A Study of the Chicago Irish Parish Community and Race Relations, 1916–1970,'' (Loyola University, Ph.D., diss., 1989), 225–26; Rev. Martin Farrell, interview with Steven Avella, September 30, 1983, AAC.

59. ''Anti-Jewish Violence Sweeps South Side,'' *National Guardian,* November 21, 1949, p.3, clipping in folder April-December 1949, box 1, CIC. On arrest records, Hirsch, *Making the Second Ghetto,* 70–84; Chicago Commission on Human Relations, ''Interracial Disturbances at 7407–7409 South Parkway and 5643 South Peoria Street,'' folder 5, box 7 ACLU.

60. Rev. Martin Farrell interview with Steven Avella, September 30, 1983, AAC. Also see Ellen Skerrit, ''The Catholic Dimension,'' in *The Irish in Chicago* (Urbana: University of Illinois Press,1987), 51–53; ACLU report, ''Anti-Negro and Anti-Semitic Violence on the South Side,'' November 25, 1949, folder April-December 1949, box 1, CIC; William Peters, ''Race War in Chicago,'' *The New Republic,* 122 (January 9, 1950), 10.

61. Homer A. Jack, ''Chicago's Violent Armistice,'' *The Nation,* 169 (December 10, 1949), 571–72; ''Irish Rule Chicago,'' *Pittsburgh Courier,* November 26, 1949; Hirsch, *Making the Second Ghetto,* 79–80.

62. People of the State of Illinois et al. v. Charles Burke et al., March 13, 1950, copy in folder 1, box 1, CIC.

63. Untitled CIC report, [n.d.], folder April-December 1949, box 1, CIC; Casimir Marcevicius noted that a number of African-American Catholics disturbed by the incidents quit attending church, in '' 'Prudence' and Our Negro Catholics,'' *Interracial Review,* 24 (January 1951), 9.

64. William Gremley, ''The Scandal of Cicero,'' *America,* 85 (August 25, 1951), 495–97; Hirsch, *Making the Second Ghetto,* 53–55.

65. Walter White, ''This is Cicero,'' *The Crisis*, 58 (August-September 1951), 434–40; Walter White to Pius XII, September 14, 1951, file ''Housing—Illinois,'' box 311, Series A, Group II, NAACP.

66. Daniel M. Cantwell, ''Postscript on the Cicero Riot,'' *Commonweal*, 54 (September 14, 1951), 543; for a more detailed account, see William Gremley, ''Social Change in Cicero,'' folder July-September 1951, box 2, CIC.

67. McNamara interview with Fr. L. W. Kozlowski, October 5, 1951, file October-December 1951, box 2, CIC; Gremley, ''Social Control in Cicero, Ill,'' 38; McNamara interview with Rev. Edward Flannery, September 25, 1951, folder July-September 1951, box 2, CIC; McNamara interview with Rev. E.J. Tony, September 25, 1951, file July-September 1951, box 2, CIC.

68. Frank London Brown, *Trumbull Park* (Chicago: Regnery, 1959) gives a fascinating, thinly veiled fictional account of the terror endured by the African-American residents. Almost thirty years later, one resident of the area was termed the ''bomber'' for his activities during the housing riots. See William Kornblum, *Blue Collar Community* (Chicago: University of Chicago Press, 1974), 73; also see Avella, This *Confident Church*, 269–72.

69. *Golden Jubilee of St. Kevin's Parish* (Chicago: 1935), n.p.

70. Rev. Dennis J. Geaney, O.S.A., ''The Chicago Story,'' *Today Magazine*, October 30, 1954, file October 1954, box 6, CIC; Rt. Rev. Thomas J. Reed to Msgr. George Casey, February 25, 1955, file ''Board Membership,'' box 8, Egan; Grant Application to Fund for the Republic [c. 1955], folder January-March 1954, box 5, CIC; CIC minutes, October 19, 1953, file October-November 1953, box 4, CIC; Lloyd Davis to Judge Kiley, October 13, 1953, file October 11-November 1953, box 4, CIC; Minutes of September 10, 1953, CIC meeting, folder August-October 10, 1953, box 4, CIC.

71. Lloyd Davis, ''Memo to CIC board,'' July 3, 1954, file June-July, 1954, box 6, CIC; ''Upon the Advice of the Vicar General of the Archdiocese of Chicago,'' [1954], folder 1953 clippings, box 5, CIC.

72. ''Meeting with Fr. McHugh,'' October 27, 1955, file October 25-November, box 9, CIC; ''Meeting with Fr. McHugh,'' August 9, 1955, file August-September 15, 1955, box 8, CIC; Mrs. Jack K. Jallings to the Catholic Interracial Council, May 21, 1954, file June-July 1954, box 6, CIC.

73. Alan Paton, ''The Negro in America Today,'' *Collier's*, 134 (October 29, 1954), 75; Joseph Beauharnais, ''Blow Your Top,'' *Daily Calumet*, [1953?] file 1953 clippings, box 5, CIC.

74. Tape in box 77, audio file, CIC.

75. ''Meeting with Fr. McHugh,'' October 27, 1955, file October 25-November, box 9, CIC; Confidential Reports to the ACLU, file 9, box 11, ACLU; untitled and undated CIC memo, file December 1956, box 14, CIC.

76. St. Clair Drake, ''Charges Churches and Schools Fail to Fight Racial Hatred Here,'' letter to the editor, *Chicago Daily News*, August 2, 1957, p.10.

77. Edward A. Marciniak, ''Books for a Sociology Library,'' *American Catholic Sociological Review*, 6 (June 1945), 105–6. Examinations in box 9, carton 2, NYCIC.

78. Fr. Hubert Roberge to Mooney, May 28, 1948, Our Lady of Victory parish file, file 18, box 9, AAD.

79. George to Tom, July 22, 1949, box 1, CIC.

80. Statistics in *Churches and Church Membership in the United States* (New York: National Council of Churches, 1956).

81. Murray H. Leiffer, *The Effective City Church* (New York: Abingdon-Cokesbury, 1949), 97, 112; Ross W. Sanderson, *The Strategy of City Church Planning* (New York: Institute of Social and Religious Research, 1932), 162–64.

82. Moses Nathaniel Delaney, ''The Interaction Between Protestant Churches and Their Social Environment in the Inner City,'' (Drew University, Ph.D. diss., 1959), 259.

83. Figures cited in Marshall Sklare, *America's Jews* (New York: Random House, 1971), 120–21.

84. Eleanor Wolf, ''The Invasion-Succession Sequence as a Self-Fulfilling Prophecy,'' *Journal of Social Issues,* 13 (1957), 47; Luigi Laurenti, *Property Values and Race: Studies in Seven Cities* (Berkeley: University of California Press, 1960), 180–84. On Boston, see Gerald Herbert Gamm, ''Neighborhood Roots: Exodus and Stability in Boston, 1870–1990'' (Harvard University, Ph.D. diss, 1994); Yona Ginsberg, *Jews in a Changing Neighborhood: The Study of Mattapan* (New York: The Free Press, 1975), 41, 51–55; Hillel Levine and Lawrence Harmon, *The Death of an American Jewish Community: A Tragedy of Good Intentions* (New York: The Free Press, 1992), 174–80, 201–2.

85. Judy Goldstein, ''The Inwood Community: Resistance to Ethnic Change-over'' (1972), paper in author's possession; Lee A. Lendt, ''A Social History of Washington Heights'' (1960), paper in author's possession; George Barrett, ''West Side Report: 'No Incidents',''*New York Times Magazine* (February 8, 1959), 68.

86. Membership list from Lawndale Clergy in Christian Action, January 25, 1962 Meeting Minutes, Lawndale file, box 2, Margaret Roach Papers, MAUND; Taylor Branch, ''The Uncivil War,'' *Esquire,* 111 (May 1989), 92–94; statistics from Alpine Wade Jefferson, ''Housing Discrimination and Community Response in North Lawndale (Chicago), Illinois, 1948–1978'' (Duke University, Ph.D. diss., 1978), 64–85; David K. Fremon, *Chicago Politics Ward by Ward* (Bloomington: Indiana University Press, 1988), 158; Hirsch, *Making the Second Ghetto,* 189–95; Audrey Fishbein, ''The Expansion of Negro Residential Areas in Chicago, 1950–1960'' (University of Chicago, A.M. thesis, 1962), 65–66.

87. Cited in Thomas J. Sugrue, ''The Origins of the Urban Crisis: Race, Industrial Decline, and Housing in Detroit, 1940–1960'' (Harvard University, Ph.D. diss., 1992), 182; William E. Hill, ''Field Report'' (1945), file ''Racial Tension, Detroit, Michigan, 1944–1946,'' box 505, series A, group II, NAACP; Lendt, 116–17.

88. Daniel J. Mallette, ''Chicago Housing and the Sin of Segregation'' (St. Mary's Seminary, Mundelein, Ill., M.A. thesis, 1956), 20–21; Dempsey J. Travis, *An Autobiography of Black Chicago,* 128.

89. Mr. Thomas J. Gordon to Stritch, October 31, 1950, Chancery Correspondence, box 2999, AAC; Lemann, *The Promised Land,* 70; Hirsch, *Making the Second Ghetto,* 4–9, 17; Rev. Martin Farrell, interview with Steven Avella, September 30, 1983, AAC.

90. On racial transition, see the various essays in *Housing Desegregation and Federal Policy,* John M. Goering, ed. (Chapel Hill: University of North Carolina Press, 1986) and Douglas S. Massey and Nancy A. Denton, *American Apartheid:*

Segregation and the Making of the Underclass (Cambridge: Harvard University Press, 1993), 37–39. For a summary of this literature, see Brian J. L. Berry, *The Open Housing Question: Race and Housing in Chicago, 1966–1976* (Cambridge, Mass: Ballinger, 1979), 401–21.

91. See the series of articles by James T. Flannery in the *Cleveland Catholic Universe-Bulletin,* especially "What Really Happens When a Neighborhood Breaks Up?" (December 14, 1956), 1, and "If Real Estate People Would Leave Us Alone" (December 21, 1956), 1; also see Dorothy Ann Blatnica, V.S.C., *"At the Altar of Their God": African-American Catholics in Cleveland, 1922–1961* (New York: Garland, 1995), 156–57.

92. Marc Fried, "Grieving for a Lost Home," in *The Urban Condition: People and Policy in the Metropolis,* Leonard J. Duhl, M.D., ed. (New York: Basic Books, 1963), 151–71; John R. Ottensmann and Michael E. Gleeson, "The Movement of Whites and Blacks into Racially Mixed Neighborhoods: Chicago, 1960–1980," *Social Science Quarterly,* 73 (September 1992), 645–62.

93. Mario T. Garcia, "Americans All: The Mexican American Generation and the Politics of Wartime Los Angeles, 1941–1945," *Social Science Quarterly,* 65 (1984), 285, 287.

94. St. Clair Drake and Horace Cayton, *Black Metropolis: A Study of Negro Life in a Northern City* (Chicago: University of Chicago Press, 1993), 601.

95. Statistics in *Local Community Fact Book: Chicago Metropolitan Area, 1960,* Evelyn M. Kitagawa and Karl E. Taeuber, eds. (Chicago: Chicago Community Inventory, University of Chicago, 1963); also see Kornblum, *Blue-Collar Community,* 73.

96. On Puerto Ricans in the Bronx, see Jill Jonnes, *We're Still Here: The Rise, Fall, and Resurrection of the South Bronx* (Boston: Atlantic Monthly Press, 1986), 110–14. On segregation more generally, Douglas S. Massey, "Residential Segregation of Spanish-Americans in United States Urbanized Areas," *Demography,* 16 (November 1979), 553–63.

97. Richard P. Taub, D. Garth Taylor, and Jan D. Dunham, *Paths of Neighborhood Change: Race and Crime in Urban America* (Chicago: University of Chicago Press, 1984), 72–73.

98. On Puerto Ricans, see Joseph Fitzpatrick, *Puerto Rican Americans: The Meaning of Migration to the Mainland* (Englewood Cliffs: Prentice-Hall, 1971), 110–11; Felix Padilla, *Puerto Rican Chicago* (Notre Dame: University of Notre Dame Press, 1987), 126–37; Joan Dee Koss, "Puerto Ricans in Philadelphia: Migration and Accommodation" (University of Pennsylvania, Ph.D. diss., 1965), 90–91.

99. Peter Skerry, *Mexican-Americans: The Ambivalent Minority* (New York: The Free Press, 1993), 34–35.

100. Reprinted in "Interracial Justice and Home Values," *Catholic Mind,* 51 (December 1953), 730–31.

101. Rev. John Quinn Corcoran, "The Disorganized Neighborhood," *Proceedings of the Forty-First Annual National Convention of the Conference of Catholic Charities* (Washington, D.C.: Ransdell Inc. Printers, 1956), 94; John Cogley, "Lost Sheep," *Commonweal,* 55 (October 19, 1951), 30.

102. Clipping enclosed with John McDermott to Rev. Robert J. Hagarty, January 24, 1964, file Jan. 17–24, box 67, CIC; "South Deering Blames Mayor for

Trouble,'' *The Daily Calumet,* January 4, 1955, p.1; Dennis Clark, galley proof of *New City* article,''Summer 1957,'' file June 15–30, box 17, CIC; ''A Report on Racial Tension and Disturbances in Levittown, Pennsylvania,'' July/October 1957, CIC files, box 1, PAHRC.

103. Keynote Address by John LaFarge, NCCIJ conference, August 25, 1958, file August 26–31, 1958, box 24, CIC; Lloyd Davis to David F. Freeman, March 11, 1955, file March 1955, box 7, CIC.

104. William Arndt, ''What a Student Can Do to Combat Racial Prejudice,'' April 21, 1952, April-May 1952 folder, box 3, CIC; Kennedy in ''Can Influence Non-Whites by Mission Spirit'' (1955), clipping in undated file, box 10, CIC.

105. Sister M. Martina Abbot, S.C., *A City Parish Grows and Changes* (Washington, D.C.: Catholic University of America Press, 1953), 81; St. Anthony of Padua parish file 15.9, AAD.

106. Janet Bower, *Older People of St. Boniface Parish: The ''Fruit Belt'' Buffalo, New York* (Buffalo: Catholic Charities of Buffalo, 1957), 7–19.

107. On friendship networks, Edward O. Laumann, *Bonds of Pluralism: The Form and Substance of Urban Social Networks* (New York: John Wiley & Sons, 1973) 106, 202; on ''class'' in Catholic neighborhoods, see Kerby Miller, ''Class, Culture, and Immigrant Group Identity in the United States: The Case of Irish-American Ethnicity,'' in *Immigration Reconsidered: History, Sociology, and Politics,* Virginia Yans McLaughlin, ed. (New York: Oxford University Press, 1990), 119–21.

108. Walter White, ''This is Cicero,'' *The Crisis,* 58 (August-September 1951), 434–40; Dietrich C. Reitzes, ''The Role of Organizational Structures: Union vs. Neighborhood in a Tense Situation,'' *Journal of Social Issues,* 9 (1953), 37–44; Joseph D. Lohman and Dietrich C. Reitzes, ''Note on Race Relations in Mass Society,'' *American Journal of Sociology,* 53 (1952), 240–46; Ira Katznelson, *City Trenches: Urban Politics and the Patterning of Class in the United States* (New York: Pantheon, 1981), 16. On the 1919 riot, *The Negro in Chicago: A Study of Race Relations and a Race Riot* (Chicago: University of Chicago Press, 1922), 399–400.

109. Faith Rich to Ed Marciniak, July 19, 1951, folder July-September 1951, box 2, CIC; Theodore Purcell, S.J., *The Worker Speaks His Mind On Company and Union* (Cambridge: Harvard University Press, 1953), 176–77, 182–84. Also see Kevin Gerard Boyle, ''Politics and Principle: The United Automobile Workers and American Labor-Liberalism, 1948–1968'' (University of Michigan, Ph.D. diss., 1990), 96.

110. Sister Mary Sheil Driscoll, G.N.S.H., ''An Experiment in the Analysis and Modification of the Racial Attitudes of a Group of High School Seniors'' (Villanova University, M.A. thesis, 1945), 27. Also see Rev. Brian A. McCullogh, O.Carm., ''The Attitude of Catholic High School Students Toward the Negro in the Light of the Church's Teaching'' (Catholic University of America, M.A. thesis, 1943), 79. Alvin E. Winder, ''White Attitudes Towards Negro-White Interaction in an Area of Changing Racial Composition'' (University of Chicago, Ph.D. diss., 1952), 209–10.

111. Lloyd Davis, ''Catholics and Eight Foundations,'' *America,* 100 (January 17, 1959), 460–61.

112. Hope Brophy, untitled history of the Archbishop's Committee on Human Relations, (April 1969), p.3, AAD.

113. "General Report of the Work Done by the Industrial Areas Foundation for the Archdiocese of Chicago: Part II: The Negro Report," [1957], file 58, SA.

114. William J. Hill, Jr., to Catholic Book Club, April 4, 1958, folder 19, box 1, JLF.

115. Mathew Ahmann to John McDermott, October 4, 1957, file October 1–15, 1957, box 19, CIC; Mrs. Constance Baker Motley to Mr. Walter White, March 11, 1954, #0173, reel 6, part 5, NAACP.

116. Much of this analysis is indebted to Gary Gerstle, *Working Class Americanism: The Politics of Labor in a Textile City, 1914–1960* (New York: Cambridge University Press, 1989), esp. 281–333.

117. Survey of Students at St. Stanislaus Kostka, January 15, 1959, file January 15–19, 1959, box 28, CIC.

118. Untitled memo on Fruitvale Study, box 85, NCCC; Rev. John Beckman to Bishop Noll, April 28, 1953, folder 1, St. Ann's parish file, Archives of the Diocese of Gary; Blatnica, *"At the Altar of Their God,"* 150.

119. Mrs. Genevieve Rafferty to Msgr. Daniel Cantwell, March 6, 1955, file 1, box 3, Msgr. Daniel Cantwell Papers, CHS; John A. McDermott, "Invitation to Leadership: The Parish and the Changing Neighborhood," *Interracial Review,* 32 (September 1959), 161.

Chapter 5

1. The best guides to the Alinsky literature are Sanford Horwitt, *Let Them Call Me Rebel: Saul Alinsky—His Life and Legacy* (New York: Alfred A. Knopf, 1989), and P. David Finks, *The Radical Vision of Saul Alinsky* (New York: Paulist Press, 1984).

2. Horwitt, *Alinsky,* 102–3.

3. Edward Duff, "The Radicals and Their Reveille," *America,* 75 (June 29, 1946), 268–70; Agnes Meyer, "Orderly Revolution," *Washington Post,* June 5, 1945, p.9; Bernard I. Kahn, "The Catholic Church and Its Relationship to the Back of the Yards Neighborhood Council" (University of Chicago, M.A. thesis, 1949), 81.

4. Horwitt, *Alinsky,* 166; Alinsky's *Reveille for Radicals* was selected as a main selection by the Catholic Book of the Month Club, as noted by Kahn, "Catholic Church," 40; John Hellman, "The Anti-Democratic Impulse in Catholicism: Jacques Maritain, Yves Simon, and Charles de Gaulle During World War II," *Journal of Church and State,* 33 (Summer 1991), 453–72; Bernard Doering, "The Philosopher and the Provocateur," *Commonweal,* 67 (June 1, 1990), 345–48; Jacques Maritain, "Of America and of the Future," *Commonweal,* 41 (April 13, 1945), 642–45.

5. Alinsky to Maritain, November 23, 1945, reprinted in *The Philosopher and the Provocateur: The Correspondence of Jacques Maritain and Saul Alinsky,* Bernard Doering, ed. (Notre Dame: University of Notre Dame Press, 1994), 21.

6. Sr. M. Elfrida, to Albert Nellum, January 3, 1959, file January 1–14, 1959, box 28, CIC; Horwitt, *Alinsky,* 316–20; Martin Millspaugh and Gurney Breckenfield, *The Human Side of Urban Renewal: A Study of the Attitude Changes Pro-*

duced by Neighborhood Rehabilitation (New York: Ives Washburn, 1960), 210–11.

7. Rt. Rev. John O'Grady, "The Changing Parish," (Washington, D.C.: National Conference of Catholic Charities, 1955, n.p.).

8. "The Pastor and the Conservation of His Parish," (Washington, D.C.: National Conference of Catholic Charities, 1952), folder 1951, undated items, box 2, CIC; Msgr. John O'Grady, "The Progress of a Year," *Proceedings of the Thirty-Fifth Annual Meeting of the National Conference of Catholic Charities* (Washington, D.C.: Ransdell Printing, 1950), 17.

9. "The Pastor and Neighborhood Conservation," *Catholic Charities Review,* 37 (January 1953), 4.

10. O'Grady, "The Changing Parish," n.p.

11. Following one request for a contribution to a Catholic anthology, Alinsky remarked that "I get so screwed up with the Catholic Church that at times I wonder whether they are more confused than I am." Alinsky to Mr. Adolph Hirsch, March 11, 1952, file 7, box 29, ESF. Alinsky even traveled to Rome to press his ideas upon Vatican representatives; see Finks, *Radical Vision,* 117–18.

12. "Urban Renewal and the Spread of Blight in American Cities," *Catholic Charities Review,* 40 (December 1956), 3–4.

13. The initial foray was made in St. Louis. To O'Grady and Alinsky's chagrin, however, officials at Catholic Charities feared that the local Citizens' Council could not be dissuaded from "devoting its energies toward the actual objective of keeping Negroes out of the community." See Msgr. John O'Grady, "Report and Application to the Emil Schwarzhaupt Foundation," May 24, 1956, folder 1, box 34, ESF; Alinsky also explored setting up an organization in heavily Polish Hamtramck, Michigan; see Lester Hunt to Alinsky, [n.d., c. 1957], file 87–89, SA; "Interim Report on the Lackawanna Project," box 34, folder 5, ESF; Report of Nicholas Von Hoffman, November 17, 1956, box 34, folder 6, ESF.

14. Carl Tjerandsen, *Education for Citizenship: A Foundation's Experience* (Santa Cruz: Emil Schwarzhaupt Foundation, 1980), 259–73.

15. Msgr. Julius Szabo, "The Role of the Church As a Leader in Local Communities—The Case History of a City, Lackawanna, New York," *Proceedings of the 43rd Meeting of the National Conference of Catholic Charities* (Washington, D.C.: Merkle Press, 1960), 100; Horwitt, *Alinsky,* 324–26; Finks, *Radical Vision,* 85–88.

16. Horwitt, *Alinsky,* 288–302; Jeff Kisseloff, *You Must Remember This: An Oral History of Manhattan from the 1890s to World War II* (New York: Harcourt Brace Jovanovich, 1989), 504.

17. Alinsky to Dan Carpenter, July 15, 1957, folder 6, box 26, ESF; Carl Tjerandsen, *Education for Citizenship,* 290.

18. "Interim Report on the Lackawanna Project," folder 5, box 34, ESF; Tjerandsen, *Education for Citizenship,* 267; Bishop Joseph Burke to Rt. Rev. Joseph E. Maguire, March 14, 1958, file 160, SA.

19. Nicholas Von Hoffman, February 9, 1957, report, file 6, box 34, ESF.

20. Tom Murphy, week of July 7–13, 1958, file 9, box 34, ESF.

21. Ed Chambers, week of May 27, 1957 and week of October 14, 1957, file 7, box 34, ESF; Tom Murphy, July 7–13, 1958, file 9, box 34, ESF; Ed Chambers, week of May 27, 1957 file 7, box 34, ESF.

22. Everett C. Parker, "How Chelsea Was Torn Apart," *The Christian Century*, 77 (February 3, 1960), 132.

23. J. Anthony Paluch to Robert Moses, October 27, 1959, housing file 4, box 117, RMP; "Title I Protests Mount in Chelsea," *New York Times*, August 24, 1959, p.23; Paul Crowell, "Mayor to Change Relocation Rule in Title I Housing," *New York Times*, August 27, 1959, p.1; "Chelsea Women Picket," *New York Times*, August 21, 1959, p.19. On Penn Station South, Robert A.M. Stern, Thomas Mellins, David Fishman, *New York 1960: Architecture and Urbanism Between the Second World War and the Bicentennial* (New York: Monacelli Press, 1995), 311.

24. Memorandum to His Eminence Francis Cardinal Spellman, September 23, 1959, housing file 4, box 117, RMP; Spellman to Moses, October 2, 1959, housing file 4, box 117, RMP. For differing perspectives on archdiocesan evaluation of the Chelsea Community Council, see "R" to Browne December 10, [1957 or 1958] folder 1958, corr. box 22, Harry Browne Papers, Columbia University Special Collections, and Dan W. Dodson, "Report on the Chelsea Community Council," VIII—5, January 10, 1961, folder 11, box 26, ESF.

25. John Moore, "Interim Report: Chelsea Community Council, April 1958," folder 9, box 26, ESF; Parker "How Chelsea Was Torn Apart," 131.

26. Dodson, "Report on the Chelsea Community Council," II, 6–10.

27. Henry J. Browne, "Groping for Relevance in an Urban Parish: St. Gregory the Great, N.Y.C.," *Cross Currents*, 21 (Fall 1971), 433–59; idem, "Parish Changes in New York's Stryker's Bay Area," *Catholic World* (December 1968), 126–27; Browne to Bishop Maguire, January 17, 1961, file 1959–1961, box 22, Browne Papers; "Crusader for Housing," *New York Times*, May 28, 1965, p.20.

28. *Conference Bulletin of the Archdiocese of New York*, 38 (December 1961), 112; Browne, "Groping for Relevance in an Urban Parish," 435. One poem circulated by Browne supporters read, "East Side, West Side, All Around the Town, Urban Renewal's Upon us, the buildings are coming down, The new ones are luxurious, So to hell with Mamie O'Rourke, She can be relocated on the outskirts of New York." See Joseph P. Lyford, *The Airtight Cage: A Study of New York's West Side* (New York: Harper & Row, 1966), 122–30.

29. Msgr. Harry Byrne, interview with author, July 24, 1991.

30. Msgr. Harry Byrne, interview with author, July 24, 1991; summary of May 23, 1962 meeting, n.d., file 1959–1961, box 22, Browne Papers.

31. Robert D. Cross, "The Changing Catholic Urban Community," *Interracial Review*, 34 (September 1961), 235; Michael E. Schiltz, "Urban Renewal and the Parish," *Interracial Review*, 34 (September 1961), 224; Rev. Robert G. Howes, "The Church and Urban Renewal: A Resume," *Interracial Review*, 34 (September 1961), 219; David B. O'Brien, *Faith and Friendship: Catholicism in the Diocese of Syracuse, 1886–1986* (Syracuse: The Catholic Diocese of Syracuse, 1987), 425.

32. *Hearings Before the United States Commission on Civil Rights: Housing* (Washington, D.C.: United States Government Printing Office, 1959), 771–72.

33. Ibid., 776.

34. Alinsky to Adolph Hirsch, May 28, 1959, folder 6, box 26, ESF; Sanford D. Horwitt, *Let Them Call Me Rebel*, 339; for an example of media coverage, see " 'Planned' Dispersion of Negroes," *Chicago Sun-Times*, May 8, 1959.

35. *Hearings Before the United States Commission on Civil Rights: Housing* (Washington, D.C.: United States Government Printing Office, 1959), 808.

36. Undated memo, folder 2, box 8, Msgr. Daniel Cantwell Papers, CHS.

37. *Hearings Before the United States Commission on Civil Rights: Housing* (Washington, D.C.: United States Government Printing Office, 1959), 778.

38. "General Report of the Work Done by the Industrial Areas Foundation for the Archdiocese of Chicago, Part I," [1957–58] file 57, SA.

39. Cardinal Meyer,"The Catholic Church and the Negro in the Archdiocese of Chicago" (Chicago, 1960), copy in AAC; Msgr. John J. Egan, interview with author, December 13, 1990.

40. Horwitt, *Alinsky,* 341; John Fish, Gordon Nelson, Walter Stuhr, Lawrence Witmer, *The Edge of the Ghetto: A Study of Church Involvement in Community Organization* (New York: Seabury Press, 1966), 23; Saul Alinsky, "The Urban Immigrant," presented at Notre Dame conference on "Roman Catholicism and the American Way of Life," February 13, 1959, file 292, SA.

41. Memo from Ed Chambers and Nicholas Von Hoffman to Msgr. John J. Egan, August 7, 1961, file 334, SA.

42. Rev. John A. McMahon, "Conservation at St. Sabina's," *Catholic Charities Review,* 37 (January 1954), 11–13; Eileen McMahon, "What Parish Are You From? A Study of the Chicago Irish Parish Community and Race Relations, 1916–1970," (Loyola University, Ph.D. diss., 1989), 253–54.

43. Daily reports of Msgr. John J. Egan, October 24, 1960, box 66, EGAN; Andrew Greeley, *Confessions of a Parish Priest: An Autobiography* (New York: Simon and Schuster, 1986), 262; Donald O'Toole to Egan, October 3, 1960, "OSC corr." file, box 39, EGAN.

44. "Souvenir Edition: Commemorating 75th Anniversary of St. Leo Parish," *Southtown Economist,* October 31, 1957, file 343, SA; *Church of St. Leo Weekly* (July 26, 1959), file 343, SA.

45. On Molloy, see Horwitt, *Alinsky,* 328–31.

46. Thelma S. Robinson to William McManus (n.d., 1960), "Negro in Chicago," box 3394, Meyer Papers, AAC; "Charge Catholic Priest Bars Negroes at St. Leo," *The New Crusader* (October 1, 1960); Horwitt, *Alinsky,* 354; Unsigned memo, December 19, 1958, file December 9–31, box 26, CIC; Fr. Kenneth Brigham, interview with Steven Avella, March 11, 1989, copy in possession of Steven Avella.

47. William Gleason, "Girls' Schools Test Integration," *Chicago American* (June 28, 1960), clipping in folder 2, box 8, Cantwell Papers; "Parish Notes," St. Leo's Parish Bulletin, May 4, 1958, September 14, 1958, August 17, 1958, in August 26–31 file, box 24, CIC; "Parish Notes," St. Leo's Parish Bulletin, July 26, 1959, file 343, SA.

48. Statement by John McDermott at Public Hearing #3, Organization of the Southwest Community, September 14, 1961, file 307, SA; McMahon, "What Parish Are You From?" 276; OSC Constitution Preamble—from OSC "Proposals for Planning and Renewal," (n.d), file 307, SA.

49. Alice Ogle, "The Power of Community Action," *Franciscan Messenger,* 19 (March 1966, n.p.); Msgr. John J. Egan, "The Archdiocese Responds," *Church in Metropolis,* 6 (Summer 1965), 16; "Priest Stirs Anti-Bias Move," *Chicago Defender* (March 21, 1961).

50. Rev. Martin Farrell, interview with Rev. Thomas Joyce, n.d, copy at Garrett-Evangelical Seminary.

51. On The Woodlawn Organization, see John Fish, *Black Power/White Control: The Struggle of The Woodlawn Organization in Chicago* (Princeton: Princeton University Press, 1973); Charles Silberman, *Crisis in Black and White* (New York: Random House, 1964), 318–48; Farrell quoted in Horwitt, 366.

52. All of this draws upon Horwitt, *Alinsky,* 390–42.

53. André Siegfried, *America Comes of Age* (New York: Harcourt, Brace, 1927), 263; also see Terrence J. McDonald, Introduction, in William R. Riordan, *Plunkitt of Tammany Hall,* Terrence J. McDonald, ed. (Boston: Bedford Books, 1994), 17.

54. Fr. Leroy E. McWilliams with Jim Bishop, *Parish Priest* (New York: McGraw-Hill, 1953), 201; on Hague and Jersey City, see Dayton David McKean's muckraking account, *The Boss: The Hague Machine in Action* (Boston: Houghton Mifflin, 1940), 121, 160–165. Also see Richard J. Connors, *A Cycle of Power: The Career of Jersey City Mayor Frank Hague* (Metuchen, N.J.: Scarecrow Press, 1971), 91; William Lemmey, "Boss Kenny of Jersey City, 1949–1972," *New Jersey History,* 98 (1980), 16.

55. "Parish Surveys" file, box 97, NCCC; "General Report of the Work Done by the Industrial Areas Foundation for the Archdiocese of Chicago, Part II [1957 approx.], file 58, SA; Jay P. Dolan et al., *Transforming Parish Ministry: The Changing Roles of Catholic Clergy, Laity, and Women Religious* (New York: Crossroad, 1989), 12; Herbert Gans, *The Urban Villagers: Group and Class in the Life of Italian-Americans* (New York: The Free Press of Glencoe, 1962), 112.

56. Vattel Elbert Daniel, "Ritual in Chicago's Southside Churches for Negroes" (University of Chicago, Ph.D. diss., 1940), 138. Also see Nicholas Lemann, *The Promised Land: The Great Black Migration and How It Changed America* (New York: Alfred Knopf, 1991), 93, 269 and William J. Grimshaw, *Bitter Fruit: Black Politics and the Chicago Machine, 1931–1991* (Chicago: University of Chicago Press, 1992), 131–35. On New York, Philip J. Murnion, *The Catholic Priest and the Changing Structure of Pastoral Ministry, 1920–1970* (New York: Arno Press, 1978), 287.

57. Charles J. V. Murphy, "The Cardinal," *Fortune,* 61 (February 1961), 154.

58. Msgr. John O'Grady, "The Housing Movement and Its Challenge to the Priest," June 28,1949, file "Speeches on Housing, 1948–1950," box 115, NCCC. Meetings for clergy interested in these issues were held in Detroit (1951) and Cleveland (1952) in conjunction with annual meetings of the National Conference of Catholic Charities. Msgr. John O'Grady, "The Priest and Housing," *Proceedings of the Thirty-Fifth Annual Meeting of the National Conference of Catholic Charities* (Washington, D.C.: Ransdell, 1950), 174; "Pastors Discuss Means to Check Slum Threat," *The New World* (December 12, 1952), 10. For an example of O'Grady's frequent testimony before Congress on housing issues, see *The Housing Act of 1954: Hearings before the Committee on Banking and Currency, House of Representatives,* 83d Congress, 2nd session (Washington, D.C.: United States Government Printing Office, 1954), 643.

59. On Boston, the church, and urban renewal see Thomas O'Connor, *Building a New Boston: Politics and Urban Renewal, 1950–1970* (Boston: Northeastern

University Press, 1993), 96–97; Gans, *The Urban Villagers,* 112; Langley Carleton Keyes, Jr., *The Rehabilitation Planning Game: A Study in the Diversity of Neighborhood* (Cambridge: MIT Press, 1969), 110–15; Peter Dreier, ''Economic Growth and Economic Justice in Boston: Populist Housing and Jobs Policies,'' in *Unequal Partnerships: The Political Economy of Urban Redevelopment in Postwar America,* Gregory Squires, ed. (New Brunswick: Rutgers University Press, 1989), 38. On the church and Chicago expressways, Steven M. Avella, *This Confident Church: Catholic Leadership and Life in Chicago, 1940–1965* (Notre Dame: University of Notre Dame Press, 1992), 215–18; on the development of the Kennedy expressway, see Ed Marciniak, *Reviving an Inner City Community: The Drama of Urban Change in East Humboldt Park in Chicago* (Chicago: Department of Political Science, Loyola University, 1977), 17–19; Stritch to Rev. Thomas Reed, August 26, 1954, ''Board Membership'' File, box 8, EGAN.

60. T. J. Shanahan to Moses, July 16, 1957, box 43, RMP; Robert Caro, *The Power Broker: Robert Moses and the Fall of New York* (New York: Vintage Books, 1975), 722–27, 738; ''General Report of the Work Done by the Industrial Areas Foundation for the Archdiocese of Chicago, Part II,'' [1957 approx.], file 57, SA.

61. Swanstrom, ''The Catholic Parish and Low-Rent Housing,'' 135; ''Catholics Stress Four Amendments,'' *New York Times,* November 7, 1938, p. 2. One 1937 project in Cleveland, Lakeview Terrace, was 50 percent Catholic. See *Dedication of St. Malachi's Memorial Church* (Cleveland, 1947), 37.

62. Msgr. John O'Grady, ''The Priest and Housing,'' 175; *General Housing Legislation: Hearings Before the Subcommittee on Housing of the Committee on Banking and Currency House of Representatives, 80th Congress, 2nd Session* (Washington, D.C.: United States Government Printing Office, 1948), 588; Thomas William Tifft, ''Toward a More Humane Social Policy: The Work and Influence of Monsignor John O'Grady'' (Catholic University of America, Ph.D. diss., 1979), 360–73.

63. Bishop Stephen Donahue to Moses, September 11, 1956, Housing file 1, box 116, RMP; Mrs. Mary K. T. Williams to Mayor Wagner, August 25, 1955, Housing Corr. file, box 116, RMP. On Manhattantown, Caro, *Power Broker,* 979–82, 1010–13.

64. Gene Gleason and Fred J. Cook, ''One Million Profit for Not Building,'' *New York World Telegram,* July 31, 1956, p.15.

65. On urban renewal generally, Arnold R. Hirsch, ''With or Without Jim Crow: Black Residential Segregation in the United States,'' in *Urban Policy in Twentieth-Century America,* Arnold R. Hirsch and Raymond Mohl, eds. (New Brunswick: Rutgers University Press, 1993), 88–90. In New York City alone, urban renewal projects displaced 15,000 families a year by 1960. Bernard Frieden, ''Policies for Rebuilding,'' in *Urban Renewal: The Record and the Controversy,* James Q. Wilson, ed. (Cambridge: MIT Press, 1966), 599.

66. Henry J. Browne, *The Parish of St. Michael, 1857–1957*(New York, 1957), 57–59.

67. John Gurda, ''The Church and the Neighborhood,'' in *Milwaukee Catholicism: Essays on Church and Community,* Steven M. Avella, ed. (Milwaukee: Knights of Columbus, 1991), 8, 25; *The Elderly of St. Rose of Lima Parish: A*

Study of the Older People in Milwaukee (Washington, D.C.: National Council of Catholic Charities, 1961), 60–61.

68. Rev. Laurence J. McGinley, S.J., oral history, in *As I Remember Fordham: Selections from the Sesquicentennial Oral History Project* (Fordham: Fordham University Press, 1991), 1–3.

69. Caro, *Power Broker,* 1010–13. On church and state, see Sen. Jacob Javits to Robert Moses, February, 8, 1957, Committee on Slum Clearance, 1957 file, box 116, RMP.

70. Moses to McGinley, June 5, 1956, housing file 1, box 116, RMP; Moses to Spellman, August 31, 1956, Housing file 1, box 116, RMP.

71. Moses to Anthony Paluch, October 20, 1959, housing file 4, box 117, RMP. A statue of Moses was placed on the Fordham campus at Lincoln Center in 1970; see Caro, *Power Broker,* 738–41, 1155. Moses includes a lavish tribute to the Fordham Jesuits in his *Public Works: A Dangerous Trade* (New York: McGraw-Hill, 1970), 517–19, 530.

72. Bruce Kuklick, *To Every Thing a Season: Shibe Park and Urban Philadelphia, 1909–1976* (Princeton: Princeton University Press, 1991), 41–45, 71, 153–54; Hannah Lees, "How Philadelphia Stopped a Race Riot," *The Reporter,* 12 (June 2, 1955), 26–29.

73. Msgr. Raymond J. Gallagher, "The Disorganized Neighborhood," *Proceedings of the Forty-First National Convention of the Conference of Catholic Charities* (Washington, D.C.: Ransdell, 1956), 88; Chester Hartman, "The Housing of Relocated Families," in *Urban Renewal: The Record and the Controversy,* James Q. Wilson ed. (Cambridge: MIT Press, 1966), 293–335. William Grigsby points out that 70 percent of displaced families were non-white, in "Housing Markets and Public Policy," in Wilson, *Urban Renewal,* 30.

74. "Record of Proceedings Held at Our Lady of Sorrows Church, November 27, 1956," box 55, EGAN.

75. On Egan, Margery Frisbie, *An Alley in Chicago: The Ministry of a City Priest* (Kansas City: Sheed & Ward, 1991).

76. Egan to Rev. Thurston N. Davis, S.J., April 2, 1958, file 402, SA; Msgr. John J. Egan, "Trojan Horse in Our Cities," *Ave Maria,* 87 (May 10, 1958), 11–13.

77. Robert Moses to Thurston Davis, *America* editor, March 27, 1958, file 402, SA; Moses to McGinley, March 27, 1958, Committee on Slum Clearance file, 1958, box 117, RMP; Msgr. Francis Lally, "Correcting Urban Renewal," *Ave Maria,* 88 (July 19, 1958), 3.

78. Avella, *This Confident Church,* 227–34; Arnold Hirsch, *Making the Second Ghetto: Race and Housing in Chicago, 1940–1960* (New York: Cambridge University Press, 1983), 135–74.

79. "Both Sides of Hyde Park-Kenwood Redevelopment," *The New World,* May 2, 1958, 7.

80. *Housing Act Amendments of 1955: Hearings before the Committee on Banking and Currency House of Representatives,* 84th Congress, 1st session (Washington, D.C.: United States Government Printing Office, 1955), 345.

81. "Statement of the Catholic Archdiocese of Chicago: Hyde Park-Kenwood Urban Renewal Plan," [1958], folder 3, Archdiocesan Conservation Committee

Papers, AAC; Hirsch, *Making the Second Ghetto,* 161–67; Msgr. John J. Egan, "More Housing . . . Less Segregation," *The New World,* May 16, 1958, 1; J. M. Kelly, "Shiny New Islands," *The New World,* April 18, 1958, 1.

82. Fr. John Gallery to various priests in the archdiocese—copy, "Board Membership" file, box 8, EGAN; Msgr. John J. Egan, interview with the author, December 13, 1990; Hirsch, *Making the Second Ghetto,* 166; editor quoted in William J. Gleason, "Interracial Commission Examines Housing," *The New World* undated clipping, folder 5, Archdiocesan Conservation Committee, AAC; Frisbie, *An Alley in Chicago,* 99; for reaction to Egan's presentation, see the files of the Archdiocesan Conservation Committee, AAC.

83. Peter H. Rossi and Robert A. Dentler, *The Politics of Urban Renewal: The Chicago Findings* (New York: The Free Press of Glencoe, 1961), 225.

84. Taub, Taylor, and Dunham, *Paths of Neighborhood Change,* 32–35; also see, *Catholics in Archdiocese of Chicago,* vol. 1, section X, copy in box 43, EGAN; Leonard Confar, *Charlestown: A Limited Study of Church and Community* (Massachusetts Council of Churches, Department of Research and Strategy, 1950), #2171, HPD; Ira Rosenwaike, *Population History of New York City* (Syracuse: Syracuse University Press, 1972), 152–53.

85. Joseph B. Schuyler, S.J., *Northern Parish: A Sociological and Pastoral Study* (Chicago: Loyola University Press, 1960), 97–107.

86. John A. McDermott, "Invitation to Leadership: The Parish and the Changing Neighborhood," *Interracial Review,* 32 (September 1959), 158.

Chapter 6

1. Kenneth Clark to George Hunton, February 4, 1959, folder 41, box 3, II, JLF; Richard Wightman Fox, *Reinhold Niebuhr: A Biography* (New York: Pantheon Books, 1985), 285.

2. News release of Fr. Hesburgh's October 27, 1959, talk at Catholic Interracial Council Communion breakfast, October 26–31 file, box 32, CIC; *Hearings Before the United States Commission on Civil Rights: Housing* (Washington, D.C.: United States Government Printing Office, 1959), 317, 425, 809, 836.

3. Data from the 1940s and 1950s suggested that 64 percent of Catholics could be classified as "urban manual workers" and Catholics were much less likely than non-Catholics to live in rural areas. See John L. Thomas, *The American Catholic Family* (Englewood Cliff: Prentice-Hall, 1956), 131–36.

4. Mathew Ahmann, "Catholics and Race," *Commonweal,* 73 (December 2, 1960), 249.

5. Mark Goldman, *City on the Lake: The Challenge of Change in Buffalo, New York* (New York: Prometheus Books, 1990), 79–80; Timothy Richard Allan, "Roman Catholicism and Inner Urban Demographic Change in Buffalo, New York, 1960–1980: A Quantitative and Interpretive Overview" (SUNY Buffalo, Ph.D. dissertation, 1992), 124; John and Helen Cort, interview with J. Anthony Lukas, n.d., file "Church and CRS," box 6, JAL.

6. Eleanor McGrogan, "One Fine Day in Warren, Ohio," *Community,* 21 (May 1962), 6–7; "Problems of Racial Changes in Parishes Discussed at Meeting in Catholic Colleges," June 2, 1958, NCWC press release in Philadelphia "Housing" file, Josephite Archives, Baltimore, Maryland.

7. Joseph Newman, "Speaking from Experience," *Community,* 21 (October

1961), 7; Joseph Newman, interview with Gerard Powers, January 7, 1980, ADC; "New Growth of an Idea," *Community*, 20 (June 1961), 8; Gerard Francis Powers, "Under One God: The Catholic Church and Race Relations in Cleveland, 1955–1970" (Princeton University, B.A. thesis, 1980), 47–54.

8. Fr. John F. Lynch to parishioners of Basilica of Our Lady of Perpetual Help, July 19, 1963, "Church and CRS" file, box 6, JAL.

9. Msgr. Daniel Cantwell to Most Rev. Cletus O'Donnell and Very Rev. Msgr. Francis W. Byrne, December 20, 1961, file December 20–28, 1961, box 47, CIC.

10. James Gordon, "A Case Study of the Roles of Clergy, Police, Press and Local Neighborhood Leaders in Race Tension on Chicago's Southside," (1963) 6, #1981, HPD.

11. John McDermott, to Msgr. Francis W. Byrne, December 21, 1961, file Dec. 20–28, box 47, CIC; Sr. Mary Nona McGreal, O.P., "A Mission Great and Difficult: Dominican Sisters in a Chicago Parish, 1891–1991," 15, copy in author's possession.

12. Walter Speedy, "A Plea for Fair Play," *Southtown Economist* (April 26, 1962); Walter Speedy to Cardinal Meyer, April, 1962, file April 11–13, box 52, CIC; CIC to Rev. William McManus, January 17, 1962, file Jan. 11–18, 1962, box 50, CIC; John A. McDermott, interview with author, December 10, 1990.

13. Lillian S. Calhoun, "Catholic Interracial Council Head, McDermott, Assesses Catholic Schools," *Chicago Defender*, February 5, 1964, p.8.

14. Note written by Daniel M. Cantwell, n.d., file "undated items," box 66, CIC; Msgr. Wolfe of Visitation quoted in *A History of the Parishes of the Archdiocese of Chicago*, vol. 2:973; also see James Gordon, "A Case Study of the Roles of Clergy, Police, Press and Local Neighborhood Leaders in Race Tension on Chicago's Southside" (1963), #1981, HPD. Sr. Mary Nona McGreal, O.P., "A Mission Great and Difficult: Dominican Sisters in a Chicago Parish, 1891–1991," 15, copy in author's possession. On the Speedy children, CIC Board Meeting, September 18, 1962, file September 14–23, 1962, box 54, CIC.

15. Joseph A. Komonchak, "What They Said Before the Council," *Commonweal*, 117 (December 7, 1990), 716; Martin A. Zielinski, "'Doing the Truth': The Catholic Interracial Council of New York, 1945–1965," (Catholic University, Ph.D. diss., 1989), 288–89.

16. Fr. Joseph Connolly, interview with Brother Thomas Spalding, January 3, 1983; "The Baltimore Story," *America*, 109 (September 14, 1963), 256; Thomas W. Spalding, *The Premier See: A History of the Archdiocese of Baltimore, 1789–1989* (Baltimore: Johns Hopkins University Press, 1989), 405–35.

17. Quoted in George V. Fornero, "The Expansion and Decline of Enrollment and Facilities of Secondary Schools in the Archdiocese of Chicago, 1955–1980: A Historical Study" (Loyola University, Ph.D. diss., 1990), 414–15.

18. The newspapers gave the clergy front-page coverage when news of the meeting was released to the press; see Dave Meade, "Integration Ordered by Cardinal Meyer," *Chicago Daily News*, January 20, 1961, p.1; Meyer to all pastors, October 28, 1961, box 46, file October 27–31, CIC; all addresses contained in *The Catholic Church and the Negro in the Archdiocese of Chicago* (Chicago, 1960), AAC.

19. Hope Brophy, interview with author, September 19, 1990.

20. Ibid.; Marie Oresti, interview with author, September 19, 1990.

21. For example, see John A. Kenny, "Our Changing Neighborhoods: Pastors Tell Their Stories," *The Voice of St. Jude,* 26 (February 1961), 15–17; Dearden to Tennelly, September 21, 1960, file 5, box 3, series 5/2, BCIM.

22. Apostolate of the Negro, Questionaires, 1962, 1–5, subseries a, series I, ACHR.

23. Rt. Rev. Harold J. Markey, no title, and Rev. John F. Finnegan, "The All-White Parish and the Negro Apostolate," Detroit Clergy Conference, September 20–22, 1960, 7–3, series II, subseries b, ACHR; "Notes on Clergy Conference," 7–3, subseries a, series II, ACHR; March 13, 1960, ACHR minutes, 7–3, series II, subseries b, ACHR.

24. Rev. Robert F. Allen to Rev. Bernard D. Labelle, October 13, 1959, 7–31, subseries b, series II, ACHR; Hope Brophy, "Planned Ad Libbing, " *Community,* 20 (June 1961), 4–6; Rev. Robert F. Allen, "The Role of a Parish in Inter-Group Relations," *Interracial Review,* 34 (May 1961), 122–23.

25. "Minutes from Meeting of Parish Human Relations Groups," March 26, 1962, 7–32, subseries b, series II, ACHR; "Report on Leadership Training Programs," August 1961-October 30, 1962, 7–32, subseries b, series II, ACHR.

26. Report on Parish Leadership Training Programs, August 1961-October 30, 1962, subseries b 7–32, series II, ACHR.

27. Elaine Latzman Moon, ed., *Untold Tales, Unsung Heroes: An Oral History of Detroit's African-American Community, 1918–1967* (Detroit: Wayne State University Press, 1994), 334–35.

28. Hope Brophy, untitled history of Archbishop's Committee on Human Relations, (1969), 11, Black Catholics file, AAD.

29. "Minutes from Meeting of Parish Human Relations Groups," March 26, 1962, 7–32, subseries b, series II, ACHR; "Report on Leadership Training Programs," August 1961-October 30, 1962, 7–32, subseries b, series II, ACHR.

30. "Report on Leadership Training Programs," August 1961-October 30, 1962, 7–32, subseries b, series II, ACHR.

31. Gerhard Lenski, *The Religious Factor: A Sociological Study of Religion's Impact on Politics, Economics and Family Life* (Garden City: Doubleday, 1961), 32–36.

32. Scott Greer, *The Emerging City: Myth and Reality* (New York: The Free Press of Glencoe, 1962), 132; Lenski, *The Religious Factor,* 32–36, 195–96, 223–25.

33. Emphasis in original. Joseph T. Leonard, S.S.J., *Theology and Race Relations* (Milwaukee: Bruce, 1963), 282.

34. C.J. McNaspy, S.J., *Let's Talk Some Sense About the Negro* (March 14, 1961), America Press pamphlet, box 55, October 24–31 file, CIC. An interesting sidelight to these discussions occurred when a white South African bishop—the only native South African in the country's hierarchy—endorsed "separate development" along lines proposed by the South African government. Quick refutations of the position as mistaken came from both his South African colleagues and American Catholic liberals. See John F. Cronin, S.S., "Is Segregation Immoral?," folder 2, box 32, Cantwell Papers, CHS.

35. Tom Gaudette to Alinsky, January 22, 1962, file 262, SA.

36. Lenski, *The Religious Factor,* 65.

37. Ahmann, "Catholics and Race"; Mathew Ahmann to Anna McGarry, box 1, series 2, NCCIJ.

38. Mathew Ahmann, "Catholics and Race," 248; " 'Freedom Riders'," *America,* 105 (May 27, 1961), 358; Zielinski, " 'Doing the Truth,' " 137, 430–31.

39. Terry Sullivan, "What Is It Like to be a Freedom Rider?" *Work,* 19 (December 1961), 5.

40. "The Peril of Pussyfooting," *Interracial Review,* 35 (June 1962), 136.

41. Mathew H. Ahmann, Preface, *The New Negro,* Mathew H. Ahmann, ed. (New York: Biblo and Tannen, 1969 [1961]), xi.

42. Diane Nash, "Inside The Sit-Ins and Freedom Rides: Testimony of a Southern Student," in ibid., 43–60.

43. Mathew Ahmann, "Freedom Riders," *Commonweal,* 74 (June 23, 1961), 329. Statement of Brooklyn Catholic Interracial Council, *Interracial Review,* 34 (July 1961), cover page; William J. Kenealy, S.J., "The Legality of the Sit-Ins," in Ahmann, *The New Negro,* 63–86; "Freedom Riders Uncover Truth Concerning Race Relations in the South," *Interracial Review,* 34 (September 1961), 235; Emery Biro and Gordon Zahn to Long Range-Planning and Executive Committee, [1959?], file December 27–31, 1959, box 33, CIC.

44. "An Open Letter to the Parishioners of St. John Brebeuf Parish, Niles, particularly those in the Oakton Manor Section," n.d., file 1959, undated items, box 36, CIC; CIC memo from John McDermott on Niles, January 10, 1962, file January 1–10, 1962, box 49, CIC; Rev. Daniel J. Mallette, "Cardinal Rugambwa Comes to Chicago," *Ave Maria,* 94 (July 22, 1961), 11–14; Rev. William F. Graney, "How to Curb Panic and Win Friends: The Skokie Story," *Interracial Review,* 35 (February 1962), 42–43; Bob Senser, "Two More Joneses Have Gone to Suburbia," *Sign,* 40 (June 1961), 20–23. Text of sermon by Fr. Sauer, February 12, 1961, file February 1–17, 1961, box 41, CIC; Bernard L. Marsh to Archbishop Meyer, February 6, 1961, file February 1–17, 1961, box 41, CIC.

45. William Hogan, interview with the author, January 4, 1991; John McDermott, interview with the author, December 10, 1991; Memo to Board of Directors from John McDermott, September 1, 1961, file August 28–31, 1961, box 45, CIC; William Hogan, "Wade-In at 'Rainbow'," *Community,* 21 (October 1961), 4–6; McDermott to Rt. Rev. Msgr. Burke, July 14, 1961, file July 1–19, box 44, CIC.

46. "Catholic Students Here Prepare Resolutions," *Chicago Sun-Times,* August 30, 1962; "Why We Went to Albany: A Report by the Catholic members of the Chicago Interreligious Delegation," file September 24–30, 1962, box 54, CIC; David Garrow, *Bearing the Cross: Martin Luther King, Jr., and the Southern Christian Leadership Conference* (New York: William Morrow, 1986), 216.

47. Msgr. John J. Egan, "You and Your Neighbors Will Decide the Future," "articles OUA" file, box 12, EGAN. Analysis in this paragraph is indebted to Robert Wuthnow, *The Restructuring of American Religion: Society and Faith Since World War II* (Princeton: Princeton University Press, 1988), 145–48.

48. "Not Words but Acts," *Commonweal,* 78 (July 26, 1963), 444; "City's Open Occupancy Hearing Set for September 18," *Archdiocesan Commission on Human Relations Newsletter,* 1 (September 1963), 1–2.

49. McDermott to Mr. Arthur Wright, June 17, 1964, box 86, CIC.

50. McDermott to James F. Maguire, S.J., n.d., file June 110, box 60, CIC.

51. Dave Meade, "Club's Racial Policy Stirs Loyola Students," *Chicago Daily News,* June 29, 1963.

52. Fr. Daniel Mallette, interview with author, August 23, 1991; Sr. M. Angelica Sen, O.S.F., "The Sister in the New City," in *The Changing Sister,* Sr. M. Charles Borromeo Muckenhirn, C.S.C., ed. (Notre Dame: Fides, 1965), 239; Sr. Angelica, O.S.F., "Teaching Teachers," *Community,* 22 (February 1963), 4–6; Sr. Angelica and Sr. Anthony Claret, "The Nun's Story: Why We Picketed," *Community,* 22 (September 1963), 4–5.

53. Sr. Angelica and Sr. Anthony Claret, "The Nun's Story," 22 (September 1963), 6–7; Dave Meade, "Open Membership at Picketed Club," *Chicago Daily News,* July 9, 1963, p.16; Thomas Webb Sexton to McDermott, August 9, 1963, file August 1–12, 1963, box 62, CIC.

54. The most avidly read contemporary account was Xavier Rynne's series of *New Yorker* pieces. See Rynne, *Vatican Council II* (New York: Farrar, Straus and Giroux, 1968). For a concise summary of the council, Michael J. Walsh, "A History of the Council," in *Modern Catholicism: Vatican II and After,* Adrian Hastings, ed. (New York: Oxford University Press, 1991), 35–48.

55. See Gerald P. Fogarty, *The Vatican and the American Hierarchy from 1870 to 1965* (Stuttgart: Anton Hiersemann, 1982), 346–403; Benny Kraut, "A Wary Collaboration: Jews, Catholics, and the Protestant Goodwill Movement," in *Between the Times: The Travail of the Protestant Establishment in America, 1900–1960,* William R. Hutchison, ed. (Cambridge: Cambridge University Press, 1989), 193–230.

56. Philip F. Berrigan, "The Challenge of Segregation," *Interracial Review,* 34 (February 1961), 30; also, Philip F. Berrigan, "The Challenge of Segregation," *Worship,* 34 (November 1960), 597–604. John LaFarge, "Some Questions as to Interdenominational Cooperation," *Theological Studies,* 3 (September 1942), 332.

57. Msgr. John O'Grady, "Unity in Community Development," *Catholic Charities Review,* 43 (January 1959), 3; Joseph B. Meegan, "Notes on Community Action for Conservation," presented to National Conference of Catholic Charities meeting, September 29, 1953, file 240, SA; Zielinski, " 'Doing the Truth'," 268.

58. Walter G. Murphy, "You Can Fight City Hall," *Sign,* 42 (November 1962), 26.

59. Quoted in Sanford Horwitt, *Let Them Call Me Rebel: Saul Alinsky—His Life and Legacy* (New York: Alfred A. Knopf, 1989), 361.

60. Andrew Greeley, "City Life and the Churches," *America,* 103 (August 27, 1960), 573–74; Gibson Winter, *The Suburban Captivity of the Churches: An Analysis of Protestant Responsibility in the Expanding Metropolis* (Garden City: Doubleday, 1961), 89–90, 164; Minutes from Meeting of Parish Human Relations groups, March 26, 1962, 7–32, subseries b, series II, ACHR.

61. Henry Clark, *The Church and Residential Desegregation: A Case Study of an Open Housing Covenant Campaign* (New Haven: College and University Press, 1965), 70–168.

62. CIC memo of August 25, 1962, file August 17–25, box 53, CIC; Dennis

Clark, ''Evasive Answers About Racial Change,'' *Interracial Review,* 33 (February 1960), 34.

63. ''Open or Closed Cities,'' *The Christian Century,* 78 (May 10, 1961), 579; Ken Pierce, ''Church Supports Hate Group,'' *Chicago Maroon,* 1961 clipping, file 1961, box 48, CIC; Everett C. Parker, ''How Chelsea Was Torn Apart,'' *The Christian Century,* 77 (February 3, 1960), 132.

64. Harvey Cox, interview with J. Anthony Lukas, n.d., box 6, file ''Church and CRS,'' JAL; Ann Orlov to Mathew Ahmann, December 6, 1963, box 12, series 10–12, NCCIJ.

65. Stanley M. Grabowski, '' 'Dirty' Hands,'' *The Homiletic and Pastoral Review,* 64 (October, 1964), 12–14; *The Tablet,* 924 (August 24, 1963), 217; *Challenge to Conscience: Report of the Metropolitan Conference on Open Occupancy* (Detroit: The Detroit Metropolitan Conference on Religion and Race, 1963); Leonard Gordon, ''Attempts to Bridge the Racial Gap: The Religious Establishment,'' *A City in Racial Crisis: The Case of Detroit Pre-and Post-1967 Riot,* Leonard Gordon, ed. (Dubuque: Wm. C. Brown, 1971), 19–20.

66. Memo to Board of Directors, August 29, 1962, file August 26–31, 1962, box 53, CIC.

67. J. M. Kelly, ''Cardinal Sums up Racial Meet Aims,'' *The New World,* January 11, 1963.

68. John LaFarge, ''Religion and Race Meeting,'' *America,* 108 (February 2, 1963), 159; Preface, in *Race: Challenge to Religion,* Mathew Ahmann, ed. (Chicago: Henry Regnery, 1963), v.

69. Rev. Dr. Martin Luther King, Jr., ''A Challenge to the Churches and Synagogues,'' *Race: Challenge to Religion,* 156, 169; Albert Cardinal Meyer, ''Interracial Justice and Love: Challenge to a Religious America,'' *Race: Challenge to Religion,* 131.

70. Dennis Clark, ''Catholic Influence on Community Practices,'' [n.d. 1963?] box 1, series 4.1, NCCIJ; John LaFarge, ''Religion and Race Meeting,'' *America,* 108 (February 2, 1963), 159.

71. J. Francis Polhaus, ''Catholic Involvement in Civil Rights Legislation,'' *Interracial Review,* 36 (October 1963), 195; Trueblood Mattingly, ''Gwynn Oak,'' *America,* 109 (August 10, 1963), 137; Thomas W. Spalding, *The Premier See,* 433–35; Fr. Joseph Connolly, interview with Brother Thomas W. Spalding, January 3, 1983.

72. On Canton, Norma Marcere to Mathew Ahmann, September 21, 1963, box 18, series 10–2, NCCIJ; Stuart Lansdowne, ''Are You Planning a Trip to Washington?'' *America,* 109 (August 24, 1963), 185. On O'Boyle and the Lewis speech, see O'Boyle to the Apostolic Delegate, September 4, 1963, ''Social Action—Civil Rights, 1963'' file, box 85, NCWC; Taylor Branch, *Parting the Waters,* 874–79; Garrow, *Bearing the Cross,* 282–83.

73. Zielinski,'' 'Doing the Truth',' 450–53; Rev. Ronald Luka, C.M.F., ''We Marched the Mass,'' *Ave Maria,* 98 (November 2, 1963), 10/28; Ned O'Gorman, ''The Freedom March,'' *Jubilee,* 11 (October 1963), 17–20; Francis E. Kearns, ''Marching for Justice,'' *Commonweal,* 78 (September 20, 1963), 553.

74. ''Red Mass Speaker Cites Lawyers' Obligations,'' *Community,* 23 (November 1963), 15.

75. ''A Sad Chapter,'' *Commonweal,* 80 (May 1, 1964), 163–164. Also see John B. Sheerin, C.S.P., ''Catholic Involvement in Civil Rights,'' *Catholic World,* 201 (April 1965), 93.

76. Harvey Cox, ''Dialogue Among Pickets,'' *Commonweal,* 79 (November 22, 1963), 245–46; ''Priests and Ministers March in Downpour for 'Race Justice','' *Boston Globe,* November 9, 1963, p. 12; J. Anthony Lukas, *Common Ground: A Turbulent Decade in the Lives of Three American Families* (New York: Alfred A. Knopf, 1985), 383.

77. ''Report on 'Priests' Meeting on Rights,' '' (1964) Cantwell file, box 3392, Meyer Papers, AAC; Richards Jacobs, ''Priests Seeking Freedom to Join in Bias Protests,'' *St. Louis Post Dispatch,* May 19, 1964, clipping in Cantwell Papers, CHS.

78. On the churches' lobbying campaign for the 1964 civil rights legislation, James F. Findlay, ''Religion and Politics in the Sixties: The Churches and the Civil Rights Act of 1964,'' *Journal of American History,* 77 (June 1990), 66–92; Mary McGrory, ''Students Behind the Civil Rights Vigil,'' *America,* 110 (May 9, 1964), 623; Gayne Richards, ''Judgement Day Along the Potomac,'' *America,* 110 (May 9, 1964), 623.

79. Fr. Daniel Mallette, interview with author, August 23, 1991; ''Report from Mississippi,'' *Justice,* 5 (August 1964), file ''Archdiocese of San Francisco,'' box 2, series 10–12, NCCIJ; also on Mississippi, Timothy Ignatius Kelly, ''The Transformation of American Catholicism: The Pittsburgh Laity and the Second Vatican Council, 1950–1980'' (Carnegie-Mellon University, 1991), 404–5. On Boston priests, Rt. Rev. George W. Casey, ''The Priest and Demonstrations,'' *Homiletic and Pastoral Review,* 64 (October 1964), 1036; Lukas, *Common Ground,* 385; Father Tom MacLeod, n.d., interview with J. Anthony Lukas, box 6, file ''Church and CRS,'' JAL.

80. ''Churches in US Hail Pope's Call,'' *New York Times,* April 11, 1963, p. 16; John LaFarge, S.J., ''Pope John on Racism,'' *Interracial Review,* 36 (April 1963), 10–11.

81. The American bishops, meeting in Rome because of the council, also issued another joint letter on racial issues. See John D. Morris, ''Full Negro Rights Urged by US Catholic Bishops,'' *New York Times,* November 17, 1963, 1/85; ''Racism and the Council,'' *America,* 109 (November 2, 1963), 507; *American Participation in the Second Vatican Council,* Monsignor Vincent A. Yzermans, ed. (New York: Sheed and Ward, 1967), 69; Bishop Robert E. Tracy, *American Bishop at the Vatican Council* (New York: McGraw-Hill, 1965), 129–35; Milton Bracker, ''US Bishops at Rome Ask Clear Race Equality Stand,'' *New York Times,* October 25, 1963, p.1.

82. Curt Gentry, *J. Edgar Hoover: The Man and The Secrets* (New York: W. W. Norton, 1991), 570–71; Paul Hofman, ''Pope and Dr. King Confer on Rights,'' *New York Times,* September 19, 1964, p.3.

83. *American Participation in the Second Vatican Council,* 238–40; Joseph A. Califano, Jr. *The Triumph & Tragedy of Lyndon Johnson: The White House Years* (New York: Simon & Schuster, 1991), 72–73.

84. Fr. Daniel Peil to Bishop Leo Pursley, August 31, 1963, Pursley correspondence, Archives of the Diocese of Fort Wayne.

Chapter 7

1. Michael Real, "Search for Understanding," *Community,* 24 (April 1965), 4; Daniel M. Cantwell, "To Witness," *Community,* 24 (May 1965), 6.

2. Gregory Nelson Hite, "The Hottest Places in Hell: Catholic Participation in the Selma Voting Rights Campaign, 1962–1965" (University of Virginia, M.A. thesis, 1994), 51–59. Also see "Love and Forgiveness," *Jubilee,* 13 (August 1965), 21–23; Charles W. Eagles, *Outside Agitator: Jon Daniels and the Civil Rights Movement in Alabama* (Chapel Hill: University of North Carolina Press, 1993), 66–68.

3. Toolen quoted in John Cogley, "The Clergy Heed a New Call," *The New York Times Magazine* (May 2, 1965), 54.

4. John McDermott, interview with Martin Zielinski, July 21, 1987; Sister Margaret Burke to Egan, November 3, 1969, file "Egan," box 101, EGAN; Sister Thomas Marguerite, "Nuns at Selma," *America,* 112 (April 3, 1965), 454; Fr. Jerome Fraser, "Seminary Professor," *Community,* 24 (April 1965), 8–9.

5. Fr. Geno Baroni, "Selma—'Golgotha without the cross,' " *National Catholic Reporter,* 1 (March 17, 1965), 1/10.

6. "Selma Police Halt Demonstration Over Minister Meeting," *Washington Post,* March 11, 1965, 1/7; *New York Times,* March 11, 1965, p.1. Also see David Garrow, *Protest At Selma: Martin Luther King, Jr., and the Voting Rights Act of 1965* (New Haven: Yale University Press, 1978), 92; "Why Sisters?" *National Catholic Reporter,* 1 (March 24, 1965), 3; NCO Priests Meeting, March 18, 1965, file "NCO Catholic Clergy Meetings," box 32, EGAN.

7. Daniel M. Cantwell, "To Witness," *Community,* 24 (May 1965), 6; Thomas, "Nuns at Selma," 454–55; Sr. Mary Peter, S.S.N.D. "Nun," *Community,* 24 (April 1965) 5; John Cogley, "The Clergy Heed a New Call," *The New York Times Magazine* (May 2, 1965), 42–43, 54.

8. Rosemary Reuther, letter to the editor, *Commonweal,* 82 (April 30, 1965), 178; "Should Priests March?" *America,* 112 (May 1, 1965), 629–630; Bill Fanning, "Our Man in Harlem," *The Catholic News* (March 18, 1965); Robyn Draper and Cele Stefanski, "Marching Has Meaning," *Community,* 24 (April 1965), 10–11.

9. Daniel Berrigan, "Selma and Sharpeville," *Commonweal,* 82 (April 9, 1965), 73.

10. On the impact of the Council, John W. O'Malley, S.J., "Developments, Reforms, and Two Great Reformations: Towards a Historical Assessment of Vatican II," *Theological Studies,* 44 (1983), 373–406; on the European origins of Catholic social thought, Peter Hebblethwaite, "The Popes and Politics: Shifting Patterns in 'Catholic Social Doctrine,' " *Daedalus* 111 (Winter 1982), 85–98.

11. O'Malley, "Developments," 398.

12. *The Documents of Vatican II,* Walter M. Abbot, S.J., ed. (New York: Guild Press, 1966), 15.

13. Ibid., 5.

14. Ibid., 201–3.

15. Karl Rahner, "Towards a Fundamental Theological Interpretation of Vatican II," *Theological Studies,* 40 (1979), 716–27. On Jesuits and Latin America, Peter McDonough, *Men Astutely Trained: A History of the Jesuits in the American*

Century (New York: The Free Press, 1992), 263–75. On American nuns, Lora Ann Quinonez, C.D.P., and Mary Daniel Turner, S.N.D.deN., *The Transformation of American Catholic Sisters* (Philadelphia: Temple University Press, 1992), 68.

16. Samuel P., Huntington argues that the surge in democratic governments during the 1970s is attributable in part to the new Catholic emphasis on the rights of individuals; see Huntington, "Religion and the Third Wave," *The National Interest* (Summer 1991), 29–42. Also see Margaret E. Crahan, "Church and State in Latin America: Assassinating Some Old and New Stereotypes," *Daedalus,* 120 (Summer 1991), 131–58; and Marcos McGrath, "The Impact of *Gaudium et Spes:* Medellin, Puebla, and Pastoral Creativity," in *The Church and Culture Since Vatican II: The Experience of North and Latin America,* Joseph Gremillion, ed. (Notre Dame: University of Notre Dame Press, 1985), 61–73. On broader issues, see Avery Dulles, S.J., *The Resilient Church: The Necessity and Limits of Adaptation* (Garden City: Doubleday, 1977), 16–17 and Dulles, *The Reshaping of Catholicism: Current Challenges in the Theology of Church* (San Francisco: Harper & Row, 1988), 19–33.

17. John B. Sheerin, C.S.P., "Catholic Involvement in Civil Rights," *Catholic World,* 201 (April 1965), 95; Rev. Joseph P. Fitzpatrick, "The School: Its Place of Witness in the Church in the Inner City," November 11, 1965, CIC file, box 1, AUS.

18. Mathew Ahmann to Boston Archdiocesan Ecumenical Commission, August 26, 1964, Boston file, box 3, series 11, NCCIJ; "Planks for a Platform," *National Catholic Reporter,* 1 (October 28, 1964), 3.

19. See the various articles and comments in *The Church in the Changing City,* Louis J. Luzbetak, S.V.D., ed. (Techny, Ill: Divine Word Publications, 1966), 1, 7, 25, 54.

20. CIC petition to Cushing, December 19, 1965, file "CIC," box 1, AUS; Anna McGarry to William Ball, October 17, 1963, file 1963, box 1, AMC; Anna McGarry, memo, January 10, 1963, "History" file, box 1, AMC.

21. Rev. Francis J. Gross, S.J., "Catholics Who Are Negroes," *Community,* 22 (December 1962), 6–7; also see Dennis Clark, "City Catholics and Segregation," in *American Catholic Horizons,* Eugene K. Culhane, S.J., ed. (New York: Doubleday, 1966), 176.

22. Msgr. John J. Egan, daily reports, January 21, 1964, box 66, EGAN; Rev. Joseph G. McGroarty, "Convert-Making Not Primary Goal of Negro Apostolate," *Brooklyn Tablet* (October, 13, 1966).

23. William J. Jacobs, "Harlem Nocturne for the Church," *Ave Maria,* 106 (October 21, 1967), 23; Sara Harris, *The Sisters: The Changing World of the American Nun* (Indianapolis: Bobbs-Merrill, 1970), 319–322; Robert McClory, "Church Loses Ground," *National Catholic Reporter,* 13 (February 4, 1977), 1; Brother Joseph M. Davis, S.M., "The Position of the Catholic Church in the Black Community," *Homiletic and Pastoral Review,* 69 (June 1969), 699–704.

24. Richard Morrisroe, interview with author, December 31, 1990; also see Eagles, *Outside Agitator,* 166.

25. Anthony J. Vader, "The Catholic Church and the Negro Community," *Chicago Studies,* 4 (Spring 1965), 29; R. Scott Appleby, "In the Church but of

the World: Parish Priests on the Edge of Vatican II,'' *US Catholic Historian*, 11 (Winter 1993), 83–100.

26. Fr. Richard Morrisroe, ''Christian Witness: Transformation in a Crucible,'' *Community*, 25 (May 1996), 10-11.

27. McDonough, *Men Astutely Trained*, (New York: The Free Press, 1992), 366–73, 386; Garry Wills, *Bare Ruined Choirs: Doubt, Prophecy, and Radical Religion* (New York: Doubleday, 1972), 196–213.

28. J. Anthony Lukas, *Common Ground: A Turbulent Decade in the Lives of Three American Families* (New York: Alfred A. Knopf, 1985), 383–85.

29. Fr. Jack Farry, interview with author, January 3, 1990; Paul Wilkes, *These Priests Stay* (New York: Simon and Schuster, 1973), 28–46.

30. Baroni to Egan, March 11, 1965, box 57b, ''Urban Renewal'' file, EGAN.

31. See ''Cardinal in Harlem,'' *America*, 109 (July 27, 1963), 86–87.

32. Spellman speech of July 11, 1963, copy in possession of Msgr. Harry Byrnes, New York City. On Catholics priests and nuns as one of the few groups of whites choosing to live in the African-American ghetto, St. Clair Drake and Horace Cayton, *Black Metropolis: A Study of Negro Life in a Northern City* (Chicago: University of Chicago Press, 1994), 197; William Brink and Lou Harris, *The Negro Revolution in America* (New York: Simon and Schuster, 1964), 133; on New York, Patrick J. Mullaney, ''The Parish and the Community,'' *Interracial Review*, 24 (December 1951), 182; Lester Gaither, ''A Negro Youth Center: Saint Charles Borromeo Youth Project, Harlem, New York, 1950–1953'' (Fordham University, M.S.S. thesis, 1954), 18–55; Willie Agatha Backus, ''A Case Study of Twenty Negro Catholic Families in Saint Charles Borromeo Parish, Harlem, Receiving Aid to Dependent Children Because of Absence of Father from the Home'' (Fordham University, M.S.W. thesis, 1952), 53; Robert I. Gannon, S.J., *The Cardinal Spellman Story* (Garden City: Doubleday, 1962), 270.

33. Charles L. Palms, C.S.P., ''A Harlem Priest Reports on Selma,'' *Catholic World*, 201 (June 1965), 175; Philip Murnion, ''Catholic Church in Harlem,'' 1964, paper in possession of Fr. Philip Murnion, New York City; Fr. Philip Murnion, interview with the author, July 23, 1991.

34. *Our Lady of Victory Church* (Brooklyn, 1968), 21; ''Clergy Group Gets Saul Alinsky to Help,'' *Catholic Free Press* (Worcester, Mass.), January 20, 1967, p. 3; Rev. John James Bird and John Francis Boyne, ''Priests and Social Workers: Priests' Use of Catholic Charities and Expectations for Social Workers in the Diocese of Brooklyn, N.Y., 1966–1967'' (Fordham University, M.S.W. thesis, 1967), 23; Joseph Judge, ''Operation Brownsville,'' *Commonweal*, 84 (May 6, 1966), 194; John Leo, ''Uplift-cum-realism in a Brooklyn Slum,'' *National Catholic Reporter*, 2 (February 16, 1966), 8; Rev. Joseph G. McGroarty, ''Convert-Making Not Primary Goal of Negro Apostolate,'' *Brooklyn Tablet* (October 13, 1966).

35. On Catholic women more generally, Mary Jo Weaver, *New Catholic Women: A Contemporary Challenge to Traditional Religious Authority* (San Francisco: Harper & Row, 1985), and James Kenneally, *The History of American Catholic Women* (New York: Crossroad, 1990).

36. James O'Gara, ''The Emerging Nun,'' *Commonweal*, 82 (April 16, 1965), 104. For a remarkable series of personal narratives by Americans sisters, see

Midwives of the Future: American Sisters Tell Their Story, Ann Patrick Ware, ed. (Kansas City: Leaven Press, 1985).

37. Msgr. John J. Egan, January 12, 1963, daily reports, box 66, EGAN; Richard M. Menges, "Urban Apostolate Nuns . . . They're Where the Action Is," *The New World* (September 16, 1966), 28; Egan to Meyer, December 27, 1963, "Meyer Correspondence" file, box 13, Egan Papers.

38. Office of Urban Affairs Minutes, August 6, 1963, Board Meeting Minutes file, box 9, EGAN.

39. Mary Cole, *Summer in the City* (New York: P.J. Kenedy & Sons, 1968), 71; "New York Nuns Reach For Relevance," *National Catholic Reporter,* 1 (November 18, 1964), 7; "Nun Marchers Out of Place—Card. Cushing," *National Catholic Reporter,* 1 (May 19, 1965), 5; Misc. file, "Activities of the Association of Urban Sisters," box 1, AUS.

40. Sr. Mary Berchmans Shea, O.S.U., "Protest Movements and Convent Life," in *The New Nuns,* Sr. Charles Borromeo [Muckenhirn], C.S.C., ed. (New York: New American Library, 1967), 57; November 12, 1964 advisory meeting, Urban Apostolate of the Sisters, file "Advisory Comm. Agenda and Minutes," box 18, EGAN.

41. Sr. M. Charles Borromeo Muckenhirn, C.S.C., "Religious Poverty," in *Vows But No Walls: An Analysis of Religious Life,* Eugene E. Grollmes, ed.(St. Louis: B. Herder, 1967), 63.

42. Sister Mary Peter Traxler, S.S.N.D., "The Ministry of Presence," in *Split-Level Lives: American Nuns Speak on Race,* Traxler, ed. (Techny, Ill.: Divine Word Publications, 1967), 3; Sr. Mary Benet, O.S.B., and Sr. Mary Francis Xavier, H.H.S., to Archbishop Cody, January 21, 1966, Urban Apostolate of the Sisters correspondence, box 18, EGAN.

43. Discussion reported in *The Church in the Changing City,* Louis J. Luzbetak, S.V.D., ed. (Techny, Ill.: Divine Word Publications, 1966), 70, 126; Richard P. McBrien, *Do We Need the Church?* (New York: Harper & Row, 1969), 214–15.

44. Luzbetak, *The Church in the Changing City,* 25, 66.

45. Mrs. Myron Greismer to "Your Excellency," May 14, 1964, "Issues/Racism" file, ADC.

46. Clara Walters, in *Brooklyn Tablet* (August 19, 1965); Mrs. John O'Connor to Catholic Interracial Council, July 18, 1961, file July 1–19, box 44, CIC; John A. Selby to John McDermott, August, 1962, file August 26–31, 1962, box 54, CIC.

47. John Leo, "Moral Influence," *National Catholic Reporter,* 1 (March 24, 1965), 5; Sr. Ernest Marie, C.S.J., "Another Demonstrator's View," *Community,* 25 (January 1966), 7.

48. Letters in *Catholic Universe Bulletin* (April 2, 1965), 5; Michael Real, "Search For Understanding," *Community,* 24 (March 1965), 5.

49. Letters to the editor of the *Catholic Universe Bulletin* (April 2, 1965), 5.

50. "Nun Demonstrators Marched Off to Jail; But Marches Continue," *The New World* (June 18, 1965), 28; Bruce Cook, "Nuns' Arrest in Chicago Brings Soul Searching in Civil Rights Ranks," *National Catholic Reporter,* 1 (June 23, 1965), 1.

51. Longtime activist Msgr. Daniel M. Cantwell sparked a lengthy debate by criticizing demonstrators for assuming that solutions to segregation in Chicago

could be developed through street demonstrations. See Cantwell, ''Letter To My Friends,'' *New City*, 4 (August 1, 1965), 5.

52. See Letters to the editor of the *New World* (June 25, 1965), 24, and (July 16, 1965), 19.

53. The argument is the central theme of both James Davison Hunter, *Culture Wars: The Struggle to Define America* (New York: Basic Books, 1991), and Robert Wuthnow, *The Restructuring of American Religion* (Princeton: Princeton University Press, 1988). Also see Christopher Lasch, *The True and Only Heaven: Progress and Its Critics* (New York: W. W. Norton, 1991), 476–532.

54. For the phrase ''community without walls,'' see Sr. Marie Augusta Neal, ''Sociology and Community Change,'' in *The Changing Sister,* Sr. M. Charles Borromeo Muckenhirn, C.S.C., ed. (Notre Dame: Fides Publishers, 1965), 16.

55. Paul Hofman, ''Pope and Dr. King Confer on Rights,'' *New York Times,* September 19, 1964, p.3.

56. Report of the Executive Director, August 1966, box 4, series 2, NCCIJ.

Chapter 8

1. Anna McGarry to William Ball, October 17, 1963, file ''1963,'' box 1, AMC.

2. Dennis Clark, ''Philadelphia—Still Closed,'' *Commonweal,* 80 (May 1, 1964), 167–70; Fr. John McNamee, interview with author, July 31, 1991; Helen Adler to Mathew Ahmann, September 1963, file ''Penn-Philadelphia,'' box 20, series 10–12, NCCIJ; John Sisson to Mathew Ahmann, September 4,1963, file ''Penn-Philadelphia,'' box 20, series 10–12, NCCIJ; William G. Weart, ''Negroes are Jeered at New Pennsylvania Home,'' *New York Times,* August 31, 1963, p.6.

3. Diary of Dennis Clark, September 9, 1963, September 28, 1963, Clark Papers, MAUND; Anna McGarry to William Ball, October 17, 1963, file ''1963,'' box 1, AMC.

4. ''1000 Leave Meeting on Tension in Folcroft,'' *Philadelphia Bulletin,* January 27, 1964; Clark, ''Philadelphia—Still Closed,'' 169.

5. Clark, ''Philadelphia—Still Closed,'' 169; Proposed Letter to Archbishop Krol by Catholic Intergroup Relations Council, n.d., file March 29–31, box 68, CIC.

6. Copy of Statement released by Catholic Intergroup Relations Council, n.d., file March 29–31, 1964, box 68, CIC.

7. ''Commission on Human Relations Established,'' *Catholic Standard and Times* (May 15, 1964), 1; Msgr. Philip Dowling, interview with the author, July 31, 1991.

8. ''Dean of Boston College Law School Calls for Stepped-up Demonstrations,'' *Catholic Herald-Citizen* (July 1, 1965); George M. Collins, ''A Taste of Freedom,'' *Boston Globe,* April 23, 1965; J. Anthony Lukas, *Common Ground: A Turbulent Decade in the Lives of Three American Families* (New York: Alfred A. Knopf, 1985), 18.

9. ''Proposal for an Experimental School and for a Parish-Community Center,'' n.d., file ''Archdiocese of Boston—Catholic and Parochial Schools,'' box 1, AUS; Misc. file, ''Activities of the Association of Urban Sisters,'' box 1, AUS; Charles A. Meconis, *With Clumsy Grace: The American Catholic Left, 1961–1975* (New York: Seabury Press, 1979), 5.

10. Lukas, *Common Ground,* 385–86; Robert B. Kenney, "NAACP Float Hit By Rubble," *Boston Globe,* March 18, 1964, p.1; Joseph McClellan, "Incomplete Induction in South Boston," *Boston Pilot,* n.d., clipping in "Church and CRS" file, box 6, JAL; Lukas interview with John and Helen Cort, n.d., file "Church and CRS," box 6, JAL.

11. Lukas, *Common Ground,* 385–86.

12. Meeting at Ruth Batson's, February 10, 1966, Wednesdays in Mississippi file, box 2, Margaret Roach Papers, MAUND.

13. CHR report on November 16, 1966, meeting of high school faculties, Commission on Human Relations File, AAB.

14. CHR report on Dorchester faculty meeting, October 13, 1966, Commission on Human Relations File, AAB; CHR report on November 16, 1966 meeting of high school faculties, Commission on Human Relations File, AAB.

15. J. Brian Sheehan, *The Boston School Integration Dispute: Social Change and Legal Maneuvers* (New York: Columbia University Press, 1984), 217–19; "Statement read by Catholic pastors of Charlestown in Feb. 1963," "Church in Charlestown" file, box 3, JAL; Lukas, *Common Ground,* 355–56.

16. Sheehan, *Boston School Integration,* 220–26, Lukas, *Common Ground,* 355–56.

17. Langley C. Keyes, Jr., *The Rehabilitation Planning Game: A Study in the Diversity of Neighborhood* (Cambridge: MIT Press, 1969), 131; Sheehan, *Boston School Integration,* 220–26; Lukas, *Common Ground,* 355–56.

18. On segregation during the 1960s, Sar A. Levitan, William B. Johnston, Robert Taggart, *Still A Dream: The Changing Status of Blacks Since 1960* (Cambridge: Harvard University Press, 1975), 151–161. Population data in *Statistical Abstract of the United States, 1974* (Washington, D.C.: U.S. Government Printing Office, 1974), 23–25.

19. Norman M. Bradburn, Seymour Sudman, Galen L. Gockel, *Racial Integration in American Neighborhoods* (Chicago: National Opinion Research Center, 1970), 327–28.

20. Ibid.; Louis Harris and Bert E. Swanson, *Black-Jewish Relations in New York City* (New York: Praeger, 1970), 54, 72–73; On Catholic support for integration, Andrew M. Greeley and Paul B. Sheatsley, "Attitudes Toward Racial Integration," *Scientific American,* 225 (December 1971), 13–19; Harold Abrahamson and C. Edward Noll, "Religion, Ethnicity and Social Change," *Review of Religious Research,* 8 (Fall 1966), esp. 22–25; Jonathan Reider, *Canarsie: The Jews and Italians of Brooklyn Against Liberalism* (Cambridge: Harvard University Press, 1985), 81–82.

21. Msgr. Daniel M. Cantwell to Fr. Griffin, June 23, 1964, file "June 21–30, 1964," box 71, CIC.

22. Annual Report of the Northwest Community Organization, 1969, pp.1–8, series I, subseries A 2–23, ACHR.

23. "Organization for the Southwest Community—an Evaluation," October 1965, "Board Meeting Agenda" file, box 9, EGAN.

24. Rt. Rev. Msgr. Bernard M. Brogan to Egan, November 4, 1966, "OAU corr" file, box 66, EGAN; Interview with Msgr. Piwowar, July 6, 196[5] and interview with Fr. Lisewski, C.S.C, July 14, 196[5], "Father McGrath" file, box 9, EGAN.

25. Market Opinion Research, "Aid to Education Study," May 24, 1968, file 13, box 36, John Cardinal Dearden Papers, MAUND; James Q. Wilson, *Thinking About Crime* (New York: Basic Books, 1983), 3–25, 34.

26. See "Ad Hoc Committee: Catholics in Support of the Civilian Complaint Review Board," *New York Daily News,* November 7, 1966, p. 67, and "Roman Catholic Priests Committee Against Civilian Review Board," *New York Daily News,* November 7, 1966, p. 102. Also see James J. Graham, "Backlash in Brooklyn," *Commonweal,* 85 (December 9, 1966), 287–91.

27. David W. Abbot, Louis H. Gold, Edward T. Rogowsky, *Police Politics and Race: The New York City Referendum on Civilian Review* (Cambridge, 1969), 13–15.

28. James L. Sundquist, *Dynamics of the Party System: Alignment and Realignment of Political Parties in the United States* (Washington, D.C.: The Brookings Institution, 1983), 382–83.

29. David Heer notes the predominance of Catholic policemen and fireman in Boston's Roslindale section in Herr, "The Role of the Working Wife in Catholic Families" (Harvard University, Ph.D. diss., 1958), 102; Eileen McMahon makes a similar observation about Catholic neighborhoods in Chicago. McMahon, "What Parish Are You From? A Study of the Chicago Irish Parish Community and Race Relations, 1916–1970," (Loyola University, Ph.D. diss., 1989), 293.

30. Eileen McMahon, "What Parish Are You From?" 309–21.

31. Daniel Peil, "We Had a Riot," sermon of July 30, 1967, copy in Pursley correspondence, Archives of the Diocese of Fort Wayne; *Transcripts of Special Hearings Concerning Civil Disorders of 1967* (South Bend, 1967), esp. 75–100, 1215–18; Fr. Francis Van Bergen, interview with Gerard Powers, December 31, 1979, ADC.

32. Nicholas Von Hoffman, "Religion Goes Into Action," *Chicago Daily News,* August 10, 1963; Austin C. Wehrwein, "Chicago Negroes Protest in Mud," *New York Times,* August 3, 1963, p. 18.

33. "Priests' Efforts Fail to Quell Cleveland Mob," *Catholic Herald Citizen* (February 8, 1964); "Priest Told to Mind Own Business and 'Pray for Us' by Youthful Mob Determined to Halt Negro Students," *Catholic Herald Citizen* (February 15, 1964), 15; James Flannery, "The Murray Hill Incident: An Eyewitness Report," *Catholic Standard and Times* (February 14, 1964); "School Board Today Faces Pickets, Sit-In," *Cleveland Plain-Dealer,* January 31, 1964, p. 8.

34. Dr. Jan Van Lier to Auxiliary Bishop Whealon, June 21, 1964, CIC file, ADC; Edward P. Gorczyca to Bishop Whealon, May 30, 1964, St. Leo parish file, ADC.

35. "Priests Try to Restore Calm to Racially Tense Section of Kensington," *Catholic Standard and Times* (October 7, 1966), 1; Peter Binzen, *White Town U.S.A.* (New York: Random House, 1970), 112, 248; Msgr. Philip J. Dowling, interview with author, July 31, 1991.

36. Msgr. Philip J. Dowling, interview with author, July 31, 1991; "Uproar at Bok," *Camden Catholic Star-Herald* (October 18, 1968); "Shared Time Seen as Aid in Cooling School Tensions," *Catholic Standard and Times* (October 18, 1968); "Post-Dismissal Incidents Blamed in School Tension," *Catholic Standard and Times* (October 18, 1968).

37. David J. Garrow, *Bearing the Cross: Martin Luther King, Jr., and the*

Southern Christian Leadership Conference (New York: William Morrow, 1986) 444, 449. The most thorough discussion of the Chicago marches is contained in James R. Ralph, Jr., *Northern Protest: Martin Luther King, Jr., Chicago, and the Civil Rights Movement* (Cambridge: Harvard University Press, 1993). On King and Chicago Catholicism, see ibid., 73–75; Arthur Southwood, "Nobel Prize Winner Honored by CIC Here," *The New World* (November 6, 1964); Fr. Daniel Mallette, interview with author, August 23, 1991; William Hogan, interview with author, January 4, 1991.

38. James H. Bowman, S.J., "Martin Luther King in Chicago," *Ave Maria*, 102 (September 25, 1965), 6–9; "Priests, Sisters, and Martin Luther King," *Community*, 25 (September 1965), 4–5.

39. Garrow, *Bearing the Cross*, 465; Fr. Daniel Mallette, interview with author, August 23, 1991; Delores McCahill, "Dr. King Confers With Cody About Civil Rights Campaign," *Chicago Sun-Times*, February 4, 1966, p. 3/52. Jesse Jackson also addressed the CIC board on December 14, 1965, see CIC board meeting minutes, December 14, 1965, box 1A, CIC.

40. "John Patrick Cody," *Chicago Defender* (June 22, 1965).

41. D.J.R. Bruckner, "Church Held Faltering in Chicago Civil Rights," *Los Angeles Times*, March 23, 1966, p. 6. On the FBI, Cody, and Cushing, see Lukas, *Common Ground*, 387–88, and David J. Garrow, *The FBI and Martin Luther King, Jr.: From "Solo" to Memphis* (New York: W.W. Norton, 1981), 176; *Investigation of the Assassination of Martin Luther King, Jr.: Hearings Before the Select Committee on Assassinations of the U.S. House of Representatives*, 95th Congress, 2nd session, vol. 6, 263–66; Bruce Cook, "How Cody Is Re-styling Chicago," *National Catholic Reporter*, 2 (February 2, 1966), 2.

42. Cody statement, July 10, 1966, "Cody—Civil rights rally" file, box 13, EGAN. The statement was written by long-time activist Msgr. John J. Egan. Msgr. John J. Egan, interview with author, December 13, 1991; Garrow, *Bearing the Cross*, 491; "Chicago Archbishop Supports Civil Rights Drive," *New York Times*, July 11, 1966, p.19.

43. I rely on Garrow, *Bearing the Cross*, 489–525, and Alan B. Anderson and George W. Pickering, *Confronting the Color Line: The Broken Promise of the Civil Rights Movement in Chicago* (Athens: University of Georgia Press, 1986); "Riot in Chicago," *America*, 115 (July 30, 1966), 117–118.

44. Michael E. Schiltz, "Catholics and the Chicago Riots," *Commonweal*, 85 (November 11, 1966), 159–63; Mary Lou Finley, "The Open Housing Marches: Chicago, Summer '66," in *Chicago 1966: Open Housing Marches, Summit Negotiations, and Operation Breadbasket*, David J. Garrow, ed. (Brooklyn: Carlson Publishing, 1989), 19–31; Donald Bloch, "Identification and Participation in Urban Neighborhoods" (University of Chicago, M.A. thesis, 1952), 72.

45. Anderson and Pickering, *Confronting the Color Line*, 221–22.

46. Garrow, *Bearing the Cross*, 499–500.

47. Bevel quoted in *Delmarva Dialog* (August 19, 1966), clipping in Josephite Archives, Baltimore, Maryland; also see Ralph, *Northern Protest*, 145–47. For the archdiocesan response, William F. Graney, "Problem Too Serious to be Left to Swastika Clutchers," *The New World* (August 19, 1966), 4; J. M. Kelly, "Will Drama of March Turn Into Strategy?" *The New World* (August 19, 1966) 4;

"Demonstrations Should End, Cody Says," *Chicago Tribune*, August 11, 1966, pp. 1/2; "Churchmen Try to Ease Race Tension in Chicago," *Catholic Star-Herald* (August 19, 1966); CIC press release, box 109, CIC.

48. Anderson and Pickering, *Confronting the Color Line*, 256, 307–8; Msgr. John J. Egan, August 15, 1966, "ABC Priests' Course," box 13, EGAN.

49. William J. Leahy to Cody, November 28, 1966, "l" file, box 2007, ACHREC; Sr. Matthias Rinderer, O.S.F., "One Sister's Chicago Education," in *The New Nuns*, Sr. M. Charles Borromeo, C.S.C., ed. (New York: New American Library, 1967), 113.

50. Robert McClory, "The Holy Terror of Saint Sabina's," *Chicago Reader*, 19 (November 17, 1989), 28; Fr. James A. Bowman, S.J., letter, *Commonweal*, 85 (December 16, 1966), 323; John A. McDermott, "A Chicago Catholic Asks: Where Does My Church Stand on Racial Justice?" *Look*, 30 (November 1, 1966), 82+; William F. Graney, "Saddest Part of Riot: 'Catholic' Know Nothings," *The New World* (August 5, 1966); Dennis Geaney, "Trouble in Chicago," *Ave Maria*, 104 (October 1, 1966), 11.

51. Dennis Geaney, O.S.A., "Trouble in Chicago," 11; John A. McDermott, "A Chicago Catholic Asks: Where Does My Church Stand on Racial Justice?," 82+; John Cogley, "Chicago Diocese Split," n.d, clipping in file "Race—Catholic," box 32, Msgr. Geno Baroni Papers, MAUND.

52. Kathleen Connolly, "The Chicago Open-Housing Conference," in *Chicago 1966: Open Housing Marches, Summit Negotiations, and Operation Breadbasket*, David J. Garrow, ed. (Brooklyn: Carlson Publishing, 1989), 82.

53. "Priests' Attempt to Salve Racial Violence Wounds," *The New World* (August 12, 1966); Connolly, "Open Housing Conference," 69.

54. Daniel M. Cantwell, letter, *Commonweal*, 85 (December 16, 1966), 321; Jerry DeMuth, "Suburb Foes of Integration Fire at Cody," *Chicago Sun-Times*, February, 13, 1967, p.18; Pat Johnson, "Speech Blasts Clergymen at Crescent Meet," *Southwest News-Herald*, February 16, 1967, p.1.

55. Copy of statement in folder 4, box 11, Cantwell Papers, CHS; Ralph, *Northern Protest*, 208; Hugh Hough, "Archdiocese Rebukes Msgr. Burke on His Housing Integration Stand," *Chicago Sun-Times*, January 22, 1967, p.3; Jerry DeMuth, "Chicago Pastor Against Racial Integration," *National Catholic Reporter*, 3 (February 1, 1967), 3.

56. Hough, "Archdiocese Rebukes Msgr. Burke," pp. 3/50.

57. Woman to Cantwell, n.d., folder 4, box 11, Cantwell Papers, CHS; Raymond F. Knauerhaze to Cody, September 15, 196[5?], file "k," box 2007, ACHREC; Marty Servick to Cantwell, January 27, 1967, folder 11, box 4, Cantwell Papers, CHS.

58. *Milwaukee Lutheran Planning Study* (Milwaukee, 1962), iv–7.

59. William Kennedy, *O Albany!* (New York: The Viking Press, 1983), 276–78.

60. William Kennedy, "Father Bonaventure Leaves Albany," *National Catholic Reporter*, 2 (June 22, 1966), 1/10; Thomas Lickona, "Another Priest, Another Ban," *Commonweal*, 83 (December 10, 1965), 298, 299.

61. "Case Histories," *Ave Maria*, 103 (January 8, 1966), 5–6; Kennedy, *O Albany!*, 168–73; Memo on O'Brien situation directed to All N.F.C.C.S. member

colleges, [November, 1965], Political Affairs Commission file, box 10, NFCCS; "Two Pacifists Start Protest," *New York Times*, December 11, 1965, p. 20; Lickona, "Another Priest," 298–99.

62. Quoted in Joseph Gallagher, *The Pain and the Privilege: Diary of a City Priest* (New York: Image Books, 1983), 300; "Shehan Defies Threat, Testifies," *National Catholic Reporter*, 2 (January 26 1966), 7; Thomas W. Spalding, *The Premier See: A History of the Archdiocese of Baltimore, 1789–1989* (Baltimore: Johns Hopkins University Press, 1989), 436.

63. Frank McLoughlin, "2 'Worker Priests' Adopt Life in the Slums," *New York Sunday News*, March 17, 1968, p. B 2/12.

64. On Daley and the CIC, Arnold R. Hirsch, "The Cook County Democratic Organization and the Dilemma of Race, 1931–1987," in *Snowbelt Cities: Metropolitan Politics in the Northeast and Midwest Since World War II*, Richard M. Bernard ed. (Bloomington: Indiana University Press, 1990), 77–78; Len O'Connor, *Clout: Mayor Daley and His City* (Chicago: Henry Regnery, 1975), 165–66. On Daley and Catholicism, Eugene Kennedy, *Himself! The Life and Times of Mayor Richard J. Daley* (New York: The Viking Press, 1978), 30–46, 197, 202–3.

65. Quoted in Gary Rivlin, *Fire and Prairie: Chicago's Harold Washington and the Politics of Race* (New York: Henry Holt, 1992), 15; Nicholas Lemann, *The Promised Land: The Great Black Migration and How It Changed America* (New York: Alfred A. Knopf, 1991), 100; "Report on 3309 South Lowe Avenue," Chicago Commission on Human Relations, file 5, box 30, Cantwell Papers, CHS; James Alan McPherson, "In My Father's House There are Many Mansions, and I'm Going to Get Me Some of Them Too!" *The Atlantic*, 229 (April 1972), 55; Mike Royko, *Boss: Richard J. Daley of Chicago* (New York: Dutton, 1971), 140–41.

66. Robert Axelrod, "Where the Votes Come From: An Analysis of Electoral Coalitions, 1952–1968," *American Political Science Review*, 72 (February 1972), 16.

67. Ibid., also see Kevin Phillips, *The Emerging Republican Majority* (Garden City: Anchor Books, 1970), 31–72.

68. Useful on these issues is Edward G. Carmines and James A. Stimson, *Issue Evolution: Race and the Transformation of American Politics* (Princeton: Princeton University Press, 1989). On "race" as the central element in the realignment of American politics, Thomas Byrne Edsall with Mary D. Edsall, *Chain Reaction: The Impact of Race, Rights and Taxes on American Politics* (New York: W.W. Norton, 1991), esp. 59–60; Jonathan Reider, "The Rise of the Silent Majority," in *The Rise and Fall of the New Deal Order, 1930–1980*, Steve Fraser and Gary Gerstle, eds. (Princeton: Princeton University Press, 1989), 243–68.

69. James W. Carey, "An Ethnic Backlash," *Commonweal*, 81 (October 16, 1964), 91–93; James O'Gara, "The X Factor," *Commonweal*, 80 (September 18, 1964), 628.

70. Phillips, *Emerging Republican Majority*, 166–67; Richard C. Haney, "Wallace and Wisconsin," *Wisconsin Magazine of History*, 61 (Summer 1978), 259–78; *America* magazine noted Wallace's success with disgust, since "his extremist positions on race are so undisguisedly opposed to Catholic teaching." "Convenient Catholicism," *America*, 110 (June 13, 1964), 815.

71. William F. Buckley, ''Remarks to the New York Police Department Holy Name Society,'' *National Review,* 17 (April 20, 1965), 324–26; John B. Judis, *William F. Buckley, Jr.: Patron Saint of the Conservatives* (New York: Simon and Schuster, 1988), 235–44.

72. ''Cardinal Too Lenient,'' CIC Boston Newsletter, 2 (January, 1966), box 12, series 10–2, NCCIJ; CIC petition to Cushing, December 19, 1965, ''CIC'' file, AUS.

73. On Catholic perceptions of ''race'' and party loyalty, see John R. Petrocik, *Party Coalitions: Realignment and the Decline of the New Deal Party System* (Chicago: University of Chicago Press, 1981), 139–47; Thomas Edsall and Mary Edsall, *Chain Reaction,* 59–60.

74. Frank Sullivan, ''Open Occupancy the Issue in 5 Southwest Side Wards,'' *Chicago Sun-Times,* February 20, 1967, p. 24; also see Ralph, *Northern Protest,* 115–16.

75. Tillman, in *Voices of Freedom: An Oral History of the Civil Rights Movement from the 1950s through the 1980s,* Henry Hampton and Steve Fayer, eds. (New York: Bantam, 1990), 312.

76. For example, see Andrew Greeley, ''Chicago Summer,'' *New City,* 5 (August 1966), 19; Nathan Glazer and Daniel Patrick Moynihan, *Beyond the Melting Pot: The Negroes, Puerto Ricans, Jews, Italians, and Irish of New York City,* 2d ed.(Cambridge: MIT Press, 1970), lxv. For Chicago archdiocesan population data and analysis, see *Catholics in Archdiocese of Chicago,* esp. vol. 1, in box 43, EGAN.

77. Rosemary Thielke, ''Muddled Like Me,'' *Ave Maria,* 101 (April 3, 1965), 5–7; ''Milwaukee Vindicates Griffin in Public Show of Fairness,'' *National Catholic Reporter,* 1 (February 17, 1965), 2; Jane Berdes, ''Old vs. New in Milwaukee,'' *National Catholic Reporter,* 1 (July 21, 1965), 1/7; Fr. Mathew Gottschalk, O.F.M., to Chicago CIC, August 23, 1963, file Milwaukee, box 22, series 10–2, NCCIJ; Catholic population estimates in *Milwaukee Lutheran Planning Study,* Walter Kloetzi, ed. (Milwaukee, 1962), esp. x-3.

78. Frank Aukofer, *City With a Chance* (Milwaukee: Bruce, 1968), 94; Rosemary Thielke, ''Muddled Like Me,'' 5–7.

79. Richard Bernard and Bill Leuders, ''The Selma of the North,'' *Milwaukee Magazine,* 11 (February 1986), 77–78; Sister Mary Dolores Rauch, ''Impact of Population Changes in the Central Area of Milwaukee Upon Catholic Parochial Schools, 1940–1970'' (University of Wisconsin-Milwaukee, M.S. thesis, 1967), 38, 54, 59, 74.

80. Bleidorn to Archbishop Cousins, October 20, 1965, folder 1, EB.

81. Ibid.

82. ''Milwaukee Priests Revise Plan to Aid Boycott of Schools,'' *New York Times,* October 18, 1965, p.27; Bleidorn to Archbishop Cousins, October 20, 1965, folder 1, EB.

83. ''Priests and Nuns Back Milwaukee School Boycott,'' *New York Times,* October 19, 1965, p.27; Aukofer, 71–72.

84. Sermon in folder 1, EB; Aukofer, *City with a Chance,* 71.

85. Thomas Sweetser, S.J., ''Rundown on a Demonstration,'' *Community,* 25 (December 1965), 5–6; Paul Wilkes, *These Priests Stay* (New York: Simon and Schuster, 1973), 107–9.

86. Ann F. Cudahy to Groppi, October 25, 1965, folder 1, box 1, JG; ''Priest Support in Race Protest,'' *New York Times,* December 11, 1965, p.22.

87. Bernard Karger to Groppi, December 6, 1965, folder 6, box 5, JG; Isabel Maguire to Groppi, October 21, 1965, folder 6, box 5, JG; ''Catholic Laymen, workers in an industrial complex,'' to Groppi, October 13, 1965, folder 3, EB.

88. ''Groppi vs. Milwaukee,'' undated clipping in folder 4, box 3, JG; St. Philip Neri parish bulletin, in September 4, 1966, letter sent to Groppi by Robert J. Feller, box 1, JG; also see ''Groppi in Black for Yule Mass,'' *National Catholic Reporter,* 3 (January 3, 1967), 7, and Karen Kelly, ''The Scene—Milwaukee,'' *Community,* 27 (October 1967), 3.

89. Aukofer, *City with a Chance,* 110–20; Richard Bernard and Bill Lueders, ''The Selma of the North,'' 74–80; Homar Bigart, ''A Militant Priest Kicks Up a Storm,'' *New York Times,* September 17, 1967, IV, 4; Richard M. Bernard, ''The Death and Life of a Midwestern Metropolis,'' in *Snowbelt Cities: Metropolitan Politics in the Northeast and Midwest Since World War II,* Richard M. Bernard, ed. (Bloomington: Indiana University Press, 1990), 170–88.

90. King to Groppi, September 4, 1967, folder 3, box 2, JG.

91. James Weighart and Dan Patrinos, '' 'Cut Off Funds to Segregated Cities,' '' *National Catholic Reporter,* 3 (September 27, 1967), 3; ''Depth of Race Hatred Illustrated in 'I am the Priest in this Picture,' '' *Texas Catholic Herald* (November 17, 1967); Aukofer, *City with a Chance,* 126–27; William D. Cohen, ''Milwaukee march 'like it is': hate, song, dreams, frustration,'' University of Minnesota *Daily* (October 3, 1967), clipping in JG; Cousins wrote an editorial for the *Catholic Herald Citizen* on Groppi—see ''Milwaukee Archbishop Backs Priest's Cause,'' *Chicago Sun-Times,* September 14, 1967.

92. ''Hub Priests Fly to Milwaukee,'' *Boston Globe,* September 16, 1967, pp.1/4; Ken Rolling, O.F.M., to Groppi, September 21, 1967, folder 4, box 3, JG; Jeremiah Murphy, ''Marching Hub Priest Says Time Has Come,'' *Boston Globe,* September 20, 1967, p. 2.

93. Ruth Burley and Richard Burley to Groppi, April 16, 1968, folder 6, box 4, JG; John and Julie McCarthy to Groppi, April 2, 1967, folder 2, box 1, JG; St. Edward's parish in Providence, April 10, 1967, folder 2, box 1, JG; Msgr. George Casey, ''Fr. Groppi Won't Stop—Thank God,'' *St. Louis Review* (September 8, 1967).

94. Patrick and Maureen Coffey in letter to editor, *St. Louis Review,* September 22, 1967, 16–17; John D. Murnant et al., to Groppi, September 21, 1967, folder 4, box 3, JG; Richard J. Noll to Archbishop Cousins and Groppi, January 10, 1967, folder 2, box 1, JG.

95. E. B. Hayes to Groppi, October 6, 1967, folder 1, box 4, JG; Bob Hoffman to Groppi, February 2, 1967, folder 2, box 1, JG; Mr. and Mrs. Cliff Lund to Groppi, June 23, 1967, folder 6, box 1, JG.

96. Mrs. E. Meyer to Groppi, September 19, 1967, folder 1, box 7, JG; Fred J. Gordon to Groppi, August 31, 1967,folder 3, box 6, JG; Mrs. A.C. Caruso to Groppi, September 22, 1967, folder 4, box 3, JG.

97. Mrs. J. N. Hipp to Groppi, September 5, 1967, folder 5, box 6, JG; Msgr. A. J. Kanckert to Groppi, May 20, 1967, folder 2, box 6, JG; ''A German-Polish Catholic Family,'' to Groppi, September 6, 1967, folder 5, box 6, JG; Mrs. Elenore R. Haubert to Groppi, July 29, 1967, folder 2, box 6, JG.

98. Mrs. John C. Fonuke to Groppi, September 3, 1967, folder 5, box 6, JG; Mrs. M. Dugan to Groppi, September 1, 1967, folder 4, box 6, JG.

99. Mrs. Louis G. Cowan to Fr. Mario Shaw, O.S.B., March 3, 1966, file Indiana, box 7, series 10–2, NCCIJ.

100. Jay P. Dolan, *The American Catholic Experience: A History From Colonial Times to the Present* (Garden City: Doubleday, 1985), 385.

101. John LaFarge, "Worship and Fellowship," *Interracial Review,* 29 (July 1956), 110. The most sophisticated overview of American Catholic spiritual life is Joseph P. Chinnici, O.F.M., *Living Stones: The History and Structure of Catholic Spiritual Life in the United States* (New York: Macmillan, 1989), esp. 179–85. The founder of the American liturgical movement, Benedictine monk Virgil Michel, also expressed interest in interracial justice. Paul Marx, O.S.B., *Virgil Michel and the Liturgical Movement* (Washington, D.C.: Catholic University of America Press, 1957), 378–79, 402; William Hogan, interview with author, January 4, 1991.

102. Rev. Daniel Peil to Bishop Leo Pursley, August 20, 1965, and August 29, 1965, in Pursley correspondence, Archives of the Diocese of Fort Wayne; Elizabeth Louise Sharum, "A Strange Fire Burning: A History of the Friendship House Movement" (Texas Tech University, Ph.D. diss., 1977), 156–60; Rt. Rev. Msgr. Joseph P. Morrison, "Towards a Christian Order in Inter-racial Relations," in *The Sacramental Way,* Mary Perkins, ed. (New York: Sheed & Ward, 1948), 353–62.

103. C.J. McNaspy, S.J., *Our Changing Liturgy* (New York: Hawthorn Books, 1966), 101–2.

104. A brilliant discussion of the vehement rejection of traditional religious rituals by Catholic liberals is Robert Orsi, " 'Have You Ever Prayed to Saint Jude?': Reflections on Fieldwork in Catholic Chicago," paper in author's possession. Also see Richard Mazziotta, "When the Saints Went Marching Out," *Commonweal,* 119 (October 23, 1992), 14–16; Lawrence S. Cunningham, "Sacred Space and Sacred Time: Reflections on Contemporary Catholicism," in *The Incarnate Imagination: Essays in Theology, The Arts and Social Sciences in Honor of Andrew Greeley,* Ingrid H. Shafer, ed. (Bowling Green: Bowling Green State University Popular Press, 1988), 248–55.

105. Dennis Clark, "Parochial Roles," in *The Parish in Crisis,* John McCudden, ed. (Techny, Ill.: Divine Word Publications, 1967), 53; Rev. Donald McIlvane, quoted in Timothy Kelly, "The Transformation of American Catholicism: The Pittsburgh Laity and the Second Vatican Council," (Carnegie-Mellon University, Ph.D. diss., 1991), 322.

106. Joseph F. Roccasalvo, S.J., "Harlem Diary," *Woodstock Letters,* 94 (1965), 427–44.

107. "Mike" to Fr. Egan, December 17, 1965, misc. file, box 9, EGAN.

108. Archdiocesan School Board Minutes, April 13, 1967, AAC; "Chicago Parents Hit Religion Books," *National Catholic Reporter,* 3 (August 16, 1967), 3. A thoughtful treatment of these matters is William D. Dinges, "Ritual Conflict as Social Conflict: Liturgical Reform in the Roman Catholic Church," *Sociological Analysis,* 48 (Summer 1987), 138–57.

109. Fr. John McNamee, interview with author, July 31, 1991. On the church renovation from the perspective of the architectural world, see Stanislaus Von

Moos, *Venturi, Rauch & Scott Brown Buildings and Projects* (New York: Rizzoli, 1987), 301–2; "Saint Francis de Sales Church," *Liturgical Arts,* 38 (August 1970), 125–26; CRS, "Electric Demolition," *Progressive Architecture,* 51 (September 1970), 92–94.

110. See especially Philip Gleason, *Keeping the Faith: American Catholicism Past and Present* (Notre Dame: University of Notre Dame Press, 1987), 82–96, and Daniel Bell, "Religion in the Sixties," *Social Research,* 37 (Autumn 1971), 461–71.

Chapter 9

1. Pedro Arrupe, S.J., "To the Members of the Society of Jesus in the United States," *Catholic Mind,* 66 (January 1968), 16–24.

2. "The Catholic Church in Boston: A Pastoral Letter of the Association of Boston Urban Priests," *Pentecost,* 1969, JAL.

3. Dearden to all pastors, August 20, 1963, "Black Catholic" file, AAD. For sympathetic portraits of Dearden, see Hiley H. Ward, "Detroit: Pangs of a Progressive Diocese," *US Catholic,* 36 (July 1971), 32–35; Robert J. McClory, "Two Cities, two Bishops, and Vatican II," *National Catholic Reporter,* 18 (April 2, 1982), 21, and idem, "President of the American Bishops: Profile of Archbishop Dearden," *Ave Maria,* 104 (December 3, 1966), 16–17. On the importance of the Vatican II experience for Dearden, see Thomas J. Reese, S.J., *A Flock of Shepherds: The National Conference of Catholic Bishops* (Kansas City: Sheed & Ward, 1992), 40–43.

4. Press Releases, Project Commitment, February 2, 1966, Spring 1967, series I, subseries B 3–39, ACHR; Hiley H. Ward, "Bias is Target of New Drive by Catholics," *Detroit Free Press,* January 1, 1966, p.1.

5. Press Releases, Project Commitment, February 2, 1966, Spring 1967, series I, subseries B 3–39, ACHR; "Heavy Attendance Greets Project Commitment Debut," *Michigan Catholic* (February 10, 1966), 1.

6. D. J. R. Bruckner, "Church Held Faltering in Chicago Civil Rights," *Los Angeles Times,* March 23, 1966, p.6; Sidney Fine, *Violence in the Model City: the Cavanaugh Administration, Race Relations, and the Detroit Riot of 1967* (Ann Arbor: University of Michigan Press, 1989), 86–87; "Launch Fund To Help Parishes in Inner City," *Michigan Catholic* (May 26, 1966, n.p.).

7. "New Coughlin Stir in Detroit," *Catholic Transcript* (June 9, 1967), 5; James Sheehan, interview with author, September 20, 1990; John Hooper, interview with author, September 21, 1990.

8. Picture on front page of *Michigan Catholic* (August 3, 1967); Alec MacNealy, interview with the author, September 21, 1990; "Local Priest Caught by Riot," *Louisville Record,* August 31, 1967, clipping in Josephite Archives, Baltimore; "Churches Were Opened to Detroit Riot Victims," *New York Catholic News* (August 3, 1967); "Clergy Efforts Ineffective in Detroit," *New Jersey Advocate* (July 27, 1967); Fr. Geno Baroni, "Detroit: City Under Fire; Ruined Homes and Lives," *The Catholic Light* (August 3, 1967), 3; Fine, *Violence in the Model City,* 297–301.

9. John F. Dearden, "Challenge to Change in the Urban Church," in *The Church and the Urban Racial Crisis,* Mathew Ahmann and Margaret Roach, eds.

(Techny, Ill.: Divine World Publications, 1967), 43–44, 50; Fine, *Violence in the Model City,* 317–19.

10. "Fifty All-White Detroit Parishes Told They Are Infected With Racism," *National Catholic Reporter,* 4 (April 3, 1965), 7.

11. John C. Haughey, "Detroit: Evolution of a Revolution," *America,* 120 (April 19, 1969), 475; William E. Schmidt, "Detroit Priest Preaches Hope Through Job Training," *New York Times,* January 1, 1991, pp. 1/7; Keith Pitcher, "Abp. Dearden Outlines How ADF Will Assist Needy in Inner City," *Michigan Catholic* (April 11, 1968), 1; "Act to Ease Racial Tensions," *Michigan Catholic* (March 14, 1968).

12. Frank Tully, letter to the editor, *Michigan Catholic* (March 14, 1968); Mr. John Lohela, letter to editor, *Michigan Catholic* (May 16, 1968); "Fifty All-White Detroit Parishes Told They Are Infected With Racism," *National Catholic Reporter,* 4 (April 3, 1968), 7; Barbara Van Rysseghem, letter to editor, *Michigan Catholic* (August 17, 1967).

13. "Loss of Aid May Close 105 Michigan Schools," *National Catholic Reporter,* 5 (December 4, 1970), 1.

14. "Now is the Time for All Good Men to . . . ," *Michigan Catholic* (May 2, 1968); Frank Tully refers to Fr. James Sheehan's questioning of school aid bill in letter to editor, *Michigan Catholic* (March 14, 1968).

15. Ward, "Detroit: Pangs of a Progressive Diocese," 33. Statistics from 1968 and 1973 editions of *The Official Catholic Directory* (New York: P. J. Kenedy and Sons, 1968), 245 and (1973) 259.

16. Thomas F. Hinsberg, memo to Dearden, November 27, 1970, file 9, box 48, Bishop Thomas Gumbleton Papers, MAUND.

17. Paul Wrobel, *Our Way: Family, Parish, and Neighborhood in a Polish-American Community* (Notre Dame: University of Notre Dame Press, 1979), 89–91; Samuel Lubell, *The Future While It Happened* (New York: W. W. Norton, 1973), 53; Thaddeus Radzialowski, "View From a Polish Ghetto," *Ethnicity,* 1 (1974), 142.

18. Raymond A. Schroth, "Detroit, 1967," *America,* 117 (August 12, 1967), 151.

19. Garry Wills, *Bare Ruined Choirs,* 49–56.

20. Rev. George H. Tavard, "Correspondence," *Commonweal,* 80 (August 7, 1964), 547.

21. John Courtney Murray, "Freedom, Authority, Community," *America,* 115 (December 3, 1966), 734–41; J. Bryan Hehir, "The Unfinished Agenda," *America,* 153 (November 30, 1985), 386–88; Gerald P. Fogarty, *The Vatican and the American Hierarchy From 1789 to 1965* (Stuttgart: Anton Hiersemann, 1982), 368–403.

22. Daniel T. Rodgers makes the point about "freedom" in *Contested Truths: Keywords in American Politics Since Independence* (New York: Basic Books, 1987), 212–22.

23. A useful summary is found in Debra Campbell, "The Struggle to Serve: From the Lay Apostolate to the Ministry Explosion," in *Transforming Parish Ministry: The Changing Roles of Catholic Clergy, Laity, and Women Religious,* Jay P. Dolan et al., eds. (New York: Crossroad, 1989), 201–80.

24. Gerard Francis Powers, "Under One God: The Catholic Church and Race Relations in Cleveland," (Princeton University, B.A. thesis, 1980), 50.

25. John J. O'Connor, "Tomorrow Is Now," at testimonial dinner, June 4, 1963, box 10, carton 2, NYCIC.

26. See Meyer to Cantwell, July 4, 1962, folder 5, box 8, Cantwell Papers, CHS.

27. McDermott to Cantwell, March 4, 1964, McDermott file, box 109, CIC; McDermott to Cantwell, March 4, 1964, McDermott file, box 109, CIC; McDermott to Clark, March 31, 1964, file March 29–31, 1964, box 68, CIC; McDermott to Ralph Duggan, June 9, 1963, file June 1–10, box 60, CIC; McDermott to Egan, October 19, 1961, file Oct. 18–21, 1961, box 46, CIC; John A. McDermott, interview with author, December 10, 1990.

28. Margaret Murphy, *How Catholic Women Have Changed* (Kansas City: Sheed & Ward, 1987), 64.

29. Some of the lay frustration is apparent in Daniel Callahan, *The New Church: Essays in Catholic Reform* (New York: Charles Scribner's Sons, 1966), 104–23.

30. "CIC President Cites Parochial Segregation," *Witness,* 1 (Chicago CIC Publication, November 1967); Edward M. Keating, *The Scandal of Silence* (New York: Random House, 1965), 59; Ed Marciniak, *Tomorrow's Christian* (Dayton: Pflaum Press, 1969), 12.

31. Hope Brophy, interview with author, September 19, 1990.

32. On this issue generally, see Sara Evans, *Personal Politics: The Roots of Women's Liberation in the Civil Rights Movement and the New Left* (New York: Knopf, 1979), 28–38.

33. Mary Daly, "A Built-In Bias," *Commonweal,* 81 (January 15, 1965), 510. Daly later voiced one of the first explicitly feminist attacks on Church structures in *The Church and the Second Sex,* 2nd ed. (New York: Harper & Row, 1975).

34. Helen Sanders, S.L., *Loretto Before and After Vatican II: 1952–1977* (Nerinx, Ken.: Sisters of Loretto, 1982), 107.

35. Dennis Geaney, "Two Notebooks on the Council," *New City,* 5 (January 1966), 6.

36. Sr. M. Benet, O.S.B., "The Sister in the Parish," in *The Parish in Crisis,* 157.

37. James H. Bowman, S.J., "Martin Luther King in Chicago," *Ave Maria,* 102 (September 25, 1965), 8–9; Maureen Fielder, "Riding the City Bus From Pittsburgh," in *Midwives of the Future: American Sisters Tell Their Story,* Ann Patrick Ware, ed. (Kansas City: Leaven Press, 1985), 37–52; Sister M. Berchmans Shea, O.S.U., "Protest Movements and the Renewal of Religious Life," *National Catholic Reporter,* 2 (April 27, 1966), 10.

38. Boston CIC newsletter, 2 (June 1966), copy in Boston file, box 12, series 10–2, NCCIJ.

39. Quoted in James Hennesey, S.J., *American Catholics: A History of the Roman Catholic Community in the United States* (New York: Oxford University Press, 1981), 312; Roger Kuhn to Mitchell P. Briggs, August 17, 1964, "LA Civil Rights" file, box 3, Commonweal Papers, MAUND; A. V. Krebs, Jr., "A Church of Silence," *Commonweal,* 80 (July 10, 1964), 467–76; Mike Davis, *City of Quartz: Excavating the Future of Los Angeles* (New York: Verso, 1991), 323–72.

40. Roger Kuhn to Mitchell P. Briggs, August 17, 1964, Emil Seliga to Mathew Ahmann, October 1, 1963, box 2, series 10–2, NCCIJ. A. V. Krebs, Jr., "A Church of Silence," *Commonweal,* 80 (July 10, 1964), 467–76, and John Leo, "The DuBay Case," *Commonweal,* 80 (July 10, 1964), 477–82 are useful summaries of the initial incidents.

41. DuBay to John Leo, January 1, 1964, "1A-LA civil rights" file, box 3, Commonweal Papers, MAUND; Press Release, Catholics United For Racial Equality, August 16, 1966, "LA Civil Rights" file, box 3, Commonweal Papers.

42. Charles W. Dahm in collaboration with Robert Ghelardi, *Power and Authority in the Catholic Church: Cardinal Cody in Chicago* (Notre Dame: University of Notre Dame Press, 1981), 28–36; Joseph Scimecca and Ronald Damiano, *Crisis at St. John's: Strike and Revolution on the Catholic Campus* (New York: Random House, 1967), 46; Rev. William H. DuBay, letter to editor, *Commonweal,* 84 (April 1, 1966), 62.

43. A provocative analysis of the birth control encyclical's impact on the American Church is contained in Andrew Greeley, William C. McCready, Kathleen McCourt, *Catholic Schools in a Declining Church* (Kansas City: Sheed and Ward, 1976), 103–54; also see Andrew M. Greeley, *The Catholic Myth,* 15–33; John Seidler and Katherine Meyer, *Conflict and Change in the Catholic Church* (New Brunswick: Rutgers University Press, 1989), 92–108.

44. Philip Murnion and Henry J. Browne, memo on James Forman demands, July 30, 1969, file 1968–69, box 24, Henry J. Browne Papers, Columbia University Special Collections.

45. On Cody, Edward R. Kantowicz, "The Beginning and the End of an Era: George William Mundelein and John Patrick Cody in Chicago," in *Patterns of Episcopal Leadership,* Gerald Fogarty, ed. (New York: Macmillan, 1989), 211; on Lucey, see Saul E. Bronder, *Social Justice and Church Authority: The Public Life of Archbishop Robert E. Lucey* (Philadelphia: Temple University Press, 1982), and David Rice, *Shattered Vows: Priests Who Leave* (New York: William Morrow, 1990), 14–24.

46. Andrew M. Greeley, *Priests in the United States: Reflections on a Survey* (Garden City: Doubleday, 1972), 102–17; Seidler and Meyer, *Conflict and Change,* 128–46.

47. Report on Negro Work and Application for Aid, August 15, 1953, folder 5, box 3, series 5/2, BCIM.

48. Report on Negro Work and Application for Aid, September 17, 1956, folder 1, box 4, series 5.2, BCIM; Report on Negro Work and Application for Aid, September 7, 1962, folder 8, box 12, series 5.2, BCIM.

49. Benedict Anderson, *Imagined Communities: Reflections on the Origins and Spread of Nationalism* (London: Verso, 1991), 14–15.

50. *The Documents of Vatican II,* Walter M. Abbott, S.J., ed. (New York: Guild Press, 1966), 264. For an introduction to these topics, see the various essays in *Modern Catholicism: Vatican II and After,* Adrian Hastings, ed. (New York: Oxford University Press, 1991).

51. Henry J. Offer, S.S.J., "Black Power—A Great Saving Grace," *American Ecclesiastical Review,* 159 (September 1968), 193–201; also see Joseph H. Fichter, "Black and Catholic, VI," *America,* 142 (March 29, 1980), 267–69.

52. Margaret Cronyn, "Centers Provide Food and Shelter," *Michigan Catholic*

(July 27, 1967), 6; Fr. Jack Farry, interview with author, January 4, 1990; Lawrence M. O'Rourke, *Geno: The Life And Mission of Geno Baroni* (Mahwah, N.J.: Paulist Press, 1991), 24. Divisions between African-American Catholics who focused on the cultural universality of the Church and those who emphasized the need for an African-American identity persisted into the 1980s. See *Faith, Hope and Charity: A Challenge to the Catholic Church in Chicago: The Development of Catholic Parishes in Predominantly Black Communities* (Chicago: Cartwright and Associates, 1989), 4/1, 4/2. I am grateful to Fr. Charles Payne, O.F.M., for giving me a copy of this report.

53. "Black Catholics' Board Affirms Bishops' Fund Rejection," December 15, 1970, "Black Catholics thru 75" file, newsroom, United States Catholic Conference.

54. Lawrence Lucas, *Black Priest/White Church: Catholics and Racism* (New York: Random House, 1970), 11–12, 235; Clements also discussed these issues on a national television news program; see "We Have Had to Accept Rejection—Can You?" *New York Times,* May 25, 1969, II, p. 35; Fr. George Clements, interview with Rev. Clarence Williams, n.d, copy in Garrett-Evangelical Library, Evanston; Claudia Peek, "Threats, Hate Letters Fail to Sway St. Cecilia Rector," *Michigan Catholic* (November 19, 1968).

55. On the meeting of the African-American priests, Lawrence Lucas, "Black Priest Assesses Caucus," *National Catholic Reporter,* 4 (May 8, 1968), 4; Cyprian Davis, O.S.B., *The History of Black Catholics in the United States* (New York: Crossroad, 1990), 257–58; Harold Schachern, "Priests Call Church Racist," *National Catholic Reporter,* 4 (April 24, 1968), 1.

56. Fr. Donald Clark in Minutes of the Seventh General Meeting of the National Conference of Catholic Bishops, November 10–14, 1969, file 15, box 22, John Cardinal Dearden Papers, MAUND. On "black" and "Catholic," see Sara Harris, *The Sisters: The Changing World of the American Nun* (Indianapolis: Bobbs-Merrill, 1970), 272–73, and Robert L. Pitts to Bishop Issenman [c. 1970], CAI-CCCA file, ADC. On the Young Christian Students, Father Vincent Giese, *You Got It All: A Personal Account of a White Priest in a Chicago Ghetto* (Huntington: Our Sunday Visitor, 1980), 28–40.

57. See "A Statement of the Black Catholic Clergy Caucus, April, 18, 1968," in *Black Theology: A Documentary History, 1966–1979,* Gayraud S. Wilmore and James H. Cone, eds. (Maryknoll: Orbis Books, 1979), 322–24.

58. Edward B. Fiske, "Boland Assailed on 2 New Vicars," *New York Times,* February 20, 1969, p.39; Edward B. Fiske, "1000 Back Priests on Racist Charge," *New York Times,* January 20, 1969, p.49; "Boland Meets Accusers; Spokesman Softens Charge," *National Catholic Reporter,* 5 (January 22, 1969), 4; Edward B. Fiske, "20 Priests Accuse Newark Pastors of 'Racial Cruelty'," *New York Times,* January 10, 1969, p. 49; Anne Buckley, "Inner City Priests Criticized, Upheld," *The Advocate* (January 16, 1969), 1; "20 Priests Brand Newark Archbishop Racist," *Catholic Standard and Times* (January 10, 1969).

59. William Hogan, interview with author, January 4, 1991; Fr. Jack Farry, interview with author, January 3, 1991; "Blacks in Controversy at St. Dorothy," *The New World* (January 10, 1969); "St. Dorothy Controversy Stirs Blacks and

Whites,'' *The New World* (January 10, 1969); James H. Bowman, ''City's Four Black Priests May Quit,'' *Chicago Daily News* (December 10, 1968), 1.

60. Lucas,*Black Priest/White Church* 120; ''Negro Pastor in Chicago Says Cody Is 'Unconsciously Racist','' *New York Times*, January 11, 1969, p.37; Francis Ward, ''Say Cody to Name Militant as Pastor,'' *National Catholic Reporter*, 5 (January 10, 1969), 1.

61. Cody to Archbishop Krol, July 15, 1969, file ''Rev. George C. Clements,'' box 2006, ACHREC; Roger T. Flaherty, ''Chicago's 7 Negro Priests Hit Church Paternalism,'' *The New World* (February 21, 1968).

62. Report of the Executive Director, August, 1968, box 6, series 1, NCCIJ; Mathew Ahmann, ''Strategies for the Future,'' in *The Church and the Urban Racial Crisis*, 233; Sister Mary Benet, O.S.B., in *Encounter*, 6 (May 1969), 1, copy in UAS file, box 2, Association of Chicago Priests Papers, MAUND.

63. ''Recommendations of the Commission on Catholic Community Action on the Proposals of the Black Task Force,'' December 1969, 1969 Commission Task Force file, ADC; Rev. John Powis, letter to editor, *Brooklyn Tablet* (July 23, 1970).

64. ''A Statement of the Black Catholic Clergy Caucus, April 18, 1968,'' 322–23; Hope Brophy, interview with author, September 19, 1991.

65. Thomas F. Campbell, ''Cleveland: The Struggle for Stability,'' in *Snowbelt Cities: Metropolitan Politics in the Northeast and Midwest Since World War II*, Richard M. Bernard, ed. (Bloomington: Indiana University Press, 1990), 112–20. On Cleveland and the Stokes campaign, see Kenneth G. Weinberg, *Black Victory: Carl Stokes and the Winning of Cleveland* (Chicago: Quadrangle Books, 1968), 104–51.

66. Report of the Executive Director, August 1967, box 5, series 1, NCCIJ.

67. NCCIJ and American Council for Nationalities Service proposal to Ford Foundation, June 20, 1967, file correspondence/memos, box 1, series 18, NCCIJ; Ralph Brody, ''Project Bridge: An Assessment of a Program to Improve black-white relations in Cleveland, Ohio,'' August 1969, 3, copy in ADC.

68. ''An Interim Report on Project Bridge Covering the Period November 1, 1967 through January 31, 1968,'' Project Bridge file, ADC.

69. Brody, ''Project Bridge,'' 6–14; ''Project Bridge Report,'' n.d., Description of Project Bridge 1968–69 file, box 1, series 18, NCCIJ.

70. ''Project Bridge Report,'' n.d., Description of Project Bridge 1968–69 file, box 1, series 18, NCCIJ; Brody, 14.

71. ''Project Bridge,'' March 1969, NCCIJ Program Reports.

72. Powers, ''Under One God,'' 133–34; Brody, ''Project Bridge,'' 58–59.

73. Brody,''Project Bridge,'' 51; Executive Committee Minutes, January 20, 1969, Bishop's Committee on Urban Affairs file, ADC; Powers, ''Under One God,'' i–ii; ''Church Tussle Heads for Courts,'' *National Catholic Reporter*, 5 (February 5, 1969), 1; Michael Gallagher, *Laws of Heaven: Catholic Activists Today* (New York: Ticknor and Fields, 1992), 12–17; Sister Helen, ''Project Bridge in Cleveland,'' in *New Works of New Nuns*, Sr. M. Peter Traxler, S.S.N.D., ed. (St. Louis: B. Herder, 1968), 162.

74. Andrew M. Greeley, *Why Can't They Be Like Us?: America's White Ethnic Groups* (New York: E. P. Dutton, 1971), 156, and Michael Novak, *The Rise of*

the Unmeltable Ethnics: Politics and Culture in the Seventies (New York: Macmillan, 1972) are good introductions to this literature. Mikulski quoted in Jack Rosenthal, "Anger at Power Structure Voiced at Urban Ethnic Parley," *New York Times,* June 17, 1970, p.49.

75. Arthur Mann, *The One and the Many: Reflections on the American Identity* (Chicago: University of Chicago Press, 1979), 5–45.

76. Quoted in ibid., 29–30. Baroni in Minutes of the Seventh General Meeting of the National Conference of Catholic Bishops, November 10–14, 1969, file 15, box 22, Dearden Papers.

77. Particularly useful here is Richard D. Alba, *Ethnic Identity: The Transformation of White America* (New Haven: Yale University Press, 1990); also see Mary C. Waters, *Ethnic Options: Choosing Identities in America* (Berkeley: University of California Press, 1990), esp. 16–51. An important essay in this literature is Herbert J. Gans, "Symbolic Ethnicity: The Future of Ethnic Groups and Cultures in America," in *On the Making of Americans: Essays in Honor of David Riesman,* Herbert J. Gans, et al., eds. (Philadelphia: University of Pennsylvania Press, 1979), 193–220.

78. Anonymous to Cardinal Cooke, 68–69 file, box 24, Browne Papers; Mann, 40–45.

79. Anthony Monahan, "Father Lawlor," *Chicago Sun-Times Midwest Magazine* (March 9, 1969), 6–14; M. W. Newman and Edmund J. Rooney, " 'Holding the Line' at Ashland Av.," *Chicago Daily News,* April 18, 1968, p.1; also see Brian J. L. Berry, *The Open Housing Question: Race and Housing in Chicago, 1966–1976* (Cambridge: Ballinger, 1979), 184–89.

80. Newman and Rooney, "Holding the Line," p.3; William Dendy, "Father Lawlor, SLCC Battle Mars Unity on South Side," *The New World* (February 9, 1968), 1.

81. Jerry DeMuth, "Lawlor Says He Just Wanted to Hold Whites," *National Catholic Reporter,* 4 (February 28, 1968), 2.

82. Lawlor, "As I see It," *Associated Block Club News,* 1 (February 1969) in Lawlor file, box 2007, ACHREC; Newman and Rooney, "Holding the Line," p.3.

83. Edmund J. Rooney, "Fr. Lawlor Coming Back to Chicago," *Chicago Daily News,* March 22, 1968, p.1; "Priest Attacks Lawlor," *The Daily Defender,* November 18, 1969, p.3.

84. "Block Clubs Plan to Picket at O'Toole," *Southtown Economist* (April 15, 1970), 1.

85. "Rally Raps Redmond," *Southtown Economist* (September 18, 1968), 1; Robert McDermott, "Father Lawlor: His Kingdom and Power," *Chicago Tribune Magazine* (February 21, 1971), 23–26; several newsletters are in the Lawlor file, box 2007, ACHREC.

86. Jerry DeMuth, "Lawlor Organizes in Suburbs, Block Clubs Now Total 160," *National Catholic Reporter,* 5 (January 8, 1969), 2; Lawlor, "The White Minority," *Southwest Associated Block Club News* (July 15, 1968), Lawlor file, box 2007, ACHREC.

87. Diane Monk, "Gage Park's Whites Sound Off in Bid to Shift School Boundary," *Chicago Sun-Times,* April, 7, 1972, p.8.

88. Lawlor, "The White Minority"; Little Flower parishioners, August 10, 1969, Lawlor file, box 2007, ACHREC.

89. "Father Lawlor Returns, Defies Cody," *Chicago American,* March 22, 1968, p.1; "Lawlor Supporters Meet," *Southtown Economist* (February 11, 1968, n.p.); Anthony Monahan, "Father Lawlor," *Chicago Sun-Times Midwest Magazine* (March 9, 1969), 6–14, all clippings in Municipal Reference Library, Lawlor file, City of Chicago.

90. Dahm, *Power and Authority,* 78–80.

91. John J. Quattrocki to Cody, February 28, 1968, Lawlor file, box 2007, ACHREC.

92. Mrs. John O'Callaghan to Msgr. Edward Egan, August 6, 1969, Lawlor file, box 2007, ACHREC; Mrs. Albert Duffy to Msgr. Edward Egan, August 20, 1969, Lawlor file, box 2007, ACHREC; Mrs. William P. Biros to Msgr. Edward Egan, October 17, 1969, Lawlor file, box 2007, ACHREC.

93. Kevin Phillips, *The Emerging Republican Majority* (Garden City: Anchor Books, 1970), 166–72; also see Garry Wills, *Nixon Agonistes: The Crisis of the Self-Made Man,* rev. ed. (New York: New American Library, 1979), 247–49.

94. One analyst termed the Catholic switch to the Republicans the most significant factor in the 1972 election. See Paul Lopatto, *Religion and the Presidential Election* (New York: Praeger, 1985), 54–62; George Gallup, Jr., and Jim Castelli, *The American Catholic People: Their Beliefs, Practices, and Values* (New York: Doubleday, 1987), 126–38; Nixon in Herbert S. Parmet, *Richard Nixon and His America* (Boston: Little, Brown, 1990), 578, 630; Rick Casey, "Nixon Pledges Aid at NCEA Meeting," *National Catholic Reporter,* 8 (April 14, 1972), 1; Rick Casey, "Nixon Chasing Catholic Voters," *National Catholic Reporter,* 8 (August 18, 1972), 1.

95. Estimates calculated from data in Diane B. Gertler, *Non-Public Schools in Large Cities, 1970–1971* (Washington, D.C.: U.S. Government Printing Office, 1974), 5, 11–12 and Thomas R. Swartz and Frank J. Bonello, "What Happened to the Catholic School Crisis?" *Journal of Church and State,* 19 (Spring 1977), 249–51. Also see the earlier estimates of Neil G. McCluskey S.J., in *Catholic Education Faces Its Future* (Garden City: Doubleday, 1968), 45, 153. In Chicago, the harried archdiocesan school board simply handed each pastor an envelope with the figure for the school drive enclosed; see George Fornero, "The Expansion and Decline of Enrollment and Facilities of Secondary Schools in the Archdiocese of Chicago, 1955–1980: A Historical Study" (Loyola University, Ph.D. diss., 1990), 49; Brother Kyrin Powers, C.F.X., "Factors Affecting the Decline in Enrollment in the Catholic Secondary Schools in the Diocese of Brooklyn" (Fordham University, Ph.D. diss., 1974), 1; Mary Perkins Ryan, *Are Parochial Schools the Answer?* (New York: Holt, Rinehart and Winston, 1964), 8.

96. Herbert Gans, *The Levittowners: Ways of Life and Politics in a New Suburban Community* (New York: Columbia University Press, 1982), 96–97; Harold M. Wattell, "Levittown: a Suburban Community," in *The Suburban Community,* William M. Dobriner, ed. (New York: G. P. Putnam's Sons, 1958), 309.

97. Figures cited in Anthony S. Bryk, Valerie E. Lee, and Peter B. Holland, *Catholic Schools and the Common Good* (Cambridge: Harvard University Press, 1993), 33.

98. William F. O'Connor, "Aid to Education," *Commonweal,* 74 (June 23, 1961), 328–29.

99. For example, Arthur Block, " 'Sister, Are You White or Black?' " *America,* 128 (January 27, 1973), 60–61; Sr. Charles Borromeo Muckenhirn, C.S.C., "Poverty and Property in Religious Life," *National Catholic Reporter,* 2 (June 1, 1966), 11. Several of the essays in *New Works for New Nuns,* Sr. M. Peter Traxler, S.S.N.D., ed. (St. Louis: B. Herder, 1968) discuss the sense of release experienced by women religious when freed from traditional educational duties; see Sr. Evangeline Meyer, S.S.N.D., "Parish Servant to a Pilgrim People," 3; Sr. Rosemary Keegan, S.C., "Climb Every Mountain," and Sr. Miriam St. John S.N.D., "A Place to Listen." Also see Michael Novak, "The New Nuns," in *The New Nuns,* Sr. M. Charles Borromeo, C.S.C., ed. (New York: New American Library, 1967), 19.

100. The author of one of the original studies, Andrew Greeley, convincingly rebuts criticisms of Catholic schools made during the 1960s and suggests critics of the schools willfully misread the data. See his *Catholic Schools in a Declining Church,* esp. 157–220, 282–329. Also see Dave Meade, "Future of Catholic Schools Here Questioned," *Chicago Daily News,* May 18, 1966, p.31; Ryan, *Are Parochial Schools the Answer?,* 55–56, 78, 159.

101. New Jersey priest quoted in James Colaianni, *The Catholic Left: The Crisis of Radicalism Within the Church* (Philadelphia: Chilton, 1968), 185; Jerry DeMuth, "Whites Quit Church Over Black Power? 'Let Them Go' Groppi Says in Chicago," *National Catholic Reporter,* 4 (October 25, 1967), 3.

102. Ann Dodson, "Negro Urges Catholics to Sell Schools," *National Catholic Reporter,* 2 (September 7, 1966), 11.

103. CIC statistics, "1969 CIC" folder, box 2, Rev. Donald McIlvane Papers, MAUND; also see Fr. Donald W. McIlvane, "Racial Balance in Catholic Schools," *Community,* 25 (May 1966), 7; Thomas Pettigrew, *Racially Separate or Together?* (New York: McGraw-Hill, 1971), 57.

104. Joseph M. Cronin, "Negroes in Catholic Schools," *Commonweal,* 85 (October 7, 1966), 14.

105. See "Catholic Schools Called Upon to Help Change any Racist Attitudes of Pupils," *Philadelphia Inquirer,* May 18, 1964, p.4; "Priests Assail Church Schools for Racism," *Philadelphia Inquirer,* May 17, 1968, p.1; *Equality* (Pittsburgh CIC newsletter), 6 (July-August 1967), 2, file Pittsburgh, series 10–2, box 20, NCCIJ.

106. "Dilworth: Too Few Non-White Pupils in Parochial Schools," *Catholic Standard and Times* (March 11, 1966). For similar criticisms in New York, see Edith Evans Ashbury, "City Plan to Use Church Schools Draws Protests," *New York Times,* August 8, 1966, pp.1/24.

107. Archdiocesan school board minutes, esp. Jan. 14, 1965, October 18, 1967, January 19, 1968, all in AAC.

108. Cody's statement to the press, March 18, 1968, Busing controversy file, box 1, Association of Chicago Priests collection, MAUND; Cody to all clergy, January 25, 1968, Busing controversy file, box 1, Association of Chicago Priests collection, MAUND.

109. "Says Catholic Schools Won't Become Havens for Segregationists," No-

vember 8, 1971, Catholic schools—integration, newsclippings file, United States Catholic Conference.

110. Henry DeZutter, "Cody's Bus Plan in Review," March 19, 1969, clipping in Busing controversy file, box 1, Association of Chicago Priests collection, MAUND; William Dendy, "Anti-Busing Group Protests at City Hall, Cardinal's Home," *The New World* (February 2, 1968); James Hannon, "Catholic and Public in Boston Integration," *Population Research and Policy Review,* 3 (1984), 222; Ronald P. Formisano, *Boston Against Busing: Race, Class, and Ethnicity in the 1960s and 1970s* (Chapel Hill: University of North Carolina Press, 1991), 50; Lukas, *Common Ground,* 400.

111. Mrs. Sophia Bakos to Cody, February 15, 1968, file "B," box 2006, ACHREC; *A History of the Parishes of the Archdiocese of Chicago,* Msgr. Harry C. Koenig, ed. (Chicago: The Archdiocese of Chicago, 1980), 1: 190. One readers' poll done by the relatively liberal journal *U.S. Catholic* found 81 percent of readers opposed to busing programs. Dan Herr, "Let's Stop Busing for Integration," *U.S. Catholic,* 37 (May 1972), 12–13.

112. Charles P. Hammock, interview with author, July 31, 1991; "Pupils Agree to Return to Class," *Catholic Standard and Times* (March 5, 1970); "Black Students Making Strong Demands to Phila. School Head," *Syracuse Catholic Sun* (March 19, 1970); "The Tough Cop," *Camden Catholic Star Herald* (June 5, 1970); Richard Murphy and John P. Corr, "Negro Students Protest Ouster of Senior, Archdiocese is Firm," *Philadelphia Inquirer,* March 4, 1970, p.1.

113. "Sociologist Claims Diocese's Schools Fail Pupils Because of Segregation," *National Catholic Reporter,* 3 (September 6, 1967), 5.

114. Bob Sakamoto, "Tradition Helps Make Mt. Carmel Top Dog," *Chicago Tribune,* September, 1, 1991, section 3, p.15; Tom Brune and James Ylisela, Jr., "The Making of Jeff Fort," *Chicago* 37 (November 1988), 204; *Report of the Chicago Riot Study Committee to the Honorable Richard J. Daley* (Chicago, 1968), 9; Maryann Kathleen Janosik, "Propagating the Faith: Catholic Educational Policy Making in Post-Vatican II Cleveland" (Case-Western Reserve, Ph.D. diss., 1989), 311–12.

115. James C. Donahue, "New Priorities in Catholic Education," *America,* 118 (April 13, 1968), 476–79.

116. "Bishop Defends Schools, Disclaims Sister's Views," *National Catholic Reporter,* 1 (April 21, 1965), 5.

117. "Sociologist Claims Diocese's Schools Fail Pupils Because of Segregation," *National Catholic Reporter,* 3 (September 6, 1967), 5.

118. Berry, *The Open Housing Question,* 234–35, 256; Henry Luskin Molotch, *Managed Integration: Dilemmas of Doing Good in the City* (Berkeley: University of California Press, 1972), 91–92; Eleanor P. Wolf and Charles N. Lebeaux, "Class and Race in the Changing City," in *Social Science and the City: A Survey of Urban Research,* Leo F. Schnore, ed. (New York: Frederick A. Praeger, 1968), 108.

119. Memo of Meeting of Bishops of Michigan with the Major Superiors, November 23, 1970, file 9, box 48, Bishop Thomas Gumbleton Papers, MAUND; *Equality* (newsletter of the Pittsburgh CIC), 6 (July-August 1967), 2, file Pittsburgh, box 20, series 10–2, NCCIJ; Rt. Rev. Msgr. George Kelly, "Just How

Integrated Are Our Catholic Schools—Part I,'' *New York Catholic News* (August 18, 1966); ''Part II'' (October 6, 1966).

120. Mary Frase Williams, *Private School Enrollment and Tuition Trends* (Washington, D.C.: Center for Educational Statistics, Office of Educational Research and Improvement, 1987), 10.

121. Sr. Mary Ancilla Leary, ''The History of Catholic Education in the Diocese of Albany'' (Catholic University, Ph.D. diss., 1957), 116.

122. Edward K. Braxton, ''Authentically black, truly Catholic,'' *Commonweal,* 112 (February 8, 1985), 73–75; Mother Patricia Barrett, R.S.C.J., ''Nun in the Inner City,'' in *The New Nuns,* Sr. M. Charles Borromeo, C.S.C., ed. (New York: New American Library, 1967), 93; Minutes of Bishop Elwell Advisory Committee, May 18, 1964, Bishop's Committee on Urban Affairs file, ADC; Chicago Archdiocesan School Board Minutes, June 10, 1965, Archdiocesan school board file, 1965, AAC.

123. St. Agnes Parish Council to Dearden, December 11, 1970, file 9, box 48, Bishop Thomas Gumbleton Papers, MAUND; ''Cardinal Hears Inner-City Views on School Closings,'' *Michigan Catholic* (December 9, 1970); ''School Sit-In at Chancery in Detroit,'' *National Catholic Reporter,* 7 (March 12, 1971), 1. On Newark, Thomas R. Brooks, ''Breakdown in Newark,'' in *The World of the Blue-Collar Worker,* Irving Howe, ed. (New York: Quadrangle Books, 1972), 107. For a statement by the National Office of Black Catholics on the value of Catholic schools to the African-American community, see ''The Collapse of Catholic Schools in the Black Community,'' *Origins,* 5 (January 8, 1976), 528–32.

124. Kathleen Riley Fields, ''Bishop Fulton J. Sheen: An American Catholic Response to the Twentieth Century'' (University of Notre Dame, Ph.D. diss., 1988), 471–76; Douglas J. Roche, *The Catholic Revolution* (New York: David McKay, 1968), 76–78; ''Sheen Donates Parish for Poor,'' *National Catholic Reporter,* 4 (March 6, 1968), 5.

125. Response within the academic community to these studies—and the implication that the success of Catholic schools was not simply due to a preselected group of students—was hostile. See James S. Coleman, ''Response to the Society of Education Award,'' *Academic Questions,* 2 (Summer 1989), 76–78. The literature on Catholic schools and ''students at risk'' is growing rapidly. A comprehensive study of Catholic secondary education and survey of the available literature is Anthony S. Bryk, Valerie E. Lee, and Peter B. Holland, *Catholic Schools and the Common Good* (Cambridge: Harvard University Press, 1993), 57–58, 218–20, 247–48. Also indispensable are James S. Coleman and Thomas Hoffer, *Public and Private High Schools: The Impact of Communities* (New York: Basic Books, 1987), and Andrew M. Greeley, *Catholic High Schools and Minority Students* (New Brunswick: Transactions Books, 1982). Also see Abigail Thernstrom, ''Out-Classed,'' *The New Republic,* 204 (May 13, 1991), 12–14; Seymour P. Lachman and Barry A. Kosmin, ''Black Catholics Get Ahead,'' *New York Times,* September 14, 1991, p.19; Susan Chira, ''Where Children Learn How to Learn: Inner-City Pupils in Catholic Schools,'' *New York Times,* November 20, 1991, A14; David Gonzalez, ''Poverty Raises Stakes for Catholic School,'' *New York Times,* April 17, 1994, pp. 1/32.

126. Much of this paragraph is drawn from Coleman and Hoffer, *Public and Private High Schools,* 211–43.

127. Each morning students were asked two questions during the morning an-

nouncements: one on a famous African-American and another on Catholic doctrine. A typical pair might be "Who was Mahalia Jackson?" and "What is the Trinity?" See David Sutor, "We're Doing It By Ourselves," *U.S. Catholic,* 37 (November 1972), 27–33, and Patricia Krizmis, "Black Pride is a Way of Life in Holy Angels School," *Chicago Tribune,* March 7, 1971, p.3.

128. Richard P. McBrien, *Do We Need the Church?* (New York: Harper and Row, 1969), 219; also see Philip Gleason, *Keeping the Faith: American Catholicism Past and Present* (Notre Dame: University of Notre Dame Press, 1987), 58–81.

129. Quoted indirectly in Andrew Greeley, *The Hesitant Pilgrim: American Catholicism After the Council* (New York: Sheed & Ward, 1966), 30.

130. Scimecca and Damiano, *Crisis at St. John's,* 3; Edward J. Ahern, "The Search for Integrity," in *Generation of the Third Eye,* Daniel Callahan, ed. (New York: Sheed & Ward, 1965), 29.

131. Layton P. Zimmer, "The People of the Underground Church," in *The Underground Church,* Malcolm Boyd, ed. (New York: Sheed and Ward, 1968), 17–18; "Experimental Parish Forms—Without Okay," *National Catholic Reporter,* 3 (September 13, 1967); George J. Hafner, "Up From the Underground," in *The Underground Church,* 125, 130; Robert E. Grossman, "The Invisible Christian," in *The Underground Church,* 210; David Kirk, "Emmaus: A Venture in Community and Communication," in *The Underground Church,* 149; James E. Groppi, "The Church and Civil Rights," in *The Underground Church,* 70–83; 3; also see Meredith B. McGuire, "An Interpretive Comparison of Elements of the Pentecostal and Underground Church Movements in American Catholicism," *Sociological Analysis,* 35 (1974) 57–65.

132. Mary Ann Glendon, *Rights Talk: The Impoverishment of Political Discourse* (New York: The Free Press, 1991) is particularly thought-provoking on contemporary political language.

133. McBrien, *Do We Need the Church?* 214–15, 219, 228–30.

134. Rev. Donald Kenna, "Chaplain's Notes," September 1966, file "race and riots—1966," box 17, Msgr. Geno Baroni Papers, MAUND.

135. Richard A. Schoenherr and Lawrence A. Young, *Full Pews and Empty Altars: Demographics of the Priest Shortage in United States Catholic Dioceses* (Madison: University of Wisconsin Press, 1993), 204–7; Helen Rose Fuchs Ebaugh, *Organizational Decline in Catholic Religious Orders in the United States* (New Brunswick: Rutgers University Press, 1993), 49. Also see Joseph M. Becker, S.J., *The Re-formed Jesuits,* vol. 1, *A History of Changes in Jesuit Formation During the Decade 1965-1975* (San Francisco: Ignatius Press, 1992), 75.

136. "An Omaha Catholic" to Archbishop Cody, April 30, 1969, Lawlor file, box 2007, AAC.

137. Dan Herr, "Stop Pushing," *The Critic,* 25 (February–March 1967), 4–6; Robert Coles, "The White Northerner," *The Atlantic Monthly,* 217 (June 1966), 56–57.

Conclusion

1. "Church of the Gesu, 1868," from *Historical Sketches of the Catholic Churches of Philadelphia* (Philadephia, c. 1895), 124–25; *The Gesu Parish, 1868–1968* (Philadelphia, 1968), 5–7.

2. Rev. Joseph Cawley, S.J., phone interview with author, July 27, 1994; "Novena of Grace Begins Wednesday," *Philadelphia Evening Bulletin,* March 1, 1925, clipping in *Bulletin* collection, Urban Archives, Temple University. Copies of various parish booklets in PAHRC. *Golden Jubilee of the Church of the Gesu: 1888–1938* (Philadelphia, 1938); Benedict Guldner, "St.Joseph's College—Philadelphia," *Woodstock Letters,* 42 (1912), 200.

3. A. J. Emerick, S.J., "The Colored Mission of Our Lady Of The Blessed Sacrament," *Woodstock Letters,* 42 (1912), 183, 187.

4. Bernard J. Newman, *Housing in Philadelphia, 1931* (Philadelphia: Philadelphia Housing Association, 1932), 8; Lenora E. Berson, *Case Study of a Riot: The Philadelphia Story* (New York: Institute of Human Relations Press, 1966), 23–24.

5. W. P. Shriver, *The Presbyterian Church in Philadelphia* (New York: Presbyterian Church, USA, 1930) 16, #2782, HPD.

6. Bernard J. Newman, *Housing in Philadelphia, 1935,* 22–23.

7. Anna McGarry, interview with Margaret Sigmund, September 29, 1976 (tape), AMC.

8. Data on parish collections in issues of Gesu's *The Parish Monthly Calendar.*

9. Karl William Henry Scholz, *Real Property Tax Base Changes in Philadelphia, 1931–1937* (Philadelphia: Institute of Local and State Governments, University of Pennsylvania, 1937), 5–6.

10. "Father James I. Maguire, S.J.," *The Parish Monthly Calendar,* 17 (September 1942), 10.

11. Ibid., 11 (August 1936), 15.

12. Ibid., 13 (September 1938), 12.

13. Ibid., 13 (February 1938), 15; 13 (August 1938), 15.

14. Ibid., 12 (February 1937), 16; 11 (August 1936), 15; 16 (July 1941), 13.

15. Fr. Thomas Love, S.J., to Cardinal Dougherty, October 18, 1940, PAHRC; *The Parish Monthly Calendar,* 12 (February 1937), 16.

16. Anna McGarry, interview with Margaret Sigmund, September 29, 1976 (tape), AMC.

17. Lenerte Roberts and Arthur Huff Fauset, "White Priests Admit Anti-Negro Drive," *Philadelphia Tribune,* February 20, 1941, p.1; Arthur Huff Fauset and Lenerte Roberts, "Churchs Push Anti-Race Drive in W. Philadelphia," *Philadelphia Tribune,* March 6, 1941, p.1; Charles S. Johnson and Associates, *To Stem This Tide: A Survey of Racial Tension Areas in the United States* (Boston and Chicago: The Pilgrim Press, 1943), 46–47; William Osborne, *Segregated Covenant,* 161–64.

18. "Memorandum on the Gesu Philadelphia Situation," March 3, 1941, folder 20, box 29, JLF; Raymond Pace Alexander to LaFarge, March 2, 1941, folder 20, box 29, JLF. Mr. Alexander was the first African-American to move onto his block in 1927. See Berson, *Case Study of a Riot,* 24.

19. William J. Walsh to Dougherty, September 23, 1939, Dougherty correspondence, PAHRC; V. A. Dever to Fr. Tenelley, October 5, 1926, folder 9, box 9, series 5.2, BCIM; Richard A. Vanbero, "Philadelphia's South Italians and the Irish Church: A History of Cultural Conflict," *The Religious Experience of Italian-Americans,* Silvano M. Tomasi, ed. (Staten Island: The American-Italian Historical Association, 1975), 50–51; James F. Connelly, *The History of the Archdiocese of Philadelphia* (Philadelphia: The Archdiocese of Philadelphia, 1976), 386–89.

20. Mary Moore to Dougherty, May 1, 1940, Dougherty correspondence, PAHRC.

21. McIntyre to LaFarge, February 26, 1941, folder 20, box 29, JLF.

22. Clarence J. Howard, "The Cardinal Came to Catto Hall," *St. Augustine's Messenger,* 19 (August 1940), 146–49.

23. Fr. William J. Walsh to Walter White, May 2, 1940, file "Catholics—1940–1955," box 167, series A, Group II, NAACP.

24. Dougherty to Fr. Thomas Love, S.J., October 19, 1940, Dougherty correspondence, PAHRC.

25. Walter J. Fox, Jr., "Brotherhood & Mrs. McGarry," *National Catholic Reporter,* 2 (June 8, 1966), 2.

26. Anna McGarry, interview with Margaret Sigmund, September 29, 1976, AMC; Walter J. Fox Jr., "Brotherhood & Mrs. McGarry," 2; Dennis J. Clark, *Erin's Heirs: Irish Bonds of Community* (Lexington: University Press of Kentucky, 1991), 83–93.

27. "Force Drive to Put Out White Priest for Racial Bias in North Philly," *Philadelphia Tribune,* March 27, 1941, p.3.

28. *The Parish Monthly Calendar,* 16 (June 1941), 14; ibid.

29. Consultors' Meeting, January 16, 1946, Consultors' records, Church of the Gesu; Johnson, *To Stem This Tide,* 46–47.

30. Sylvester and Althea McCook, interview with author and Fr. George Bur, S.J, August 10, 1994.

31. Consultors' Meeting, August 14, 1942, Consultors' records, Church of the Gesu; Consultors' Meeting February 20, 1946, Consultors' records, Church of the Gesu.

32. Sylvester and Althea McCook, interview with author and Fr. George Bur, S.J, August 10, 1994; Consultors' Meeting, May 30, 1945, Consultors' Records, Church of the Gesu; John P. Smith to LaFarge, August 18, 1944, folder 20, box 29, JLF; "ISOccasions," *Institute of Social Order Bulletin,* 1 (December 1943), 8.

33. *Philadelphia's Negro Population: Facts on Housing* (Philadelphia: Commission on Human Relations, 1953), 50–52.

34. Rev. William Michelman, S.J., interview with author, August 11, 1994; John S. Monagan, *Horace: Priest of the Poor* (Washington, D.C.: Georgetown University Press, 1985), 106–11.

35. Rev. James Gormley, S.J., *Saint Joseph's Preparatory School* (Philadelphia, 1976); Rowland T. Moriarty, "It's the Old 'Prep Spirit' Aid Offers Swamp St. Joe's," *Philadelphia Evening Bulletin,* February 1, 1966, in *Bulletin* clipping collection, Urban Archives, Temple University; Consultors' meeting, April 27, 1951, Consultors' Records, Church of the Gesu.

36. *The Gesu Parish, 1868–1968* (Philadelphia, 1968).

37. *North Philadelphia DataBook* (Philadelphia: Philadelphia City Planning Commission, 1986), 9, 11; "Once Thriving Gesu Church Notes 75th Year, Stands Out as Island in North Phila. Blight," *Philadelphia Evening Bulletin,* December 8, 1963, clipping in *Bulletin* collection, Urban Archives, Temple University.

38. Acel Moore, "A School That Works Gets Big-Name Help," *Philadelphia Inquirer,* September 12, 1991, A15.

39. Roger S. Ahlbrandt, Jr., *Neighborhoods, People and Community* (New

York: Plenum Press, 1984), 26, 63–65, 187; Peter Steinfels, ''Future of Faith Worries Catholic Leaders,'' *New York Times,* June 1, 1994, pp. A1/B8; Andrew M. Greeley, *The Catholic Myth: The Behavior and Beliefs of American Catholics* (New York: Scribner, 1990), 24, 126–43.

40. J. Anthony Lukas, *Common Ground: A Turbulent Decade in the Lives of Three American Families* (New York: Alfred A. Knopf, 1985), 399. For a similar claim about Catholicism in Brooklyn, Joe Sexton and Jennifer Steinhauer, ''A Growing Rage in Bay Ridge,'' *New York Times,* November 29, 1994, p. B/8.

41. The term ''de facto congregationalism'' comes from R. Stephen Warner, ''The Place of the Congregation in the Contemporary American Religious Configuration,'' in *American Congregations,* vol. 2, *New Perspectives in the Study of Congregations,* James P. Wind and James W. Lewis, eds. (Chicago: University of Chicago Press, 1994), 77–80.

42. Linda Loyd, ''A Tearful Farewell for Church,'' *Philadelphia Inquirer,* June 28, 1993, B1; ''7 Schools to Close in Phila., Chester,'' *Philadelphia Inquirer,* April 3, 1993, A1. Also see, ''Cardinal of Detroit Orders 30 Parishes in the City to Close,'' *New York Times,* January 9, 1989, A/11; Isabel Wilkerson, ''Catholic Parish Closings Bring Tears in Chicago,'' *New York Times,* July 9, 1990, A/10; Peter Steinfels, ''Brooklyn Bishop is Facing Diverse Cultures and Tensions,'' *New York Times,* February 22, 1990, B 1/5.

43. Two statements by the African-American bishops are especially notable. ''What We Have Seen and Heard,'' *Origins,* 14 (October 18, 1984), 273–87 and ''Brothers & Sisters to Us,'' *Origins,* 9 (November 29, 1979), 381–89.

44. Barry A. Kosmin and Seymour P. Lachman, *One Nation Under God: Religion in Contemporary American Society* (New York: Harmony Books, 1993), 127-31; ''The Pope's Address to Black Catholics,'' *Origins,* 17 (September 24, 1987), 251–52.

45. Reynolds Farley and Walter R. Allen, *The Color Line and the Quality of Life in America* (New York: Oxford University Press, 1989), 117–18.

46. Paul Elie, ''Hangin' With the Romeboys,'' *The New Republic,* 206 (May 11, 1992), 18–26. On a remarkable Catholic parish in Brooklyn, Martin Gottlieb, ''Church Mirrors Neighborhood's Hope,'' *New York Times,* August 15, 1993, p.38.

47. Much of the initial energy along these lines came from Msgr. John Egan's Catholic Committee on Urban Ministry. See John A. Coleman, ''Serving the Servants: The Catholic Committee on Urban Ministry,'' *America,* 138 (March 11, 1978), 182–85. On the Alinsky legacy, see Sanford Horwitt, *Let Them Call Me Rebel,* 544–48; Robert Fisher, *Let the People Decide: Neighborhood Organizing In America* (Boston: Twayne, 1984), 121–52; Ernie Cortes, ''Reflections on the Catholic Tradition of Family Rights,'' in *One Hundred Years of Catholic Social Thought,* ed. John A. Coleman (Maryknoll: Orbis Books, 1991), 155–71; Donald C. Reitzes and Dietrich C. Reitzes, *The Alinsky Legacy: Alive and Kicking* (Greenwich: JAI Press, 1987); Peter Skerry, *Mexican-Americans: The Ambivalent Minority* (New York: The Free Press, 1993), esp. 162–201. On the congruence between Catholic social thought and Alinsky-style organizations, Charles E. Curran, *Critical Concerns in Moral Theology* (Notre Dame: University of Notre Dame Press, 1984), 171–199.

48. See Timothy Garton Ash, *The Uses of Adversity: Essays on the Fate of*

Central Europe (New York: Random House, 1989), 193–95, 270–79; José Casanova, *Public Religions in the Modern World* (Chicago: University of Chicago Press, 1994), 114–34.

49. Robert N. Bellah, Richard Madsen, William M. Sullivan, Ann Swidler, Steven M. Tipton, *The Good Society* (New York: Alfred A. Knopf, 1991), esp. 281–82; also see the initial work of the same group of authors, *Habits of the Heart: Individualism and Commitment in American Life* (Berkeley: University of California Press, 1985). On social capital, see Robert D. Putnam, "Bowling Alone: Democracy in America at the End of the Twentieth Century," *Journal of Democracy*, 6 (January 1995), 65–78. For an overview on these matters, see the essays contained in *Catholicism and Liberalism: Contributions to American Public Philosophy*, R. Bruce Douglass and David Hollenbach, eds. (Cambridge: Cambridge University Press, 1994).

50. This notion is broadly indebted to Alasdair MacIntyre, *After Virtue* (Notre Dame: University of Notre Dame Press, 1984), esp. 251–52.

INDEX

(*Illustrations are indicated by italic page numbers.*)

C

D